Self-Assessment: 650 BOFs for MRCP(UK) and MRCP(I) Part I

Second Edition

This page was intentionally left blank

Second Edition

Self-Assessment: 650 BOFs for MRCP(UK) and MRCP(I) Part I

Osama S. M. Amin

MD, FRCP(Edin), FRCP(Glasg), FRCPI, FRCP(Lond), FACP, FAHA, FCCP

**Clinical Associate Professor of Neurology
International Medical University,
Kuala Lumpur, Malaysia**

BMA LIBRARY
BRITISH MEDICAL ASSOCIATION

Copyright © 2017. Osama S. M. Amin.

Osama S. M. Amin
Clinical School, International Medical University,
Jalan Rasah, 70300 Seremban, Negeri Sembilan,
Malaysia
Email: dr.osama.amin@gmail.com, osamashukir@imu.edu.my

First Edition: 2009.
Second Edition: 2017
ISBN: 978-1-365-61609-9

Distributed by Lulu Press, Inc. Northern Carolina, USA.

Dedication

To my lovely family:

Sarah, Awan, and Naz

Acknowledgements

I woul like to thank my dear patients; their real clinical scenarios were used to formualte and generate these questions.

A special gratitude goes to my other half, wife Sarah, for her endless support and encouragement, and of course her extreme patience.

Osama

Preface

"No human being is constituted to know the truth, the whole truth, and nothing but the truth; and even the best of men must be content with fragments, with partial glimpses, never the full fruition", Sir William Osler (1849-1919).

First of all, this is the 2nd edition of my previous book, "Self-Assessment for MRCP(UK) part 1", which was published in 2009. The book has undergone several changes and updates. Membership diplomas of the Royal Colleges of Physicians of the United Kingdom, MRCP(UK), and Ireland, MRCP(I), have several things in common. Their part I examination at a glance: one-day examination; two-three hour papers (UK) or one-three hour paper (Ireland); 200 (UK) or 100 (Ireland) best of five questions; no photographic materials; and sat in an examination hall. The best initial step is to read accredited textbooks. After then, self-assess yourself. Read this book, chapter after chapter, try to find out your mistakes and gaps in knowledge, and then re-read what you have missed.

In writing this book, I tried my best to include the commonest examination themes. You may encounter negative or positive stems (i.e., one wrong stem or one correct stem out of five stems). The questions' objective is to teach, i.e., a rapid review of every subject and theme. This book differs from my book *"Get Through MRCP part I; BOFs"*, which was published in the year 2008 by the Royal Society of Medicine Press in London. The latter concentrates mainly on the diagnosis and management, i.e., what is the diagnosis, what is the next best step, what feature is consistent with your preliminary diagnosis, and so on?

Undoubtedly, if you are well-prepared, you will pass the examination very easily. No need to panic when you hear your colleagues' past experiences. Lack of preparation is the single most common reason for failure. Remember, practice makes perfect. Read and self-assess, that's it!

For a more comprehensive self-assessment, try my *"Get Through MRCP part I"*, *"Neurology: Self-Assessment for MRCP(UK) and MRCP(I)"*, and *"Mock Papers for MRCPI part I, 2nd Edition"* books.

Good luck with your career and exams!

Osama S. M. Amin
January 2017

Recommended Reading

1.   Walker B, Colledge NR, Ralston S, Penmen I. *Davidson's Principles and Practice of Medicine, 26* *26th edition*. London: Churchill Livingstone; 2014.

2.   Kumar P, Clark ML. *Kumar and Clark's Clinical Medicine, 8th edition*. Philadelphia: Saunders Ltd.; 2012.

3.   Longo D, Fauci A, Kasper D, Hauser S, Jameson J, Loscalzo J. *Harrison's principles of Internal Medicine, 18th edition*. New York: McGraw-Hill Professional, 2011.

4.   Dale DC, Federman DD, *ACP Medicine, 3rd edition*. Philadelphia: BC Decker Inc.; 2007.

5.   Barrett KA, Barman SM, Boitano S, Brooks H. *Ganong's Review of Medical Physiology, 24th edition*. New York: McGraw-Hill Education / Medical; 2012.

6.   Lang TA, Secic M (eds.). *How to Report Statistics in Medicine, Annotated Guidelines for Authors, Editors, and Reviewers, 2nd edition*. Philadelphia: The American College of Physicians, 2006.

7.   Rosene-Montella K, Keely EJ, Lee RV, Barbour LA (eds.). *Medical Care of the Pregnant Patient, 2nd edition*. Philadelphia: The American College of Physicians; 2007.

Table of Contents:

This page was intentionally left blank

Chapter 1
Cardiology

Questions

1) A 65-year-old man visits the physician's office because of exertional chest pain and frequent palpitations. He is a heavy smoker but does not drink alcohol. The patient's 12-lead ECG reveals a prominent R-wave in lead V_1. All of the following are causes of prominent R-wave in lead V_1, *except*?
   a) Posterior wall myocardial infarction
   b) Right bundle branch block
   c) Mirror image dextrocardia
   d) Wolf-Parkinson-White syndrome type A
   e) Wolf-Parkinson-White syndrome type B

2) A 60-year-old man, who has a history of long-standing type II diabetes and hypertension, has been referred by his general practitioner for further evaluation of chest pain. You do 12-lead ECG and this reveals ST-segment elevation in leads V_5, V_6, and I. Which one of the following may result in ST-segment elevation?
   a) Left ventricular hypertrophy
   b) Right ventricular hypertrophy
   c) Digoxin effect
   d) Early repolarization after an episode of angina
   e) Sub-endocardial myocardial infarction

3) A 23-year-old male is brought to the Acute and Emergency Department with a 2-hour history of palpitation and dizziness. His previous notes reveal similar attacks. The patient's QRS complexes during sinus rhythms demonstrate delta waves. The commonest type of arrhythmia in Wolff-Parkinson-White syndrome is?
   a) Ventricular tachycardia
   b) Ventricular fibrillation
   c) Ventricular premature complexes
   d) Atrial ectopic beats
   e) AV nodal re-entrant tachycardia

4) A 15-year-old boy has been experiencing progressive exertional breathlessness over the past several months. He has a Turner's syndrome-like phenotype. You do chromosomal studies and secure the diagnosis of Noonan's syndrome. The following are seen in Noonan's syndrome, *except*?
   a) Mental retardation
   b) Pulmonary artery stenosis
   c) Branch pulmonary artery stenosis
   d) Hypertrophic cardiomyopathy
   e) Mitral stenosis

5) A 24-year-old woman presents with transient loss of consciousness which turns out to be due to torsades de pointes ventricular tachycardia. Her 12-lead ECG, in sinus rhythm, shows a QTc interval of 0.56 second. All of the following may result in QT-interval prolongation, *except*?
   a) Hypokalemia
   b) Hypocalcemia
   c) Hypomagnesemia
   d) Hypermagnesemia
   e) Rheumatic carditis

6) A 20-year-old man visits the cardiology outpatients' clinic to consult you about his brother who has hypertrophic cardiomyopathy and the likelihood of himself being affected by this disease. With respect to hypertrophic cardiomyopathy, which one of the following is the *correct* statement?
   a) Caused by a mutation in sodium channels
   b) 25% of cases have a positive family history of the same disease
   c) 30% of cases are associated with left ventricular outflow obstruction
   d) Almost all patients are symptomatic
   e) The commonest subtype of ventricular hypertrophy in the Western World is the apical one

7) A 47-year-old uremic man presents with central chest pain that is aggravated by swallowing and taking a deep breath. The patient is incompliant with his hemodialysis program. You consider acute pericarditis. His 12-lead ECG solidifies your clinical impression. Regarding acute pericarditis, which one is the *false* statement?
   a) The commonest cause is the idiopathic form
   b) The majority of cases progress to pericardial tamponade
   c) In uremic pericarditis, there is a high risk of pericardial tamponade formation
   d) The resulting pericardial rub is usually transient
   e) Tuberculous pericarditis rarely presents as an acute process

8) A 69-year-old man presents with severe substernal chest pain. You examine the patient and do some tests. Your diagnosis is acute myocardial infarction and you consider thrombolytic therapy. Which one of the following represents an absolute contraindication to thrombolytic therapy in acute ST-segment elevation myocardial infarction?

a) A 70-year-old hypertensive male with acute chest pain, new early diastolic murmur in the left lower sternum, and ST-elevation in lead II, III and, aVF
b) A 40-year-old woman with 7 missed periods and progressive abdominal distension
c) A 34-year-old diabetic male has recently undergone retinal LASER therapy and presents with chest pain, shock, raised JVP and clear lung bases
d) ST-segment elevation in lead V1-3, blood pressure of 180/100 mmHg, which is easily controlled with antihypertensive medications
e) A 60-year-old male who is a known case of ischemic heart disease and presents with a new left bundle branch block and central chest pain for 2 hours

9) A 66-year-old man takes many daily medications for his blood pressure control. He admits to having had orthostasis symptoms over the past few weeks. You tell him that his postural hypotension is due to dilatation of his blood vessels. All of the following cardiovascular medications have a predominantly vanodilating effect, *except?*
a) Hydralazine
b) Lisinopril
c) Prostacyclin
d) Isosorbide dinitrate
e) Prazosin

10) A 73-year-old diabetic man complains of ischemic chest pain for 2 hours. You admit him and diagnose unstable angina. You further assess him whether he has any high risk factors. "High" risk factors in unstable angina are considered to be all of the following, *except?*
a) Development of heart failure
b) Prolonged chest pain
c) ST-segment elevation
d) High cardiac troponin-I blood level
e) Normal 12-lead ECG

11) A 65-year-old man develops right-sided heart failure because of right coronary artery occlusion. You review the vascular anatomy of the heart. The following statements are correct, *except?*
a) The right coronary artery supplies the AV node in 90% of cases
b) The right coronary artery supplies the SA node in 60% of cases

c) The main stem of the left coronary artery can be easily dilated during percutaneous coronary intervention in atheromatous narrowing

d) Prinzmetal angina usually occurs at night

e) Prinzmetal angina may occur with/without coronary atheroma

12) A 50-year-old man takes amiodarone tablets because of permanent atrial fibrillation. He has impaired left ventricular systolic function. Which one is the *false* statement about this cardiac medication?

a) About 40% of its formulation is iodine

b) Potentiates the effect of warfarin

c) Potentiates the effect of digoxin

d) Corneal deposits of the medication are usually irreversible

e) Prolongs the plateau phase of the cardiac action potential

13) You ask your junior house officer to examine the JVP of this 33-year-old male patient, who is found to have raised JVP. The list of causes of raised JVP includes all of the following, *except?*

a) Mediastinal lymphoma

b) Mediastinal irradiation 2 years ago

c) ST-segment elevation in the right V_4 lead

d) Liver cirrhosis

e) Atrialization of the right ventricle

14) A 20-year-old woman presents with resistant hypertension. You a systolic bruit in the inter-scapular area. With respect to coarctation of the aorta, which one of the following is the *incorrect* statement?

a) In adults, it usually causes and presents as hypertension

b) Bicuspid aortic valve is a common association

c) Rib notching is usually seen from the 3rd rib to the 9th rib during infancy

d) Aortic dissection is a recognized complication

e) Subarachnoid hemorrhage is a risk

15) A 66-year-old man has dilated cardiomyopathy. You assess him and treat him accordingly. The following medications improve survival figure in chronic congestive heart failure, *except?*

a) Bisoprolol

b) Metoprolol

c) Carvedilol

d) Spironolactone

e) Atenolol

16) A 34-year-old woman presents with fever and skin rash. She was diagnosed with combined mitral valve disease few years ago. Examination reveals a new aortic regurgitation and Osler's nodes. Regarding infective endocarditis, which one is the *wrong* statement?
  a) *Streptococcus viridans* is the cause in 50% of cases of native valve endocarditis
  b) *Staphylococcus* species may be the cause in 50% of prosthetic valve endocarditis
  c) The commonest cause of culture-negative endocarditis is partial treatment with antibiotics
  d) Three negative serial blood cultures usually exclude infective endocarditis in the appropriate clinical setting
  e) The sensitivity of transesophageal echocardiography for the detection of vegetations is 95% versus 65% figure of the transthoracic approach

17) A 22-year-old man consults you prior to a scheduled dental extraction. He has combined mitral and aortic regurgitations due to rheumatic etiology. You educate him regarding the prophylaxis of infective endocarditis. The following cardiac lesions are associated with a moderate to high risk of infective endocarditis, *except?*
  a) Ventricular septal defect
  b) Hypertrophic cardiomyopathy
  c) Combined mitral valve disease
  d) Mitral valve prolapse without significant mitral regurgitation
  e) Lone mitral stenosis

18) A 30-year-old woman was admitted to the general medical ward and had been receiving treatment for infective endocarditis. She had severe rheumatic reflux. With respect to the treatment of infective endocarditis, which one of the following statements is the *incorrect* one?
  a) Persistent fever despite appropriate antibacterial antibiotic therapy may reflect a non-bacterial cause
  b) Cidal drugs, rather than statics, should be used
  c) Blood culture results should be obtained before starting any treatment
  d) The intravenous route of antibiotics is always preferred
  e) Persistent fever may imply drug fever

19) A 37-year-old female presents with progressive exercise intolerance and dry cough. After examining her, you think that she has mitral valve disease. You arrange for echocardiography. The latter reveals severe mitral stenosis. Regarding lone mitral stenosis, all of the following are true, *except?*

a) Malar flush may be observed
b) There is a low cardiac output state
c) Rarely congenital in origin
d) A 3rd heart sound is commonly heard
e) Approximately 20% of cases remain in sinus rhythm

20) A 63-year-old man presents with exertional breathlessness. Examination reveals systolic ejection murmur of grade III/VI at the aortic region. The murmur radiates to the right carotid. With regard to aortic stenosis, all of the following are true, *except*?
a) Carotid shudder is common
b) Syncope calls for surgical intervention
c) It is sub-valvular in William's syndrome
d) Marked left ventricular hypertrophy is a risk factor for angina and arrhythmia
e) Aortic valve replacement is the usual mode of treatment

21) A 34-year-old man presents with palpitations and orthopnea. You detect bilateral pitting ankle edema. You final diagnose congestive heart failure and you discuss the treatment plan with the patient. Regarding the objectives of medical treatment of heart failure, all of the following statements are correct, *except*?
a) Reduction of afterload
b) Optimization of preload
c) Augmentation of cardiac contractility
d) Controlling the heart rate
e) Aggressive diuresis

22) A 59-year-old man, who has long-standing poorly controlled hypertension, presents with exercise intolerance. Transthoracic echocardiography reveals concentric left ventricular hypertrophy and moderate diastolic dysfunction. He asks what his diastolic heart failure type entails. Regarding diastolic heart failure, which one of following is the *correct* statement?
a) Aortic regurgitation usually causes diastolic dysfunction
b) The treatment is not similar to that of left ventricular systolic dysfunction
c) Myocardial ischemia starts with systolic dysfunction and then results in diastolic dysfunction
d) Tachycardia is needed to compensate for the dysfunction
e) ACE inhibitors have been shown to improve the mortality figure

23) A 19-year-old male patient has been referred for further assessment of high blood pressure. He denies family history of such condition. He is on no medications and takes no illicit drugs. He neither smokes nor drinks alcohol. After a thorough history taking and examination, you consider secondary causes of his hypertension. Secondary hypertension is suspected in the presence of all of the following, *except?*
   a) Rapid development
   b) Poor response to standard medical therapy
   c) Age blow 20 years or above 50 years
   d) Mooning of the face and abdominal striae
   e) Strong family history of hypertension

24) A 55-year-old man visits you as part of his regular check-up of his hypertension and diabetes. He takes many daily medications and herbal remedies. He enquires about his long-term outlook. Regarding long-standing hypertension, choose the most *correct* statement?
   a) It has a weak adverse effect on the coronary arteries
   b) It is a strong risk factor for the development of intra-cerebral hemorrhage
   c) In diabetic nephropathy with persistent proteinuria, the target blood pressure should be ideally below 150/90 mmHg
   d) ACE inhibitors should be routinely prescribed if there are no contraindications
   e) Diuretics are usually not effective in elderly population

25) A 67-year-old man presents with a 3-hour history of central chest pain that is associated with nausea and breathlessness. The list of risk factors for ischemic heart disease does *not* include which one of the following?
   a) Hyperhomocysteinemia
   b) Obesity
   c) Sedentary life style
   d) High blood fibrinogen
   e) Diet rich in vegetables

26) A 76-year-old man presents with many complaints. He has long-standing poorly controlled hypertension. Complications of long-standing hypertension include all of the following, *except?*
   a) Diastolic heart failure
   b) Coronary artery disease
   c) Stroke
   d) Peripheral vascular disease

e) Infective endocarditis

27) A 60-year-old female presents with mitral stenosis-like picture. Her echocardiography reveals a mass within the heart. You repeat the examination and review the echocardiography findings. You finally diagnose cardiac myxoma. Regarding atrial myxoma, which one is the *false* statement?
  a) May present as an infective endocarditis-like picture
  b) May have a vasculitis-like presentation
  c) High risk of local recurrence following surgical removal in sporadic cases
  d) High risk of local recurrence following surgical removal in familial cases
  e) Familiar cases are usually associated with right-sided myxoma

28) A 44-year-old man receives a beta-blocker for his essential hypertension. All of the following beta-blockers are non-cardioselective, *except?*
  a) Metoprolol
  b) Nadolol
  c) Propranolol
  d) Pindolol
  e) Timolol

29) An undergraduate medical student asks your help to examine the JVP of a patient who has congestive heart failure. You demonstrate the examination sequences and discuss the various abnormalities with him. Which of the following you have *not* told him?
  a) A rapid *x*-descent and steep *y*-descent occur in pericardial constriction
  b) A rapid *x*-descent and blunted *y*-descent are found in pericardial tamponade
  c) A large *v*-wave is seen in tricuspid regurgitation
  d) Hepatojugular reflux is significant if the JVP elevation sustains upon continuous pressure on the abdomen
  e) Regular cannon *a* waves indicate complete heart block

30) You are reviewing an ECG strip of one of your medical ward patient. His PR interval is 0.04 second. Short PR interval occurs in?
  a) Bradycardia states
  b) Treatment with propranolol
  c) Ischemic heart disease
  d) Treatment with theophylline
  e) Hypothyroidism

31) One of your junior house officers asks about the characteristics of Q waves in ECGs. Deep and permanent Q-wave may be seen in?
   a) Hyperacute phase of acute anterior wall myocardial infarction
   b) Wolff-Parkinson-White syndrome
   c) Established sub-endocardial infarction
   d) Established acute phase of isolated posterior wall myocardial infarction
   e) During Prinzmetal angina

32) A 16-year-old male has coarctation of the aorta. He has surfed the internet and read about his disease. He is afraid that his anomaly may have additional fearful associations. All of the following are commonly associated with coarctation, *except?*
   a) Epstein's anomaly
   b) Bicuspid aortic valve
   c) Patent ductus arteriosus
   d) Ventricular septal defect
   e) Berry aneurysm at the circle of Willis

33) A 66-year-old man is brought to the Emergency Room because of few hours' crushing chest pain. You suspect acute myocardial infarction. With respect to medical treatment of acute myocardial infarction, all of the following are true, *except?*
   a) Aspirin reduces the mortality figure by 30%
   b) Thrombolysis reduces the mortality figure by 25-50%
   c) Beta blockers prominently reduce the susceptibility to arrhythmias
   d) Intravenous nitroglycerin has no effect on the overall mortality figure
   e) Morphine is better to be given by a subcutaneous route

34) A 55-year-old man, who sustained acute anterior wall myocardial infarction 1 week ago, will be discharged home today. You educate him about his medications and tell him about the importance of regular visits. Regarding secondary prophylaxis of myocardial infarction, which one is the *wrong* statement?
   a) ACE inhibitors have a favorable effect because of many independent mechanisms
   b) Bata blockers should always be considered if no contraindication is present
   c) Statins have been shown to produce regression of atherosclerotic plaques
   d) Cardiac rehabilitation programs should not be forgotten

e) Digoxin has a favorable effect on myocardial protection and mortality

35) A 66-year-old man presents with attacks of severe substernal chest pain upon exertion. These pains are relieved by rest and sublingual nitroglycerin. You consider chronic stable angina and you order exercise ECG testing. Contraindication to this mode of cardiac testing encompasses all of the following, *except?*
   a) Severe uncontrolled hypertension
   b) Fever or any febrile illness
   c) Severe aortic stenosis
   d) Acute pericarditis
   e) Subjective feeling of weakness

36) A 61-year-old man is due to do exercise ECG testing using a treadmill. The following findings reflect a strong positive exercise ECG testing, *except?*
   a) Deep ST-segment depression
   b) Wide-spread ECG changes
   c) Prolonged ST-segment depression after finishing the test
   d) ST-segment elevation
   e) Mild to moderate ST-T segment abnormalities in a middle-aged woman with chest pain and no risk factors for ischemic heart disease

37) Your colleague consults you about this 22-year-old young man, who has abnormal exercise ECG. You interview the patient and examine him. You review the ECG trace. Finally, you tell the patient that he has false positive results. Causes of false positive exercise ECG testing include all of the following, *except?*
   a) Current treatment with digoxin
   b) Underlying hypertrophic cardiomyopathy
   c) Marked left ventricular hypertrophy
   d) Early development of wide-spread ECG changes during stage I of Bruce protocol
   e) Presence of mitral valve prolapse

38) One of your junior house officers asks you about the causes of electromechanical dissociation of the heart, because one of his patients developed this. All of the following can cause electromechanical dissociation, *except?*
   a) Pulmonary thromboembolism
   b) Tension pneumothorax
   c) Profound shock due to hemorrhage
   d) Hypokalemia

e) Treatment with intravenous xylocaine

39) You educate this 65-year-old man about the use of ACE inhibitors following his myocardial infarction. He will discharged today. Which one of the following is the *incorrect* statement regarding the use of ACE inhibiters in post-myocardial infarction?
a) They decrease ventricular remodeling and aneurysmal formation
b) Prevent the onset of overt heart failure in asymptomatic patients
c) Reduce hospitalization rates
d) Should be given to all patients if they are tolerated and there are no contraindications
e) Contraindicated if there is coexistent chronic renal failure

40) A 55-year-old man is not compliant with his antihypertensive medication because this medications makes him urinate a lot during work. Regarding the treatment of hypertension with diuretics, all of the following are true, *except?*
a) Loop diuretics are always preferred because of their powerful effect
b) Diuretics used in the treatment of hypertension uncommonly produce acid-base and electrolyte disturbances
c) As anti-hypertensive medications, they work by an independent mechanism of their diuretic action
d) First-line agents in elderly patients
e) They may worsen serum lipid profile

41) A 30-year-old female presents with progressive exercise intolerance and exertional chest pain. Further work-up has revealed primary pulmonary hypertension. With respect to this disease, which one is the *false* statement?
a) An association with HIV infection has been documented
b) Medial hypertrophy and fibrinoid necrosis occur in all branches of the pulmonary arterial tree
c) Physical signs are usually unimpressive until right-sided heart failure develops
d) Five out of every 50 cases are familial
e) Most patients die within 2-3 years of diagnosis

42) A 66-year-old man, who admits to smoking heavily for 45 years, presents with bilateral pitting leg edema and cyanosis. After conducting proper examination and doing some investigations, you diagnose cor pulmonale. Regarding cor pulmonale, which one is the *wrong* statement?
a) May be acute or chronic
b) The first physical sign is usually raised JVP
c) May be caused by post-polio syndrome

d) The end-result of many chronic debilitating lung diseases
e) Defined as right-sided heart failure

43) You have been asked to examine a 70-year-old man with COPD. He presents with excessive day time somnolence and poor concentration. The patient's lips are a little bit blue in color and his outstretched hands are shaky. There is a bounding radial pulse. You intend to examine further to confirm your preliminary diagnosis. What would examine?
a) Chest for new changes
b) Eye fundus
c) The respiratory rate for hypoventilation
d) Heart for heart failure
e) Neck for any swelling

44) A 65-year-old hemiplegic man presents with shortness of breath and bloody sputum. His right calf is swollen and tender. A preliminary diagnosis of pulmonary thromboembolism is made. Which one is the *false* statement with respect to pulmonary thromboembolism?
a) It has diverse clinical manifestations, ranging from totally asymptomatic to sudden death
b) Bed-side echocardiography is a useful diagnostic tool
c) Symptomatic pulmonary thromboembolism occurs in 1% of postoperative patients
d) Autopsy incidence of post-operative thromboembolism is 10-25%
e) Up to 15% of pulmonary thromboemboli come from leg veins

45) A 55-year-old man presents to the Emergency Room with sudden severe retrosternal chest pain, 8 days after undergoing right-sided total hip replacement. You do chest X-rays and you examine them carefully. Chest X-ray findings in pulmonary thromboembolism encompass all of the following, *except?*
a) May be entirely normal
b) Subtle changes may be present, such as regional oligemia
c) Peripheral wedge-shaped opacities are very commonly seen
d) Elevation of a hemi-diaphragm is a well-recognized sign
e) Lung abscess formation has been documented

46) A 40-year-old female presents with dyspnea and tachypnea 5 days after giving birth to a healthy-looking full-term girl. You do ECG for her in addition to other investigations. Regarding ECG changes in acute pulmonary thromboembolism, which one of the following is uncommonly seen?

a) Sinus tachycardia
b) New right bundle branch block
c) Acute atrial fibrillation
d) $S_1Q_3T_3$ pattern
e) T-wave inversion

47) A 30-year-old man receives therapy for pulmonary thromboembolism, which has developed 7 days after a road traffic accident. Regarding the treatment of pulmonary thromboembolism, which one is the *incorrect* statement?
   a) Emergency embolectomy is rarely needed
   b) Thrombolytic agents should be used in massive pulmonary thromboembolism proved by chest CT or pulmonary angiography
   c) Heparin therapy has been shown to decrease mediator-induced pulmonary vasoconstriction
   d) Heparin therapy does not reduce the overall mortality in pulmonary thromboembolism
   e) Oral anticoagulants have no place in the management of life-threatening thromboembolic disease

48) A 66-year-old man, who underwent left knee joint replacement 9 days ago, presents with right-sided pleuritic chest pain. Chest X-ray film reveals an elevated right hemi-diaphragm and small basal liner opacities. Pulmonary thromboembolism is your final diagnosis. General measures in acute pulmonary thromboembolic events include all of the following, *except*?
   a) Opiates may be necessary to relieve pain and distress
   b) The use of inotropes results in prominent symptomatic benefit in massive embolization
   c) Diuretics and vasodilators should be avoided at all times
   d) O_2 should be given to all hypoxemic patients
   e) Resuscitation by external cardiac massage may be successful in moribund patients by dislodging and breaking up a large central embolus

49) A 30-year-old pregnant woman visits the physician's office to ask about the possibility of having a child with a heart anomaly. With respect to the incidence of various congenital heart diseases, all of the following anomalies and their corresponding incidence are correct, *except*?
   a. VSD-30%
   b. Atrial septal defects-10%
   c. Patent ductus arteriosus-10%

    d.   Tetralogy of Fallot-16%

    e.   Coarctation of aorta-7%

50) A 6-year-old child has been referred to the cardiology department. The referral letter states that the child has a congenital heart anomaly. Regarding congenital heart diseases, all of the following are correct, *except?*

    a)   Aortic regurgitation may complicate VSD due to loss of support of the right coronary cusp

    b)   Tetralogy of Fallot patients are immune against Eisenmenger syndrome

    c)   ASD primum usually presents in adulthood with cardiac arrhythmia

    d)   Preventing the closure of patent ductus arteriosus is useful in certain congenital heart diseases

    e)   Transposition of great vessels is the commonest cause of cardiac cyanosis at birth

51) A 20-year-old female has primary amenorrhea, webbed neck, and wide carrying angles. She presents with hypertension and a cardiac murmur. You consider Turner's syndrome. Turner's syndrome may be associated with all of the following, *except?*

    a)   ASD-triphalangeal thumb

    b)   Infant of a mother with bipolar disorder-Ebstein anomaly

    c)   Maternal rubella-infant patent ductus arteriosus

    d)   Epileptic mother-interrupted aortic arch in the neonate

    e)   Noonan's syndrome-mitral regurgitation

52) A 39-year-old man presents with head nodding and palpitations. Examination reveals early blowing diastolic murmur down the left sternal border. You diagnose aortic regurgitation. The following are important clinical clues to the underlying etiology of this type of valvular reflux, *except?*

    a)   Bilateral small irregular pupils

    b)   Long-standing low back pain, bamboo spine, and upper lung lobes fibrosis

    c)   Plucked chicken skin in both cubital fossae

    d)   Webbed neck

    e)   Cloudy cornea

53) A 30-year-old female presents with progressive dyspnea and hemoptysis. Examination reveals apical, mid-diastolic, rumbling murmur. Her echocardiographic study confirms mitral stenosis. With respect to this valvular narrowing, all of the following are true, *except?*

a) Giant *v*-waves may be seen on JVP examination
b) The development of acute pulmonary edema in pregnancy calls for emergency valvotomy
c) Trans-mitral diastolic gradient of more than 15 mmHg indicates mild degree of stenosis
d) An episode of pulmonary edema without a precipitating cause is an indication for surgical treatment
e) Although secondary pulmonary hypertension is common, but Graham-Steell murmur is rare

54) You examine a 5-year-old boy and find a pansystolic murmur in the left lower sternal border, which can be heard all over the precordium. You do investigations to uncover the origin of this non-benign murmur. ECG findings in uncomplicated congenital heart disease include all of the following, *except*?
a) ASD secondum-partial right bundle branch block
b) ASD primum-left axis deviation
c) VSD-biventricular hypertrophy
d) Ebstein anomaly-left ventricular hypertrophy
e) Coarctation of aorta-left ventricular hypertrophy

55) A 4-year-old child has pansystolic murmur and productive cough. His chest is full of crackles. You do chest X-ray examination. With respect to the findings in chest X-ray film of patients with congenital heart disease, which one is the *correct* statement?
a) It is rarely abnormal in long-standing cases
b) Boot-shaped heart is usually suggestive of ASD
c) The sign of 9 is seen in aortic coarctation
d) Large hili are observed in Eisenmenger syndrome
e) Patent ductus arteriosus is seen as a linear opacity in the left upper hemithorax

56) A 36-year-old man has recurrent attacks of troublesome supraventricular tachycardia, for which he has been prescribed daily amiodarone. This medication can cause?
a) Pulmonary edema
b) Whitening of the skin
c) Leukocytosis
d) Retinal deposits and secondary degeneration
e) Hypothyroidism

57) A 23-year-old man, who has rheumatic mitral regurgitation, presents with frequent palpitations and exercise intolerance. After examining him carefully, you find many signs which reflect severe mitral reflux. You have *not* detected which one of the following?
- a) 3rd heart sound
- b) Long duration of the murmur
- c) Signs of heart failure
- d) Presence of thrill
- e) 4th heart sound

58) A 45-year-old man, who was reasonably well and healthy, complains of palpitation over the past 3 weeks. His 12-lead ECG reveals fast atrial fibrillation. Transthoracic echocardiography is normal apart from irregularity of the cardiac beats. You prescribe a medication to control his heart rate. What else you would prescribe?
- a) Warfarin
- b) Heparin
- c) Aspirin
- d) Dabigatran
- e) No need for another medication

59) A 66-year-old man presents with central chest pain that is crushing in nature and radiating to the left shoulder for the past 4 hours. He has hypertension and type II diabetes. His past notes show high blood lipids. Which one of the following is *not* given as part of his medical treatment?
- a) Alteplase
- b) Aspirin
- c) Metoprolol
- d) Morphine
- e) Diazepam

60) A 65-year-old man presents with hemodynamic collapse and a rapid feeble pulse. One week ago, his general practitioner prescribed atenolol to treat frequent ventricular ectopics. His last ECG which was done 2 days ago revealed a regular heart rate of 36 beats per minute. What is the likely cause of his new presentation?
- a) Acute myocardial infarction
- b) Complete heart block-associated Stokes-Adams attack
- c) Bradycardia-associated torsade de pointes ventricular tachycardia
- d) Cardiac tamponade
- e) Conversion disorder

This page was intentionally left blank

Cardiology

Answers

1) e.

The other causes of prominent R -wave in lead V_1 are hypertrophic cardiomyopathy and right ventricular hypertrophy. Isolated posterior wall myocardial infarction is rare in clinical practice and is usually associated with inferior wall myocardial infarction; therefore, one should look at leads II, III, and aVF (i.e., for right coronary artery occlusion). Mirror image dextrocardia and wrong lead connections (so-called limb lead reversal) are usually forgotten causes of prominent R wave in lead V_1. Rarely, it may be a normal variant.

2) d.

Early repolarization after an episode of angina is associated with ST-segment elevation while the other stems result in ST segment depression. Causes of ST-segment elevation:

1- Full thickness myocardial infarction (ST-segment elevation myocardial infarction or STEMI).
2- Early repolarization after an attack of angina.
3- Acute pericarditis and acute myocarditis.
4- Ventricular aneurysm.
5- Transiently during cardiac and coronary angiography.
6- During Prinzmetal angina or acute Takotsubo cardiomyopathy.
7- Left bundle branch block (in leads V_{1-3} only).
8- Myocardial tumors and traumas.
9- Hyperkalemia (only leads V_{1-2}).
10- Brugada syndrome (in leads V_{1-3} with right bundle branch block pattern).

NB: Takotsubo cardiomyopathy is characterized by transient apical left ventricular dysfunction which mimics myocardial infarction, but in the absence of significant coronary artery disease.

3) e.

Ventricular arrhythmias in Wolff-Parkinson-White (WPW) syndrome are highly atypical and suggest either an alternative diagnosis or a co-existent pathology. The commonest tachycardia is orthodromic (uncommonly antidromic) AV nodal re-entrant tachycardia. A special concern is the development of atrial fibrillation (AF), which may degenerate into ventricular fibrillation. Atrial fibrillation occurs in 10-30% of WPW syndrome patients.

Spontaneous AF is more common in patients with anterograde conduction through the accessory pathway, and is uncommon in the rare patients with WPW who have only concealed retrograde conduction through the accessory pathway. Patients with accessory pathways that have short antegrade refractory periods are more liable to develop AF. AF originates within the atria, independent of the accessory pathway, but the accessory pathway functions as another route for the conduction of atrial impulses into the ventricles.

4) e.

Left-sided cardiac lesions are not part of Noonan's syndrome and their presence suggests an acquired defect, e.g., mitral stenosis following rheumatic fever. The pulmonic valve is typically dysplastic in Noonan's syndrome.

Noonan syndrome is a relatively common autosomal dominant disorder, with an estimated incidence of one in 1000 to 2500 live births. It is characterized by dysmorphic features, proportionate short stature, and heart disease, most commonly pulmonic stenosis and hypertrophic cardiomyopathy. In addition, frequently associated with this syndrome are webbed neck, chest deformity, cryptorchidism, mental retardation, and bleeding diatheses. More than 50% of cases are due to a mutation in the *PTPN11* gene on chromosome 12, which encodes the non-receptor protein tyrosine phosphatase SHP-2.

5) d.

The long QT syndrome (LQTS) is a disorder of myocardial *repolarization* that is characterized by a prolonged QT interval on the electrocardiogram. This syndrome is associated with an increased risk of a characteristic the life-threatening cardiac arrhythmia, torsade de pointes. The primary symptoms in patients with LQTS include palpitations, syncope, seizures, and sudden cardiac death. It may be caused by:

1- Inherited syndromes (congenital long QT syndromes): Jervell-Lange-Nielsen syndrome (which is autosomal recessive and is associated with sensori-neural deafness; its course is highly malignant) and Romano-Ward syndrome (which is autosomal dominant and is a pure cardiac phenotype; its course is generally benign when compared with Jervell-Lange-Nielsen). Most cases result from mutations in various components of the cardiac *potassium* channels causing delayed phase 3 of cardiac action potential; only LQT3 subtype results from sodium channel mutation causing prolongation of the phase 2 of cardiac action potential.

2- Electrolytes disturbance: hypokalemia, hypomagnesemia, and hypocalcemia.
3- Mitral valve prolapse.
4-Rheumatic carditis
5- Drugs and medications (class Ic and III anti-arrhythmic medications).
6- Bradycardia-associated (any cause of bradycardia): this is the rationale behind using isoprenaline infusion to treat such cases. However, this is contraindicated in congenital syndromes because of their already increased sympathetic drive.

6) c.

Hypertrophic cardiomyopathy results from sarcomeric contractile proteins gene mutations. Many types of mutations have been detected, and certain gene mutations predict a poor prognosis. About 50% of cases have a positive family history (25% of idiopathic dilated cardiomyopathy patients display a positive family history). The presence of left ventricular outflow obstructive element does not predict a gloomy outcome; this form of dynamic obstruction occurs in about $1/3^{rd}$ of cases. Many cases are totally asymptomatic and are detected by doing echocardiography for some reason or another. On the other extreme, some patients present with sudden death. The asymmetric septal type is the commonest one, but the apical variety is the predominant one in the Far East. There are certain variants that do not have cardiac hypertrophy, at all. The commonest mutations occur in:

1- Beta myosin heavy chain gene: this is associated with elaborate ventricular hypertrophy.
2- Troponin gene: there is little or no ventricular hypertrophy, but patients demonstrate abnormal vascular responses (mainly hypotension) upon doing exercise, and there is a high risk of sudden death.
3- Myosin -binding protein C gene: usually manifests later in life, and is often associated with prominent cardiac dysrhythmia and systemic hypertension.

There is no correlation between the degree of left ventricular (LV) outflow obstruction and symptoms. Some patients with severe LV outflow tract obstruction remain asymptomatic for many years; at the other extreme, cardiac arrest or sudden death may be the initial presentation. However, many patients with hypertrophic cardiomyopathy develop one or more of the following symptoms: dyspnea on exertion, orthopnea and paroxysmal nocturnal dyspnea, chest pain, presyncope and syncope, palpitation, postural lightheadedness, fatigue, and bipedal pitting edema (which is actually rare).

These symptoms can be induced by a variety of mechanisms which may include: LV outflow tract obstruction at rest, LV outflow tract obstruction that is present only with provocation (such as exertion or straining), impaired myocardial function in the absence of obstruction, arrhythmias (or conduction delay), and impaired filling due to diastolic dysfunction.

7) b.

The commonest type of pericarditis is the idiopathic variety. In patients with acute pericarditis in whom no cause is identified (so- called idiopathic pericarditis), the etiology is frequently presumed to be viral or autoimmune. Autoimmune factors may be particularly important in patients with recurrent acute pericarditis. The presence of pericarditis should be suspected in the following clinical settings: persistent high fever in patients with pericardial effusion; unexplained new radiographic cardiomegaly; unexplained hemodynamic deterioration after myocardial infarction, cardiac surgery, or cardiac diagnostic or interventional procedures. The disease rarely progresses to tamponade formation, and most of cases follow a benign course. Uremic pericarditis has a risk of hemorrhagic transformation. The occurrence pericarditis in dialysis patients indicates an inadequate dialysis regimen; it is an indication to do dialysis in end-stage renal disease patients (who are not receiving any form of renal replacement therapy). The resulting rub is usually transient and fluctuating. Generally, disappearance of the rub reflects a transient phenomenon (which is very common in clinical practice), resolution of the inflammatory process, or fluid collection in the pericardial sac (effusion or tamponade). Tuberculous pericarditis usually presents either as progressive collection of large volume of fluid or as a chronic constrictive picture.

8) a.

Involvement of the right coronary ostium in type A aortic dissection results in right coronary artery occlusion and inferior wall myocardial infarction; thrombolysis is contraindicated. Type A dissection may also cause aortic regurgitation and pericardial tamponade. Proliferative diabetic retinopathy, pregnancy, and easily controllable hypertension are a *relative* contraindication for r-tPA administration. Stem "e" strongly calls for using r-tPA.

9) a.

Hydralazine and minoxidil have a pure *arteriolar* dilating effect. The other stems have combined arteriolar and vanodilating effects.

10) e.

High risk unstable angina is reflected by the presence of *any* one of the following:

1- Post-infarct angina.
2- Recurrent chest pain at rest.
3- Development of heart failure (clinical and/or by echocardiographic detection).
4- Occurrence of cardiac dysrhythmias.
5- Presence of transient ST-segment elevation.
6- Presence of ST-segment depression.
7- Persistence of deep T-wave inversion.
8- Serum cardiac troponin level greater than 0.1 µg/L.

11) c.

Stems "a" and "b" are true and mirror the predominance of the right coronary arterial (RCA) system in supplying the upper part of the conduction system; therefore, RCA occlusion results in AV block in inferior wall myocardial infarction. Stem "c" is false; percutaneous coronary dilatation of the main stem of the left coronary artery should only be done by highly experienced operators; otherwise, coronary grafting surgery should be contemplating.

12) d.

Amiodarone has multiple effects on myocardial depolarization and repolarization that make it an extremely effective anti-arrhythmic drug. Its primary effect is to block the *potassium* channels, but it can also block sodium and calcium channels and the beta and alpha adrenergic receptors. The approved clinical use of amiodarone has been limited to refractory ventricular arrhythmias because of relatively high incidence of side effects that can range in severity from mild to potentially lethal. The corneal deposits are usually reversible. The other well-recognized side effects are pulmonary fibrosis, slate-grey skin pigmentation, and hepatotoxicity. Amiodarone potentiates the effects of warfarin and digoxin (a common polypharmacy in cardiac disease patients). About 40% of the drug is iodine; this may incite thyroid reactions (hypo- or hyperthyroidism). It is very lipophilic and is concentrated in adipose tissue, cardiac and skeletal muscle, and the thyroid. Elimination from the body occurs with a half-life of about 120 days. Amiodarone toxicity can therefore occur well after drug withdrawal.

13) d.

Fixed non-pulsatile elevation in JVP suggests superior vena cava obstruction, which may result from mediastinal masses with secondary compression. Mediastinal irradiation carries a risk of inciting chronic pericardial inflammation with constriction (the JVP is elevated with steep x and y descents). The ST-segment elevation in the right chest leads (e.g., the V4R) reflects right ventricular infarction (resulting in hypotension, raised JVP, and clear lung bases). Atrialization of the right ventricle occurs in Ebstein anomaly (with tricuspid regurgitation resulting in raised JVP and prominent v wave). The effective circulating blood volume is contracted in cirrhosis patients and this would result in clinical hypovolemia and low JVP.

14) c.

Coarctation usually presents as hypertension in adults (or heart failure in infancy). The mechanical obstruction to the blood flow is largely responsible for the elevation of blood pressure in the upper extremities. In addition, renal hypoperfusion may lead to enhanced renin secretion and subsequent volume expansion. Volume expansion produces a further elevation in blood pressure restoring renal perfusion and renin secretion toward normal. Girls with Turner's syndrome are at increased risk with a prevalence rate of greater than 10%. Bicuspid aortic valve is a very common association; it may be found in 50% of cases. The classical rib notching is usually obvious after the age of 6 years (from the 3 rd rib down to the 9 th one). Coarctation patients are at risk of aortic dissection; this should always be looked for in the presence of severe tearing chest pain. Hemorrhagic stroke in coarctation patients may result from hypertensive (spontaneous) intra-cerebral hemorrhage or subarachnoid bleed from ruptured cerebral (Berry) aneurysm (this usually happens between the ages of 10-30 years). In addition to the increased frequency of intracranial aneurysms, dilated collateral arteries within the spinal canal may accompany coarctation. These vessels can compress the spinal cord or can rupture, causing a clinical picture of subarachnoid hemorrhage. Infective endocarditis is rare in coarctation, and the risk is virtually eliminated by successful surgical repair. However, associated cardiac abnormalities, especially bicuspid aortic valve, carry independent risks. Both lesions are considered to have an intermediate risk of infective endocarditis.

15) e.

Apart from atenolol, the other drugs that improve survival figure in chronic congestive heart failure are ACE inhibitors. Diuretics (apart from spironolactone) and digoxin do not improve the survival figure in chronic congestive heart failure. Digoxin does, however, reduce hospitalization rates.

16) d.

Blood culture-negative infective endocarditis (IE) is defined as endocarditis without etiology following inoculation of at least three independent blood samples in a standard blood culture system with negative cultures after seven days of incubation and subculturing. Cultures remain negative in 2 -7% of patients with IE even when the utmost care is taken in obtaining the proper number and volume of blood cultures and patients with prior antibiotic treatment are excluded; the frequency is higher in patients who have already been treated with antibiotics. Causes of culture-negative endocarditis:
1- Prior antibiotic treatment (the commonest cause).
2- Infection with a fastidious organism (e.g., HACEK group) or an organism that is difficult to be cultured (e.g., Brucella species).
3- Fungal endocarditis and Q fever.
4- Marantic and Libman-Sacks endocarditis (these are non-infective causes).
5- Poor lab techniques and errors in collecting the blood samples (Staph. epidermidis colonies are frequently reported by the lab as a normal skin commensal).

17) d.

The following states have *high* risk of developing infective endocarditis (IE) and need antimicrobial prophylaxis when indicated: prosthetic heart valves, (including bioprosthetic and homograft valves); a prior history of IE; complex cyanotic congenital heart diseases (such as single ventricle states, transposition of the great arteries, and tetralogy of Fallot); surgically constructed systemic or pulmonary conduits.

Moderate risk conditions encompass the following:
1- Most other congenital cardiac malformations (not listed above), particularly bicuspid aortic valves (with the exception of isolated secundum atrial septal defects and six months or more after surgically repaired atrial septal defect, ventricular septal defect, or patent ductus arteriosus).

2- Acquired valvular dysfunction (e.g., aortic or mitral stenosis or regurgitation) and patients who have undergone valve repair.

3- Hypertrophic cardiomyopathy with latent or resting obstruction

4- Mitral valve prolapse with valvular regurgitation on auscultation and/or thickened leaflets on echocardiography

5- Intracardiac defects that have been repaired within the preceding six months or that are associated with significant hemodynamic instability.

The following conditions are associated with a *low* risk of IE. The value of antimicrobial prophylaxis prior to dental and surgical procedures in these settings is generally considered to be negligible and endocarditis prophylaxis is not recommended:

1- Physiologic, functional, or innocent heart murmurs.

2- Isolated secundum atrial septal defect 6 or more months after successful surgical or percutaneous repair of an atrial septal defect, ventricular septal defect, or patent ductus arteriosus.

3- Mitral valve prolapse without mitral regurgitation or leaflet thickening on echocardiography, although additional clinical judgment may be required.

4- Aortic valve sclerosis, defined by focal areas of increased echogenicity and thickening of the valve leaflets without any restriction of motion and with a peak velocity less than 2.0 m/sec.

5- Physiologic mitral regurgitation on echocardiography with structurally normal valves and no murmur.

6- Mild or hemodynamically insignificant tricuspid regurgitation.

7- Coronary artery disease (including previous coronary artery bypass graft surgery).

8- Previous rheumatic fever or Kawasaki disease without valvular dysfunction.

9- Cardiac pacemakers (intravascular and epicardial) and implanted defibrillators.

10- Coronary artery stent implantation.

18) c.

Generally, we should *guess* the "best treatment" for infective endocarditis based on the patient's risk factors and history, e.g., think of Staphylococci or fungi as causes of tricuspid valve endocarditis in intravenous drug addicts. The occurrence of persistent fever despite giving optimal medical therapy may indicate failure of medical treatment and persistent infection (which is an indication for surgery). The other causes of persistent fever in those patients are:

1- Resistant organism, e.g. Staph aureus to benzyl penicillin.
2- A non-bacterial cause, such as fungal organisms.
3- Cannula-associated superficial thrombophlebitis.
4- Drug fever.
5- Coexistent pathology, e.g., wound infection, bed sore, or a visceral abscess that is difficult to be treated (however, this abscess might actually be the cause of the endocarditis).

19) d.

In general, a 3rd heart sound indicates rapid early ventricular filling and/or high blood flow across the mitral valve (or tricuspid valve in right sided 3rd heart sound) *or* myocardial systolic dysfunction. Isolated mitral stenosis results in a sluggish blood flow across the diseased valve and does not impair the left ventricular myocardial function. The presence of a 3rd heart sound in mitral stenosis patients reflects an associated significant mitral regurgitation (and the latter is the predominant hemodynamic lesion) or a coexistent left ventricular dysfunction from other causes.

20) c.

Aortic stenosis (AS) is supra-valvular in William's syndrome (with mental retardation, hypercalcemia, and elfin faces). Sub-valvular aortic stenosis can result from a variety of fixed lesions. These include a thin membrane (the most common lesion), thick fibromuscular ridge, diffuse tunnel-like obstruction, abnormal mitral valve attachments, and occasionally, accessory endocardial cushion tissue. There are at least two anatomic forms of supra -valvular AS. The majority of patients (60 to 75%) have an hourglass deformity consisting of a discrete constriction of a thickened ascending aorta at the superior aspect of the sinuses of Valsalva (the hourglass deformity). More diffuse narrowing for a variable distance along the ascending aorta is seen in the remaining patients. Supra-valvular AS is also common in patients with homozygous familial hypercholesterolemia and occurs infrequently in heterozygote individuals. A thrill is commonly felt at the neck (carotid shudder). The presence of symptoms should call for surgical intervention regardless of the severity of stenosis. The cause of the stenosis is influenced by the age of the patient; young patients display a congenital or bicuspid etiology, while the calcific degenerative AS is seen in elderly patients. Aortic stenosis is a risk factor for infective endocarditis (before and after valve replacement).

21) e.

The absence of leg edema after prescribing diuretics as a mode of treatment in congestive heart failure indicates over-diuresis; hypovolemia results and pre-renal failure ensues rapidly.

22) b.

The objectives of treating diastolic heart failure differ from those of congestive heart failure states with systolic dysfunction. The objective of treatment of diastolic dysfunction, in general, are:

1- Control heart rate: this would give adequate time for the ventricles to fill properly; therefore, tachycardia has a deleterious effect.
2- Control edemas: using diuretics; this would resolve pulmonary congestion and improve exertional breathlessness.
3- Control hypertension.
4- Identify any underlying cause and remove it (if possible): e.g., aortic stenosis (aortic stenosis, not regurgitation, is a cause of diastolic dysfunction).

23) e.

Secondary hypertension may complicate long-standing essential hypertension, e.g., the development of atherosclerotic renal artery stenosis. Renal artery stenosis is both, a cause of hypertension and a result of long-standing hypertension; the same applies to chronic renal failure. Strong family history of hypertension is usually indicative of the essential variety rather than secondary causes.

24) b.

Poorly controlled hypertension is a risk factor for intracerebral hemorrhage. Diuretics and non-dihydropyridine calcium channel blockers are the first-line agents in elderly people with essential hypertension. The target blood pressure in diabetic patients *with* nephropathy and persistent proteinuria is <130/80 mmHg.

25) e.

Stem e is a protective one. The other risk factors are smoking, hypercholesterolemia, and positive family history of ischemic heart disease. Diabetes mellitus is considered a coronary artery disease equivalent when assessing coronary artery risks, i.e., the target LDL-cholesterol should be <100 mg/dl.

26) e.

Isolated systolic heart failure is highly unusual in long-standing hypertension; hypertension results in diastolic heart failure or combined diastolic/systolic heart failure. Hypertension is not a risk factor for infective endocarditis.

27) c.

Recurrence of cardiac myxoma following successful surgical excision usually occurs in familial cases. The commonest site is the fossa ovalis at the left side of the inter -atrial septum (70%). Atypical sites of growth (especially in the right side of the heart) usually indicate familial predisposition. Cardiac myxoma may present with fever, malaise, weight loss, anemia, raised ESR and C-reactive protein (malignancy and vasculitis are the differential diagnoses), and cardiac murmur with embolic phenomena (differentiating cardiac myxoma from infective endocarditis may be impossible without doing echocardiography). Cardiac myxoma is the commonest primary cardiac tumor, which mainly occurs in women. The Carney complex is an inherited, autosomal dominant disorder characterized by multiple tumors, including atrial and extra-cardiac myxomas, schwannomas, and various endocrine tumors; the cardiac myxomas generally are diagnosed at an earlier age than sporadic myxomas and have a higher tendency to recur.

28) a.

Metoprolol, acebutolol, and atenolol are cardio-selective beta blockers. Selectivity (or cardioselectivity) refers to the ability of a drug to preferentially block the beta-1 receptors. Certain beta blockers (such as pindolol and acebutolol) possess partial agonist activity (also called intrinsic sympathomimetic activity or ISA). These drugs cause a slight to moderate activation of the beta receptor even as they prevent access of natural and synthetic catecholamines to the receptor sites. The result is a weak stimulation of the receptor.

Mild to moderate ISA in a beta blocker does not interfere with efficacy. They may, however, cause less reduction in heart rate, less depression of atrioventricular conduction, and less negative inotropy than beta blockers without this property. Labetalol and carvedilol have an additional alpha adrenoceptor blocking activity.

29) e.

The JVP is a "window" through which we may see what is happening in the heart. In heart block, cannon "a" waves tend to be irregular (as the atria contract against the closed tricuspid valve at irregular intervals); if these waves are regular, think of cardiac nodal rhythms, slow ventricular tachycardia, 2:1 atrioventricular block, and bigeminy. Regular cannon waves may occur in 1st degree atrioventricular block with a markedly prolonged PR interval and atrial systole occurring during the preceding ventricular systole.

30) d.

Short PR-interval occurs in:
1- Any tachycardia state (theophylline may cause tachycardia; exercise, anxiety, hyperthyroidism,...etc.).
2- Congenital short PR interval (WPW and Long-Ganong-Levine syndromes).
3- Ventricular ectopic occurring immediately after a sinus "p" wave.
4- Nodal rhythms.

31) b.

WPW syndrome has many diverse and atypical ECG manifestations, e.g., pseudo-myocardial infarction pattern and pseudo -LVH pattern. Posterior wall myocardial infarction has a prominent R wave in lead V₁. The other causes of pathological Q wave are hypertrophic cardiomyopathy and errors in the leading and calibration of ECG machine.

32) a.

Ebstein anomaly is not commonly associated with aortic coarctation; it may be associated with maternal lithium ingestion during pregnancy.

33) e.

Morphine in the acute myocardial infarction setting should be given intravenously, as there is a risk of hematoma formation if the subcutaneous route is used (these patients would be on a thrombolytic therapy or heparin). Aspirin per se enhances the effect of the thrombolytic therapy and improves the mortality figure.

34) e.

Digoxin has no place in the secondary prophylactic programs of myocardial infarctions (unless there is a clear-cut indication for its use, e.g., atrial fibrillation). Aspirin (or other anti-platelets), ACE inhibitors, beta-blockers, and statins are the required medications to achieve secondary prophylaxis against myocardial infarction.

35) e.

Stem "e" may interfere with the performance and interpretation of the test but it is not a contraindication to do exercise ECG testing; poor functional performance during this mode of testing may reflect poor cardiac function, however. There are many contraindications to do exercise ECG testing, and these are:

1- Any significant left ventricular outflow obstruction, e.g., severe aortic stenosis and hypertrophic cardiomyopathy.
2- Suspected or documented left coronary artery main stem stenosis.
3- Untreated congestive heart failure.
4- Poorly controlled severe systemic hypertension.
5- Suspected aortic dissection.
6- During the course of acute coronary syndromes, e.g. unstable angina.
7- Acquired complete heart block.
8- Any febrile illness in general.
9- Acute pericarditis or myocarditis.

36) e.

The presence of abnormal exercise ECG in stem "e" patient most likely represents false positive result rather than being true ischemic changes.
This false positive result would misdirect the physician to proceed to coronary interventions.

In addition to the other stems, the development of ECG changes at an early stage of Bruce protocol (i.e., low threshold for ischemia) indicates a positive test and a high risk for subsequent coronary events. Exercise ECG testing can confirm the diagnosis of ischemic heart disease, identify high risk patients, and may predict the prognosis.

37) d.

Stem "d" indicates a strong positive exercise ECG testing (a high-risk for subsequent coronary events); coronary angiography is indicated. The other causes of false positive results are electrolyte disturbances (e.g., hypokalemia), anemia, beta-blockers therapy, left bundle branch block, and hyperventilation.

38) e.

This is a pulseless electrical activity (PEA), which was previously termed electromechanical dissociation. Causes of PEA are hypoxia, hypokalemia/hyperkalemia, hypovolemia, hypothermia, tension pneumothorax, cardiac tamponade, toxic/therapeutic disturbances, and thromboembolic/mechanical obstruction.

39) e.

ACE inhibitors in stem "e" have been shown to be effective and of great benefit. Do not be afraid of the renal effect of ACE inhibitors in this situation; in fact, they are the drug of choice in uremia with hypertension as well as post-myocardial infarction secondary prophylaxis.

40) a.

Thiazides are the first line agents in hypertension in elderly population. Loop diuretics are preferred in those with coexistent heart failure. Long-term complications of diuretic use are acid/base and electrolyte disturbances (hypochloremic alkalosis, hypokalemia, hyponatremia), hyperlipidemia (may worsen lipid profile), hyperuricemia (flare-up of gout may occur), hyperglycemia (diabetic control may become difficult), and photosensitization. Thiazides are contraindicated in patients who are allergic to sulfa (this group includes sulfonamides, dapsone, sulfonylureas, and acetazolamide).
They are contraindicated in severe renal failure (virtually ineffective). Diuretics used in the treatment of hypertension uncommonly produce acid-base and electrolyte disturbances, however.

41) d.

One out of 10 cases of primary pulmonary hypertension is familial and an autosomal dominant mode of inheritance has been suggested. The definitive treatment is heart-lung transplantation; medical treatment is only weakly effective and mainly used as a bridge to surgery. Medical therapy may include prostacyclin infusion, nitrous oxide inhalation (directly to the pulmonary vascular bed), warfarin (as *in situ* thrombosis occurs in some cases), and nifedipine (as a pulmonary vasodilator). Medial hypertrophy and fibrinoid necrosis are detected in all branches of the pulmonary arterial tree and these would result in wide-spread pulmonary vascular obstruction.

42) e.

Cor pulmonale is the dilatation and/or hypertrophy of the right ventricular chamber in response to an increase in the pulmonary vascular resistance that is not necessarily associated with heart failure; a common misbelieve is that cor pulmonale means right-sided heart failure. Raised JVP is the first sign to appear in general, even before leg edema. Post-polio syndrome may weaken respiratory muscles and result in chronic hypoxia and hypercapnia.

43) b.

This COPD patient has features of chronic CO_2 narcosis; those patients have cerebral vasodilatation and raised intracranial pressure (and papilledema). The patient's hands may reveal:
1- Flapping tremor.
2- Cyanosis.
3- Warmness.
4— Rapid bounding pulse.

44) e.

Up to 70-80% of emboli to the lungs come from the venous system of the legs, only 10-15% of them originate from the pelvic veins. Rare causes of lung embolization are amniotic fluid, placenta, air, fat, tumor (especially choriocarcinoma), and parasites (especially schistosomes) as well as septic emboli from right-sided endocarditis.

45) c.

The chest plain X-ray film could be entirely normal-looking (12%), especially in acute massive pulmonary thromboembolism. However, subtle changes of regional oligemia may be seen, or prominence of one of both pulmonary arteries may be noticed. Horizontal atelectasis can be seen in 69% of patients (usually in the lower zones), while non-specific parenchymal abnormalities may occur in 58% of cases. Some degree of pleural effusion (which might be bloody) is detected in 47% of patients. Cardiomegaly is common on the plain posterior-anterior films, especially in massive emboli. Although highly characteristic in the appropriate clinical settings, peripheral wedge-shaped opacities (of pulmonary infarctions) are uncommon. Because of segmental collapse, the hemi-diaphragm may be elevated. Secondary infection of the infracted area can end up with abscess formation.

46) d.

Although $S_1Q_3T_3$ pattern is highly suggestive of acute pulmonary emboli if the clinical features are compatible, but in clinical practice, it is uncommon. T-wave inversion in leads V_1 and V_2 can be very useful instead (reflecting the occurrence of right ventricular strain). The ECG, apart from sinus tachycardia, might be totally normal, however.

47) d.

Heparin therapy substantially reduces mortality by preventing further embolic events and reduces mediator -induced pulmonary vasoconstriction and bronchospasm from thrombin activation and platelet aggregation. Oral anticoagulants do not act immediately.

48) b.

The use of inotropic agents offers little benefit, as the hypoxic dilated right ventricle is already maximally stimulated by endogenous catecholamines. Transcatheter suction embolectomy has been applied as a mode of therapy (with good results when compared with stem "e").

49) d.

The incidence of common cardiac anomalies are tetralogy of Fallot (6%), pulmonic stenosis (7%), aortic stenosis (6%), complete transposition of great vessels (4%).

50) c.

Long-standing VSD can result in loss of support of the right coronary cusp and aortic regurgitation follows. Because of the right ventricular outflow obstruction, Fallot's patients are protected from the future development of Eisenmenger syndrome (their lung is already oligemic). ASD secundum patients usually presents with cardiac arrhythmias in adulthood (mostly atrial fibrillation). ASD primum, one the other hand, because of the associated mitral and/or tricuspid regurgitation, tends to present much earlier. Cardiac cyanosis at birth is rarely due to tetralogy of Fallot (transposition of great vessels is the usual culprit).

51) e.

Lutembacher syndrome encompasses ASD secundum and tri-phalangeal thumb. Lithium ingestion in pregnancy has been linked to the 3% incidence of Ebstein anomaly in the fetus. Maternal rubella infection may result in fetal ASD or patent ductus arteriosus. Valproate sodium can cause double aortic arch or interrupted aortic arch. Left-sided cardiac lesions are incompatible with Noonan's syndrome; on the contrary, right-sided cardiac lesions are very common, e.g., pulmonic stenosis.

52) e.

Argyll-Robertson pupil in the context of aortic regurgitation may suggest syphilitic aortitis. About 4% of ankylosing spondylitis patients develops aortic regurgitation (after a median of 15 years). The plucked-chicken skin in both cubital fossae of pseudoxanthoma elasticum is a good clue as an underlying cause of aortic reflux. Neither Turner syndrome nor Noonan syndrome result in isolated aortic reflux. The cloudy cornea of mucopolysaccharidosis may be seen with the aid of slit-lamp; aortic regurgitation is common in those children.

53) c.

Secondary pulmonary hypertension and tricuspid regurgitation may occur; giant *v* waves of the latter lesion may be noticed on the JVP examination. Sudden deterioration due to atrial fibrillation (which does not respond to medical treatment) is another indication of surgical intervention, as well as limitation of physical activity despite optimal medical treatment. A trans-mitral diastolic gradient of 15 mmHg reflects severe stenosis.

54) d.

ECG findings of ASD secundum patients usually reveal right axis deviation (and right bundle branch block, which may be partial), while primum defects impart left axis deviation (the PR interval may be prolonged and there is right bundle branch block). The volume overload of VSD produces left ventricular hypertrophy and ends up gradually with bi-ventricular hypertrophy. In Ebstein anomaly, there are tall peaked P waves in standard lead II, right bundle branch block with small amplitude QRS complexes, WPW syndrome type B (i.e., the QRS complex is negative in the right pre-cordial leads), and paroxysmal supra-ventricular tachycardia.

55) d.

Many changes can be seen on the chest X-ray films several years after birth in patients with congenital heart disease; by this time, we may notice cardiomegaly, enlarged pulmonary artery,...etc. Boot-shaped heart is usually suggestive of tetralogy of Fallot. The reverse sign of 3 (or the sign of E) is seen on lateral barium swallow films, and rib notching is seen from the 3rd to the 9th rib by the age of 6 years in coarctation patients; we may also notice left ventricular hypertrophy and cardiomegaly. Dilatation of the proximal pulmonary arteries with peripheral pruning of the pulmonary vasculature indicates severe pulmonary hypertension, e.g., in Eisenmenger syndrome. The ductus arteriosus can be delineated upon doing angiography; it cannot be simply seen on the plain chest film.

56) e.

Idiopathic pulmonary fibrosis-like picture may occur in patients on long-term amiodarone therapy. There are several different types of pulmonary toxicity that are causes by amiodarone.

Chronic interstitial pneumonitis is the most common; the other manifestations include organizing pneumonia, acute respiratory distress syndrome, and a solitary pulmonary mass. A nonproductive cough and dyspnea are present in 50 to 75% of affected individuals at presentation. Pleuritic pain, weight loss, fever (33 to 50% of cases), and malaise can also occur. The physical examination often reveals bilateral inspiratory crackles, while clubbing is *not* seen. The slate-grey skin pigmentation and photosensitization may occur. This bluish-grey skin pigmentation occurs in 1-3% of patients on chronic amiodarone therapy and appears to be due to the deposition of lipofuscin in the dermis. There may be a tissue threshold for amiodarone in individual patients above which skin discoloration appears and below which it fades. Thus, patients disturbed by skin pigmentation who are taking large doses (>400 mg/day) may notice improvement in skin discoloration by reducing the dose. Leukopenia may occur. Ocular complications are corneal deposits, lens opacities, and optic neuropathy. The presence of corneal microdeposits is not considered a contraindication to the use of amiodarone therapy, since visual acuity is rarely affected. In any patient with visual symptoms who is taking amiodarone, other common factors (a change in refractive correction, progression of age-related cataract, or increased intraocular pressure) should be considered before attributing the change to the drug. Two mechanisms are responsible for thyroid dysfunction because of amiodarone:

Intrinsic drug effects: amiodarone inhibits outer ring 5'-monodeiodination of T_4, thus decreasing T_3 production; reverse T_3 accumulates since it is not metabolized to T_2. Amiodarone, and particularly the metabolite desethylamiodarone, blocks T_3-receptor binding to nuclear receptors and decreases expression of some thyroid hormone-related genes. Amiodarone may have a direct toxic effect on thyroid follicular cells, which results in a destructive thyroiditis.

Effects due to iodine: iodine is a substrate for thyroid hormone synthesis. It is actively transported into thyroid follicular cells and organified onto tyrosyl residues in thyroglobulin.

The normal autoregulation of iodine prevents normal individuals from becoming hyperthyroid after exposure to an iodine load (e.g., radiocontrast). When intrathyroidal iodine concentrations reach a critical high level, iodine transport and thyroid hormone synthesis are transiently inhibited until intrathyroidal iodine stores return to normal levels (the Wolff-Chaikoff effect).

NB: a transient rise in serum aminotransferase concentrations occurs in approximately 25% (range 15 to 50 percent) of patients soon after amiodarone is begun. The patients are usually asymptomatic, but the drug should be discontinued if there is more than a two-fold elevation. Symptomatic hepatitis occurs in less than 3 percent of patients; potential complications include cirrhosis and hepatic failure. As a result, it is recommended that liver function tests be monitored at baseline and every six months.

57) e.

Fourth heart sounds (S₄) does not occur in mitral regurgitation. Mitral mid-diastolic flow murmur is another sign of severity. S₄ occurs after the P wave on the electrocardiogram and coincides with atrial systole and "*a*" waves of the atrial pressure pulse, and with the apical impulse, while third heart sound (S₃) occurs as passive ventricular filling begins after actual relaxation is completed. It appears to be related to a sudden limitation of the movement during ventricular filling along its long axis. It coincides with the *y* descent of the atrial pressure pulse and the end of the rapid filling phase of the apical impulse, occurring usually 0.14 to 0.16 second after the second heart sound.

58) c.

Lone atrial fibrillation (AF) was initially defined in the era before modern echocardiography to identify a cohort of young patients with AF who had no clinical evidence of cardiovascular disease and were at low risk for thromboembolism. Patients could have paroxysmal, persistent, or chronic AF. With the availability of echocardiography, the definition includes no echocardiographic evidence of cardiac or pulmonary disease. In some risk models of patients with AF, the presence of diabetes is considered a risk factor for thromboembolism. By conventional definitions, such patients could still have lone AF. Lone AF has generally been applied to patients under 60 years of age. However, some studies have included those up to the age of 65 years or even older who appear to be at increased thromboembolic risk compared to younger patients. Lone atrial fibrillation can be treated with aspirin alone; no need for acute or chronic anticoagulation, as the risk of thromboembolic phenomena is very low.

59) e.

The use of diazepam as a "sedative" should be discouraged in acute myocardial infarction patients, as it may cause confusion.

60) c.

The overall picture is suggestive of bradycardia-associated torsades de pointes ventricular tachycardia. The rapid feeble pulse is against complete heart block, and there is no suggestion of an acute ischemic event.

Chapter 2
Clinical Hematology and Oncology

Questions

1) A 13-year-old boy is brought by his parents to see you because of their child's poor school records. He is pale and has spoon-shaped nails. Blood film reveals hypochromic microcytic anemia. There are low serum ferritin and transferrin saturation of 11%. Differential diagnosis of iron deficiency anemia includes all of the following, *except*?
   a) Aluminum toxicity
   b) Beta thalassemia syndromes
   c) Alpha thalassemia syndromes
   d) Lead poisoning
   e) Aplastic anemia

2) A 44-year-old man presents with progressive pallor. He is malnourished. He denies upper or lower gastrointestinal bleeding, and he is on no medications including illicit drugs. The MCV is 105 fL. The blood film shows macrocytosis, but there are no megaloblastic changes. Macrocytosis without megaloblastic bone marrow changes may result from all of the following, *except*?
   a) Myelodysplastic syndromes
   b) Aplastic anemia
   c) Multiple myeloma
   d) Chronic myeloid leukemia
   e) Fish tapeworm infestation

3) A 33-year-old woman complains of undue fatigability upon exertion. She was reasonably well and healthy and there is an unremarkable past medical history. She neither drinks nor smokes, and she denies drug ingestion. You detect an MVC of 103 fL. Which one of the following does *not* cause macrocytosis *with* normal bone marrow histology?
   a) Liver disease
   b) Neonatal
   c) Pregnancy
   d) Post-hemorrhage
   e) Myelomatosis

4) You are reading a blood film report of one of your patients. The report states that both microcytic and macrocytic RBCs are present. This dimorphic blood picture is not found in?
   a) A patient with long-standing iron efficiency anemia who has received blood transfusion because of symptomatic coronary artery disease
   b) A patient with long-standing iron deficiency anemia who has been receiving iron therapy since 4 weeks
   c) Celiac disease patient who is non-compliant with gluten-free diet

    d)   Bone marrow study of a patient showing characteristic rings in the maturing erythroid series

    e)   A patient with long-term epilepsy who receives phenytoin

5) A 6-year-old African boy presents with severe right lower tibial pain for 1 day. He has frontal bossing and jaundice. The spleen is not palpable. His mother says that he developed a stroke 2 years ago, and his older brother is suffering from the same condition. His hemoglobin is 8 g/dl and there are many sickle cells on the peripheral blood film. HbS level is 85%. Regarding sickle cell anemia, all of the following are true, *except?*

    a)   Despite the degree of anemia in the chronic stable state, it is rarely symptomatic

    b)   In contrast to beta thalassemia major, blood transfusion is rarely indicated

    c)   The objective of blood transfusion is to raise the hemoglobin A level to 85-90% of the total hemoglobins and the PCV should be between 30-36%

    d)   Pulmonary infarction-chest syndrome is the commonest cause of death in adults

    e)   Stroke in general afflicts 70% of adults

6) An 8-month-old infant is brought by his mother to consult you about her son's progressive pallor and yellow sclerae. She states that his feeding is well, but one of his cousins has a blood disease. His hemoglobin is 6 g/dl, serum indirect bilirubin is 3 mg/dl, and serum LDH is raised. Hemoglobin electrophoresis reveals HbF of 90%. Regarding beta thalassemia, which one of the following is the *incorrect* statement?

    a)   Hb A$_2$ may be normal or even low in beta thalassemia minor

    b)   Regular blood transfusion is the mainstay of treatment of beta thalassemia major

    c)   Beta thalassemia minor usually comes into light after failure of oral iron therapy for mild hypochromic anemia

    d)   Heart failure in beta thalassemia major results only from iron overload

    e)   Target cells are seen on peripheral blood films

7) A 47-year-old man is brought to the Emergency Room by the ambulance. The paramedics say that he is too pale and his urine is very dark. His hemoglobin is 4 g/dl. You find hemoglobinemia, hemoglobinuria, high serum LDH, and very low serum haptoglobin.

Your preliminary diagnosis is an intravascular hemolysis and you are trying to discover its underlying cause. Causes of intravascular hemolysis include all of the following, *except*?
   a) Falciparum malaria
   b) Clostridia septicemia
   c) Immediate major ABO incompatibility reaction
   d) Paroxysmal nocturnal hemoglobinuria
   e) Hemolytic disease of the newborn

8) One of your colleagues asks you about the causes and features of microangiopathic hemolytic anemia because one of their patients has hemolysis, skin rash, and renal impairment. Regarding hemolytic uremic syndrome (HUS) and thrombotic thrombocytopenic purpura (TTP), which one is the *wrong* statement?
   a) Neurological manifestations are more prominent in TTP than in HUS
   b) Renal impairment is more prominent in HUS than in TTP
   c) Skin rash is mainly seen in TTP rather than HUS
   d) Prolonged PT and aPTT occur in most cases of HUS and TTP
   e) Vincristine may be used in the treatment

9) As part of routine check-up for a 37-year-old woman, you detect an eosinophil count of 1200 cell/mm^3. She denies symptoms or being on medications. Her stool is free of parasites, and her past medical history is unremarkable. Blood eosinophilia is *not* be caused by?
   a) Psoriasis
   b) Trichinosis
   c) Long-treatment with prednisolone
   d) Hodgkin's disease
   e) Chronic myeloid leukemia

10) You discuss monocytosis with undergraduate medical students because one of your patients has chronic myelomonocytic leukemia. Monocytosis may be induced by all of the following conditions, *except*?
   a) Bone marrow recovery after chemotherapy
   b) Sarcoidosis
   c) Myelodysplastic syndrome
   d) Kala azar
   e) Hairy cell leukemia

11) A 28-year-old epileptic man presents with painful mouth ulcers and fever. His absolute neutrophil count is 210 cells/mm^3.

You suspect carbamazepine-induced neutropenia. Neutropenia is *not* the result of which one of the following?
a) Racial
b) Associated viral infections
c) Hypersplenism
d) Treatment with carbimazole
e) Leptospirosis

12) A 16-year-old female presents with heavy menses and tiredness for 1 month. Examination reveals cervical and axillary lymphadenopathy, palpable spleen, and wide-spread skin ecchymosis. Her white cell count is 42000 cells/mm^3 and there are 94% blasts. Bone marrow studies confirm the diagnosis of acute lymphoblastic leukemia L2-subtype. Poor prognostic factors in acute lymphoblastic leukemia include all of the following, *except*?
a) Male sex
b) Age less than 1 year or more than 10 years
c) CNS involvement at the time of diagnosis
d) The common ALL type
e) Very high blood leukocyte count at the time of diagnosis

13) A 48-year-old woman presents with fever, malaise, and epistaxis for 3 weeks. Examination reveals multiple mouth ulcers and pallor. Her white cell count is 8550 cells/mm^3 and there are many blasts containing Auer rods. Further bone marrow studies disclose a diagnosis of acute myeloid leukemia (AML) M2-subtype. The following subtypes of AML are correctly matched with their cytogenic abnormality, *except*?
a) M6 subtype- inv 16
b) M3 subtype- t(15,17)
c) M5 subtype- t(8,21)
d) M1 subtype- t(1,15)
e) M2 subtype- t(2,14)

14) A 33-year-old man is referred to you for further management of disseminated intravascular coagulation (DIC). His blood film reveals many blasts .Bone marrow examination is consistent with promyelocytic subtype of acute myeloid leukemia (AML M3-subtype). Regarding AML M3-subtype, all of the following are *correct*, except?
a) The presence of t(15,17) confers a good prognosis
b) The genetic defect lies in RARA gene
c) Treatment with ATRA enhances further differentiation of promyeloblasts

d) It is almost always of a hypo-granular type
e) Evidence of DIC is present in up to 80% of patients at the time of diagnosis

15) A 60-year-old man presents with a 3-month history of fatigue. Examination reveals a large spleen. His blood count discloses a white cell count of 62000 cells/mm^3, and you find full myeloid series at different stages of maturation. Which one of the following with regard to chronic myeloid leukemia is *incorrect*?
a) The absence of Philadelphia chromosome portends a poor prognosis
b) Treatment with imatinib mesylate produces an 80% rate of cytogenetic remission
c) Basophilia is usually the first sign of accelerated crisis
d) Associated vitamin B$_{12}$ deficiency is very common
e) LAP (leukocyte alkaline phosphatase) score is low

16) A 61-year-old man developed recurrent chest infections over the past 5 months. Examination revealed cervical and axillary lymphadenopathy, a huge spleen, and a palpable liver edge of five fingers below the right costal margin. He is pale. His serum immunoglobulins level is low and the white cell count is 51800 cells/mm^3. Serum LDH is high and there is widening of the mediastinum on the chest X-ray film. You consider chronic lymphocytic leukemia. Regarding chronic lymphocytic leukemia, choose the *wrong* statement?
a) Trisomy 12 is the commonest cytogenetic abnormality
b) Smudge cells are seen on peripheral blood film together with absolute lymphocytosis of mature-looking lymphocytes
c) Bullous skin rash may occur
d) Monoclonal band is detected in some patients on doing serum protein electrophoresis
e) Hyperuricemia is common, as is clinical gout

17) A 16-year-old schoolboy presents with pyrexia of unknown origin for 3 months. Examination reveals pallor and asymmetric cervical lymph node enragement of a firm rubbery consistency. There is no hepatosplenomegaly. His hemoglobin is 9.5 g/dl and eosinophilia is found on blood counts. Lymph node biopsy reveals Reed-Sternberg cells. With regard to Hodgkin's disease, which one of the following is *not* true?
a) Serum LDH is raised and is a useful guide to the bulk of the disease
b) Usually there is anemia, neutrophilia, thrombocytosis and, in some, eosinophilia at the time of diagnosis
c) Residual masses after treatment are common

d) Lymphocyte-depleted type usually has poor prognosis
e) There is an association with long-standing CMV infection

18) A 60-year-old woman presents with dehydration and azotemia over a matter of 3 weeks. Her serum calcium is high and you find anemia and panhypogammaglobulinemia. You do bone marrow aspirate and it reveals hypercellularity with 70% plasma cells. There is a prominent M-band on serum protein electrophoresis. Regarding multiple myeloma, all of the following are true, except?
a) The ESR is usually very high
b) 1% of myelomas is non-secretory
c) Human Herpes virus type 8 infection is a recognized association
d) Isotope bone scanning is usually normal
e) Serum alkaline phosphatase is usually raised at the time of diagnosis

19) Your junior house officer asks you about the significance of M-bands on serum protein electrophoresis and whether it indicates myeloma all the time or not. Features that favor monoclonal gammopathy of undetermined significance over multiple myeloma are all of the following, *except*?
a) Low level of the serum paraprotein
b) The paraprotein remains stable over many months of observation period
c) No bone disease
d) No renal disease
e) Hypogammaglobulinemia

20) A 56-year-old male, who has moderately severe COPD, donates 2 pints of blood every 4 month because of secondary polycythemia and a hematocrit of 66%. Secondary polycythemia can result from all of the following, *except*?
a) Adaptation to high altitude
b) Pickwickian syndrome
c) Hemoglobin-M disease
d) Primary hepatocelluar carcinoma
e) Polycythemia rubra vera

21) A 60-year-old man was diagnosed cecal carcinoma after developing iron deficiency anemia (IDA). His platelets count is 600 000/mm^3. You ascribe this thrombocytosis to his long-standing IDA. Causes of reactive thrombocytosis include all of the following, *except*?
a) Post-splenectomy states
b) Aftermath of severe hemorrhage

c) Following major surgical operations
d) Post-partum states
e) Polycythemia rubra vera

22) A 69-year-old male has chronic anemia, for which he has been receiving regular blood transfusion over the past 6 months. His blood film reveals dysplastic changes in all cell lines. Bone marrow study confirms a diagnosis of refractory anemia with ring sideroblasts. Regarding myelodysplastic syndromes (MDS), which one is the *incorrect* statement?
   a) Increasingly seen in elderly, but no age is exempt
   b) Cytopenia is very common and dysplastic changes in the bone marrow are seen
   c) 40% of cases progress to AML
   d) 5q⁻ syndrome has a relatively good prognosis and is mainly observed in woman
   e) The commonest subtype is refractory anemia

23) A 32-year-old female presents with bilateral deep venous thrombosis of the legs. She displays a bad obstetrical history. She takes medications for superior sagittal sinus-associated pseudotumor cerebri. IgG anti-phospholipid antibody titer is positive. Regarding antiphospholipid syndrome, all of the following are true, *except?*
   a) The patient's aPTT is prolonged and does not correct with addition of normal plasma
   b) Addison disease can be a complication
   c) Thrombocytopenia and immune hemolytic anemia may occur
   d) Chorea, migraine, and transverse myelopathy may develop
   e) There is a bleeding diathesis

24) A 15-year-old female has been experiencing heavy menses since menarche. Her older brother and sister have a blood disease. Bleeding time is prolonged but she displays a normal platelets count. The aPTT is 80 seconds (control 28--40 sec). Regarding von Willebrand disease, all of the following are correct, *except?*
   a) The commonest inherited mild bleeding tendency
   b) DDAVP is contraindicated in type II B disease
   c) Bleeding time is a useful screening tool
   d) Some cases are associated with normal amount of von Willebrand factor antigen
   e) The disease always has an autosomal dominant form of inheritance

25) A 5-year-old child presents with recurrent joint swelling which involved the knees, elbows, ankles, and wrists over the past 4 years. No skin bruises are seen. The parents says that no gum or nosebleed occurs and they deny child physical abuse. His deceased uncle has a similar problem. Regarding hemophilia A, choose the *wrong* statement?
   a) $1/3^{rd}$ of cases are kept in the community by fresh new mutations
   b) The PT and the bleeding time are normal
   c) Preferably treated by recombinant human factor VIII concentrate
   d) The prolonged aPTT does correct with the addition of normal plasma
   e) The normal activity of factor VIII is 200-300%

26) A 30-year-old female presents with generalized bleeding tendency for the past 8 weeks. She was reasonably well and healthy, and she took no medications. She denies preceding flu or an upper respiratory tract infection. You examine the patient thoroughly and you run some blood tests. You find a platelets count of $15000/mm^{3.}$ You diagnose idiopathic thrombocytopenic purpura. Regarding idiopathic thrombocytopenic purpura (ITP), all of the following are true *except?*
   a) The presence of splenomegaly may be a clue to chronic lymphocytic leukemia
   b) In adults, it usually has a chronic fluctuating course rather than an acute monophasic illness
   c) About $2/3^{rd}$ of cases respond to steroid therapy initially
   d) Bone marrow study has a limited role in the diagnosis
   e) In adults, the disease is usually preceded by an upper respiratory tract infection

27) A 62-year-old male, who has been diagnosed recently with advanced multiple myeloma, is referred for hemodialysis because of renal insufficiency. Renal impairment in multiple myeloma may result from all of the following, *except?*
   a) Hyperuricemic nephropathy
   b) Hypercalcemia and dehydration
   c) Amyloidosis and light chain deposition
   d) Severe and recurrent urinary tract infections
   e) Treatment with high doses of glucocorticoids

28) A 63-year-old woman developed a low-threshold fracture of the left humerus. Her ESR is 88 mm/hour and serum calcium is 12.6 mg/dl. Bone marrow biopsy reveals multiple myeloma. With respect to multiple myeloma, which one of the following is *not* considered a bad prognostic factor?

a) Severe hypoalbuminemia
b) Plasma cell leukemia
c) Low serum beta 2 microglobulin level
d) Intractable renal failure
e) Hemoglobin <7 g/dl

29) A 58-year-old man developed azotemia 4 weeks ago. His ESR is 91 mm/hour. You also find anemia, panhypogammaglobulinemia, and 24-hour urinary protein of 1.2 g/d. Urine albumin dipstick testing is negative. You consider multiple myeloma. Which one of the following statements about multiple myeloma is not true?
a) High fluid intake is an important part of the management
b) The commonest paraprotein secreted is IgG
c) Recurrent chest infections imply immune paresis rather than bone marrow failure
d) Headache, blurred vision, and epistaxis result from hyperviscosity of IgA paraproteinema
e) Raised serum alkaline phosphatase reflects the severity of the disease at the time of diagnosis

30) A 56-year-old man is referred for further evaluation because of huge spleen and fatigue. One of the differential diagnoses is hairy cell leukemia. With respect to hairy cell leukemia, choose the *incorrect* statement?
a) Severe neutropenia and monocytopenia are common and characteristic
b) It is a rare indolent B-cell lymphoproliferative disorder
c) The malignant B-cells characteristically carry CD25 and CD103 antigen
d) The response to cladribine is favorable and long-term remission is the rule in many cases
e) Lymph node enlargement is very common

31) A 65-year-old man presents with progressive pallor. He denies blood loss. Past medical history includes long-standing hypertension, for which he takes atenolol tablets. Hemoglobin is 7.8 g/dl and MCV is 105 fl. Blood film and bone marrow studies show dysplastic changes but there is no frank malignancy. You diagnose myelodysplastic syndrome. With respect to myelodysplastic syndromes (MDS), all of the following statement are correct, *except?*
a) The medical treatment is still unsatisfactory
b) Chromosomal analysis frequently reveals abnormalities in chromosomes 15 and 17
c) The disease is probably more common than acute leukemia

d) Secondary MDS is mainly seen in patients who have received chemotherapy or radio-therapy
e) Macrocytosis is characteristic

32) A 32-year-old woman, who was reasonably well and healthy until one month ago, she started to develop progressive pallor. She denies heavy menses. Her past medical history is unremarkable and there is no family history of note. Investigations reveal anemia, reticulocytosis, indirect hyperbilirubinemia, and raised serum LDH. Coombs test is positive and a search for a secondary cause of hemolysis fruitless. Acquired immune-hemolytic anemia (AIHA) is characterized by all of the following, *except?*
a) The commonest cause is the idiopathic form
b) The presence of smudge cells on blood film may indicate a secondary etiology
c) Direct Coombs test detects IgE and complements but not IgA antibodies
d) Direct Coombs test requires at least 200 antibodies to be attached to the surface of RBCs to be positive
e) There is an association with HIV infection

33) A 17-year-old male presents with severe colicky abdominal pain for the past few hours. Subsequently, gallstones are found and he undergoes cholecystectomy. The patient is pale and gives a history of recurrent leg ulcers and a positive family history of a blood disease. Blood film reveals spherocytes and polychromasia. With respect to hereditary spherocytosis, which one of the following is the *wrong* statement?
a) The commonest abnormalities are seen in beta spectrins and ankyrins of the red cell membrane
b) Pigment gallstones occurs in up to 50% of cases
c) Recurrent severe crisis is an indication for doing splenectomy
d) About 75% of cases come from fresh new mutations
e) Coombs test is negative

34) A 7-year-old child presents with severe and progressive pallor after an apparently mild upper respiratory tract infection, for which he has received a particular medication. Investigations reveal hemoglobin of 4.7 g/dl, reticulocytosis, raised serum LDH, indirect hyperbilirubinemia, hemoglobinemia, and a very low level of serum haptoglobin. Eight weeks later, direct assessment of the red cell G6PD enzyme shows a very low level. Regarding G6PD deficiency, all of the following statement are true, *except?*

a) The prematurely released reticulocytes may have a normal level of the enzyme glucose-6-phosphotase dehydrogenase
b) Bite cells and blister cells are seen on peripheral blood film during attacks of hemolysis
c) Heinz bodies can be seen only after staining the blood film with supravital stain
d) May present as neonatal jaundice
e) The African type of deficiency is very severe

35) A 61-year-old woman presents with pallor and easy fatigability. She has a history of primary atrophic hypothyroidism, for which she takes daily thyroxin tablets. Her up-to-date serum TSH is within the treatment target. Examination reveals pallor, jaundice, grey hair, and patches of vitiligo. Blood film show macrocytosis and hyper-segmented neutrophils. Serum vitamin B_{12} level is very low. Megaloblastic anemia is characterized by all of the following, *except?*

a) The blood film cannot differentiate between vitamin B_{12} and folic acid deficiencies
b) Neurological manifestations favor vitamin B_{12} over folic acid deficiency
c) Serum folate is less useful than red cell folate in securing the diagnosis of folate deficiency
d) The response to hematinic replacement may precipitate severe hyperkalemia
e) The response to hematinic replacement becomes obvious within minutes

36) Your colleague asks you about the lab feature of red cell hemolysis because he sees many cases every day in the hematology outpatient clinic. Features that point out towards red blood cells hemolysis are all of the following, *except?*

a) Raised serum LDH
b) Indirect hyperbilirubinemia
c) Reticulocytopenia
d) Hemoglobinemia
e) Positive urinary hemosiderin

37) A 58-year-old man presents with exertional shortness of breath. Examination reveals pallor. He states that he notices bloody stool every time he uses the toilet. Proctoscopy detects large hemorrhoids. Blood film shows hypochromic microcytic red blood cells. When assessing iron status in iron deficiency anemia (IDA), certain facts should be kept in mind. Which one of the following is the *incorrect* statement?

a) Serum ferritin is raised in acute phase responses
b) Plasma iron is raised in liver diseases and hemolysis
c) Serum transferrin levels are low in liver disease and acute phase responses
d) Transferrin saturation less than 40% is highly suggestive of iron deficiency anemia
e) Increased serum soluble transferring receptors level is very helpful in the diagnosis of iron deficiency anemia

38) A 41-year-old man is diagnosed aplastic anemia. He receives 3 pints of packed RBCs. He asks if blood transfusion carries any risks and complications. With respect to severe allergic reaction associated with blood transfusions, all of the following statements are correct, *except?*
a) When you notice it, stop the infusion immediately
b) Give O_2
c) Advise future saline-washed blood components, if indicated
d) Return the blood unit intact to the blood bank with any other used or unused units
e) Bronchospasm, angioedema and hypotension are rarely observed

39) A 36-year-old man receives blood transfusion following a car accident which has resulted in a massive blood loss. Within few minutes of the transfusion, he develops fever, rigor, and flank pain. You suspect major ABO incompatibility reaction. Regarding major ABO incompatibility reactions, all of the following statements are correct, *except?*
a) Immediately, stop the infusion, take down the unit and the giving set and return it to the blood bank
b) Commence intravenous saline infusion and measure urine output
c) Inform the hospital transfusion department immediately
d) Treat DIC if present
e) Ensure urine output of at least 10 ml/minute

40) A 24-year-old female visits the physician's office consult him about taking oral contraceptive pills (OCPs). She states that her older brother has a tendency to thrombosis and she heard that OCPs may precipitate such events. Which one of the following is *not* an indication for thrombophilia screening?
a) Leg deep venous thrombosis in an elderly patient following hip surgery
b) Portal vein thrombosis
c) Combined arterial and venous thromboses
d) Recurrent unexplained venous thrombosis

e)   Family history of venous thrombosis

41) A 35-year-old man presents with a 1-week history of recurrent nosebleed and mouth ulcers. Examination reveals fever, wide-spread petechial rash, hemorrhagic bullae in the mouth with multiple ulcers, and pallor. No organomegaly or lymphadenopathy are detected. He denies infectious, drug, or toxic exposure, and there is no family history of note. Blood film shows pancytopenia but there are no abnormal cells. Bone marrow aspirate is difficult and a dry tap is obtained. You intend to do bone marrow biopsy. What is the likely diagnosis?
   a)   Acute myeloid leukemia
   b)   Aplastic anemia
   c)   Acute lymphoblastic leukemia
   d)   Advanced Hodgkin's disease
   e)   Early non-Hodgkin's lymphoma

42) A 54-year-old man visits the Emergency Room. He developed severe non-traumatic epistaxis before 2 hours. Examination shows cervical and axillary lymph node enlargement. The spleen is enlarged 10 cm below the left costal margin and there is a palpable liver edge of 4 cm below the right costal margin. Investigations show white cell count 51000 cells/mm^3, predominantly of mature-looking lymphocytes, and there are some smudged cells. The platelets count is 11000 cells/mm^3. Raised serum LDH is found. What does the man have?
   a)   Chronic lymphocytic leukemia
   b)   Advanced non-Hodgkin's lymphoma
   c)   Aplastic anemia
   d)   Hairy cell leukemia
   e)   Acute myeloid leukemia

43) A 31-year-old woman presents with a 3-day history of bluish-red spots over both legs. She denies other symptoms or being on medications. Examination is unremarkable apart from legs petechiae. Investigations reveal white cells of 5100 cells/mm^3 (but no abnormal cells were seen), hemoglobin of 13 g/dl, and platelets of 14000/mm^3. Bone marrow aspirate shows megakaryocytic hyperplasia. What does the woman have?
   a)   Aplastic anemia
   b)   Idiopathic thrombocytopenic purpura
   c)   SLE
   d)   Chronic myeloid leukemia
   e)   Hemophilia B

44) A 75-year-old man presents a 4-moth history of fatigue and early satiety. Examination reveals an enlarged spleen of 10 cm below the left costal margin. Blood counts show leukocytosis of 42000 cells/mm^3, with full spectrum of granular series. Which one of the following is not a suitable treatment option in this man?
a) Bone marrow transplantation
b) Interferon alpha
c) Imatinib mesylate
d) Hydroxyurea
e) Chlorambucil

45) A 9-year-old girl presents has been experiencing fever and joint pains over the past 1 month. Examination reveals hepatosplenomegaly as well as cervical and axillary lymph node enlargement. Further work-up confirms a diagnosis of acute lymphoblastic leukemia L$_2$-subtype. Which one of the following is *not* part of the overall management plan of this type of leukemia?
a) Induction chemotherapy
b) Bone marrow transplantation
c) Interferon alpha injections
d) Platelets transfusion
e) Hydration and allopurinol

46) A 17-year-old girl has just finished her induction chemotherapy for M$_2$ acute myeloid leukemia. She is in remission for the time being. What is the most appropriate next step?
a) Maintenance chemotherapy
b) Craniospinal irradiation
c) Bone marrow transplantation
d) Regular intrathecal methotrexate
e) Re-induction chemotherapy

47) A 70-year-old man visits the physician's office because of a 6-month history of fatigue. Blood film shows lymphocytosis of 49000/mm^3 (with many smudge cells). There are normal hemoglobin and platelets. You find neither organomegaly nor lymph node enlargement. What is your best next step?
a) Bone marrow transplantation
b) Observation
c) Start CHOP chemotherapy regimen
d) Oral melphalan
e) Oral chlorambucil

48) A 56-year-old male has had fatigue and early satiety over the past several weeks. There is a huge spleen but you detect no lymph node enlargement. You do investigations to find out the cause. These are neutropenia and monocytopenia. Flow cytometry detects CD25 and CD103 on the surface of B-cells. The neutrophil alkaline phosphatase score is very high. What does the man have?
   a)   Chronic lymphocytic leukemia
   b)   High grade non-Hodgkin's lymphoma
   c)   Hairy cell leukemia
   d)   Prolymphocytic leukemia
   e)   Aplastic anemia

49) A 71-year-old man visits the outpatients' clinic. He has developed progressive pallor over few months. His hemoglobin has reached a level of 6.7 g/dl after receiving many packed RBCs transfusions within the past 6 months. He denies blood loss. Upper and lower gastrointestinal endoscopies are unremarkable. MCV is 106 fl. The bone marrow is hypercelluar and there are prominent dysplastic changes. Blast cells comprise 4% of the total bone marrow cell population. What does the man have?
   a)   Acute myeloid leukemia
   b)   Acute lymphoblastic leukemia
   c)   Myelodysplastic syndrome
   d)   Bone marrow malignant secondary tumors infiltration
   e)   Aplastic anemia

50) A 54-year-old man has a hemoglobin level of 9 g/dl. He was diagnosed with chronic osteomyelitis few years ago. Stool examination is negative for occult blood. Upper and lower gastrointestinal endoscopies fail to detect any abnormality. Bone marrow study is unremarkable. Serum ferritin is within its normal reference range, and you find low serum iron level. His MCV is 79 fl. What is the likely cause of this anemia?
   a)   Iron deficiency anemia
   b)   Megaloblastic anemia
   c)   Acquired sideroblastic anemia
   d)   Anemia of chronic diseases
   e)   Aplastic anemia

Clinical Hematology and Oncology

Answers

1) e.

Aluminum toxicity should always be looked for in uremic patients on long-term hemodialysis; resistance to erythropoietin therapy may be the clue. Alpha and beta thalassemia syndromes, including the minor traits, result in hypochromic microcytic anemia. Lead poisoning and sideroblastic anemia can cause a dimorphic blood picture. The list of causes of microcytosis and hypochromia is relatively narrow; in addition, it is a common clinical problem. Aplastic anemia causes normocytic or macrocytic anemia. Copper deficiency and zinc poisoning are rare causes of microcytic anemia (MVC <80 fL).

2) e.

The term macrocytosis indicates an increase in the MCV (>100 fL) while megaloblastic changes are the characteristic pathological features that are observed in the cells, mainly in the nucleus. Fish tapeworm infestation causes vitamin B_{12} deficiency and megaloblastic anemia. The first four stems result in characteristic changes in the bone marrow. Normal MVC is 80 -100 MCV; higher values (macrocytosis) occur in ethanol abuse, folic acid deficiency, vitamin B_{12} deficiency, myelodysplastic syndromes, acute myeloid leukemia (e.g., erythroleukemia), reticulocytosis, hemolytic anemia, response to blood loss, response to appropriate hematinic (e.g., iron, B_{12}, and folate therapies), drug-induced anemia (e.g., hydroxyurea, zidovudine, and chemotherapeutic agents), and liver disease.

3) e.

Stem "e" causes characteristic changes in the bone marrow (of increased plasma cells number and bone marrow hypercellularity) in addition to peripheral blood macrocytosis.

4) e.

Dimorphic blood picture is a commonly encountered in clinical medicine. It has four causes, in general:
1. Patients with pre-existent macrocytosis or microcytosis who receives blood transfusion.
2. During the way of recovery from a hematinic deficiency (i.e., after replacing the deficient hematinic).
3. Combined deficiency states (simultaneously or in succession), .e.g., iron and folate deficiencies in patients celiac disease.

4. Primary sideroblastic anemia where there are 2 clones; one clone is normal and the other clone is the abnormal one which results in the production of macrocytes.

In stem "e", phenytoin causes macrocytosis because of phenytoin-induced folate deficiency.

5) e.

In sickle cell disease, the resulting anemia is uncommonly symptomatic, despite the degree of anemia in the chronic stable state. It is usually tolerated because of the rightward shift in the oxyhemoglobin dissociation curve (which facilitates oxygen offloading to peripheral tissues). Regular blood transfusions are uncommonly indicated, except in those who have developed recurrent life-threatening crises (such as recurrent stroke). The objective is to reduce the HbS

(and therefore, occlusive complications) and not to raise the patient's PCV (if the latter is high, it may induce stagnation and further sickling). Up to 10% of children will develop some form of stroke. Stroke in adults is uncommon. Acute-chest syndrome is the usual cause of death in adults. Three significant predictors of an adverse outcome were identified in sickle cell anemia, and these are: dactylitis before the age of one year; hemoglobin concentration <7 g/dL; and leukocytosis in the absence of infection. The peripheral blood smear may reveal sickled red cells, polychromasia (indicative of reticulocytosis), and Howell-Jolly bodies (reflecting hyposplenism secondary to repeated splenic infarctions). The red cells are normochromic unless there is coexistent thalassemia or iron deficiency. If the age-adjusted MCV is not elevated, the possibility of sickle cell-beta thalassemia, coexistent alpha thalassemia, or iron deficiency should be considered.

6) d.

HbA2 is usually elevated in beta thalassemia trait patients; however, there may be a co-existent iron deficiency anemia resulting in a low or normal HbA2. Together with iron chelation therapy and daily folic acid, regular blood transfusion is the mainstay of treatment of beta thalassemia major. Beta thalassemia minor usually comes into light after a failure of oral iron therapy for a mild hypochromic anemia. This is confirmed by doing hemoglobin electrophoresis afterwards. Cardiac iron accumulation together with the high output state of severe anemia may result in heart failure. Target cells on peripheral blood films are part of the disease's profound dyserythropoiesis.

Profound microcytic anemia accompanied by bizarre red cell morphology is the hallmark of thalassemia major. The hemoglobin level may be as low as 3 to 4 g/dL. Red cell morphology is dramatic in most patients, with extreme hypochromia and poikilocytosis, predominance of microcytes, tear drop and target cells, and the visibility (even in routine stains) of clumped inclusion bodies (representing precipitates of alpha globin within the red cell) . These precipitates (Heinz bodies) can be more readily identified after the use of methyl violet or other supravital stains.

7) e.

Hemolytic disease of the newborn causes mainly extra-vascular type of hemolysis. If intravascular hemolysis is suspected, the following tests are of value: measurement of the plasma hemoglobin concentration, measurement of free hemoglobin in the urine supernatant, and testing for hemosiderin in the urine sediment 7 days after the incident (allowing time for hemosiderin-containing tubular cells to be shed into urine). Recognizing hemolysis (intra- or extra-vascular) is not difficult in the classic patient, who may have many or all of the following: new onset of pallor and anemia, jaundice with increased indirect bilirubin concentration, gallstones, splenomegaly, presence of circulating spherocytic red cells, increased serum LDH, reduced (or absent) level of serum haptoglobin, positive direct anti-globulin test (Coomb's test), and increased reticulocyte percentage or absolute reticulocyte number (indicating the bone marrow's response to the anemia). Abnormalities on the peripheral blood smear suggesting hemolysis include: spherocytes, fragmented red cells, bite cells, acanthocytes, and teardrop red cells.

8) d.

The presence of global brain dysfunction with fever, skin rash, and renal impairment should always prompt the physician to search for TTP. For unknown reason, renal impairment is more prominent in HUS rather than in TTP. Purpuric skin rashes or frank ecchymosis are more common and more compatible with HUS rather than TTP. Impaired coagulation studies (i.e., prolongation of PT and aPPT) indicate DIC; these are normal in both HUS and TTP. Vincristine may prematurely release platelets from the bone marrow; it can be used in the management plan.

9) c.

Eosinophilia may also occur in eczema and other skin diseases, Filariasis, Ascariasis, and eosinophilic granuloma. Glucocorticoids use results in neutrophilia and eosinopenia (the reverse occurs in Addison's disease, i.e., neutropenia and eosinophilia). Hodgkin's disease is usually overlooked as a cause of high peripheral blood eosinophil count. Although accepted upper limits of normal blood eosinophil numbers vary somewhat, a value above 600 eosinophils/mm^3 of blood is abnormal in the vast majority of cases. The degree of eosinophilia can be categorized into mild (600-1500 cells/mm^3), moderate (1500-5000 cells/mm^3) or severe (>5000 cells/mm^3). Modest to marked eosinophilia of unknown cause occasionally is observed in HIV-infected patients. The eosinophilia is usually due to one of the following conditions rather than being directly induced by HIV-1:

1. Leukopenia may lead to an increased eosinophil percentage without absolute eosinophilia, and therapy with GM-CSF can stimulate actual eosinophilia.
2. Reactions to medications.
3. Adrenal insufficiency due to cytomegalovirus and other infections with consequent eosinophilia.
4. Eosinophilic folliculitis, a common dermatologic disorder in HIV-positive patients.
5. Marked hyper-eosinophilia may develop in association with the hyper-immunoglobulin E syndrome or exfoliative dermatitis.

10) e.

Moncytosis may be encountered in: bone marrow recovery from radiotherapy; inflammatory bowel disease; SLE; myelomonocytic leukemia; tuberculosis; typhus; brucellosis; SBE; malaria; and trypanosomiasis. Stem "e" causes monocytopenia. Monocytes normally comprise <10% of the total circulating white blood cells, with the absolute monocyte count being <800/mm^3 in normal adults.

11) e.

Neutropenia could have a racial basis, as in Arabs and Africans. Viral infections usually result in neutropenia. The low white cell count in hypersplenism usually coincides with low platelets. The list of medications and drugs that cause neutropenia is very long, e.g., carbimazole and penicillamine. Leptospirosis characteristically causes neutrophilia, which is a useful clue to differentiate this infection from viral hepatitis.

The absolute neutrophil count (ANC) is equal to the product of the white blood cell count (WBC) and the fraction of polymorphonuclear cells (PMNs) and band forms noted on the differential analysis:

$$ANC = WBC \; (cells/mm^3) \; x \; percent \; (PMNs + bands) \div 100$$

Neutrophilic metamyelocytes and younger forms are not included in this calculation. Neutropenia is defined as an absolute neutrophil count (ANC) of less than $1500/mm^3$. This definition of neutropenia is applicable for all ages and ethnic groups except newborn infants who have an elevated ANC in the first few days of life, and certain populations of blacks and Yemenite Jews who normally have slightly lower WBC and ANC. Neutropenia is often categorized as mild, moderate or severe, based upon the level of ANC. Mild neutropenia corresponds to an absolute neutrophil count between 1000 and $1500/mm^3$, moderate between 500 and $1000/mm^3$, and severe with less than $500/mm^3$. The risk of infection begins to increase at an ANC below $1000/mm^3$.

12) d.

The common ALL type has a good prognosis. The T-cell type usually presents with a mediastinal mass and a very high leukocyte count and has a gloomy prognosis. In addition, t(9,22) cytogenetic abnormality has a very poor prognosis (unlike CML, where its presence confers a good outcome).

13) b.

Knowing the cytogentics of AML is very important, not only for the prognosis but also to decide the future treatment options (such as bone marrow transplantation in those with poor cytogenetics). The M2 subtype has t(8,21), while t(15,17) occurs in M3 subtype. Inversion of chromosome 16 (inv16) is found in ALL. The t(1,19) mutation occurs in pre-B ALL while t(8,14) is seen in ALL-L3 (Burkitt's) subtype.

14) d.

The presence of t(15,17) confers a good prognosis in AML-M3 subtype. Retinoic acid receptor alpha (RARA) defect is responsible for the hallmark of the disease, and that is failure of maturation of the bone marrow's promyelocytes into mature white cells.

The maturation arrest can be bypassed with the use ATRA (all- trans retinoic acid); this would cause a rapid development of neutrophilia which is commonly seen after the introduction of this medication. The leukemic cells are usually hypergranular. DIC occurs in 80% of cases and is usually asymptomatic.

15) d.

Philadelphia chromosome is found in 5% of childhood ALL, 25% of adult ALL, and 1% of adult AML. Philadelphia chromosome is a derivative chromosome 22 created by a reciprocal translocation between the region containing the BCR gene on the long arm of chromosome 22 (22q) and the segment that contains the ABL gene on the long arm of chromosome 9 (9q). Due to its inhibitory effect on the BCL-ABL fusion gene product, treatment with imatinib mesylate produces an 80% rate of cytogenetic remission. Peripheral blood basophilia usually marks the beginning of the accelerated phase. Serum vitamin B_{12} is raised due to high vitamin B_{12}-binding protein (which is produced by the CML cells). CML characteristically lowers the LAP score; the other causes of low LAP score are paroxysmal nocturnal hemoglobinuria and some cases of aplastic anemia and myelodysplastic syndromes.

16) e.

In CLL, trisomy 12 is the commonest cytogenetic abnormality and 3q abnormalities are common as well; both portend a poor prognosis. Warm immune-hemolytic anemia and immune thrombocytopenic purpura may occur and mark the disease as stage C. Serum LDH is usually raised, as in other lymphoproliferative disorders. The doubling time is usually long and the cell turnover is low; therefore, hyperuricemia in the chronic stable phase is unusual (unless other causes are operative, e.g., renal impairment. Note the following in CLL:

1. Coombs test is positive in 35% of cases; however, overt immune hemolytic anemia occurs only in 11% of patients.
2. Autoimmune thrombocytopenia complicates 2-3% of cases.
3. Agranulocytosis develops in 0.5% of patients. It should be remembered that, in patients with total white blood cell counts >100,000/mm^3, the absolute neutrophil count may be normal (i.e., >1500/mm^3) in spite of a neutrophil percentage as low as 1 or 2%.
4. Hypogammaglobulinemia occurs in 8% at the time of diagnosis; a figure that rises to 33% late in the course.

5. Approximately 5% of patients display serum M-paraprotein.
6. Pure red cell aplasia is rare and usually occurs early in the course.

17) e.

As is the case in many hematological malignancies, serum LDH is commonly raised and is a useful guide to the bulk and burden of the disease. The ESR is usually elevated and indicates an active disease. Residual masses after successful treatment is common, and hence, at times, it becomes difficult to decide whether these represent lymphomatous masses or not; PET scan may be used to differentiate between them. The lymphocytic-predominant variety has a good prognosis. An association with EBV infection has been suggested in Hodgkin's disease.

18) e.

Myeloma's high ESR is responsible only for 3% of cases of ESR>100 mm/hour. One percent of myelomas is non-secretory; on the other hand, there may be no serum paraprotein, but only urinary Bence-John's protein (albumin stick-negative proteinuria). Human Herpes virus type 8 infection is a recognized association; the infected stem cells may be responsible for the secretion of high level of IL -6. Isotope bone scan is usually normal, as there is an overall inhibitory effect on bone osteoblasts. Only in the presence of a recent fracture, serum alkaline phosphatase is elevated.

19) e.

Immune paresis is favors multiple myeloma over monoclonal gammopathy of undetermined significance (MGUS), as do bone and renal diseases. MGUS is defined by the following three criteria:
1. The presence of a serum monoclonal protein (M-protein, whether IgA, IgG, or IgM) at a concentration of less than 3 g/dL.
2. Fewer than 10% plasma cells in the bone marrow.
3. The absence of lytic bone lesions, anemia, hypercalcemia, and renal insufficiency related to the plasma cell proliferative process.

In addition, MGUS is characterized by:
1. A predilection for the development of multiple myeloma or a related malignancy at the rate of 1% per year.
2. No evidence for a neoplastic plasma cell proliferative disorder or a B cell lympho-proliferative disorder.

3. MGUS has been reported in association with several non-malignant disorders. It is not clear whether these associations are pathogenetically related or merely represent coincidental associations, given the relatively frequent occurrence of MGUS in the general population above the age of 50.
4. MGUS is asymptomatic.

20) e.

Secondary polycythemia may also be found in neonates, smokers, COPD, and morbid obesity as well as high affinity hemoglobins. Anemia is more common than polycythemia in renal cell carcinoma. Polycythemia vera is a primary type of polycythemia. The following are germline and somatic mutational causes of polycythemia: polycythemia rubra vera, activating mutations of the erythropoietin receptors, Chuvash polycythemia, idiopathic familial polycythemia, methemoglobinemia, high oxygen affinity hemoglobins, and absent (or reduced) 2,3-DPG mutase. Chuvash polycythemia is a form of congenital polycythemia that is endemic in the Chuvash population of the Russian Federation (mid-Volga region of European Russia). It causes thrombotic and hemorrhagic vascular complications that lead to early mortality, usually before the age of 40 years. The serum erythropoietin concentration in polycythemic patients is elevated or inappropriately normal for the given elevation of the hematocrit. However, the erythrocyte progenitors are hypersensitive to erythropoietin, perhaps due to an up-regulation of erythroid erythropoietin, a finding that blurs the distinction between primary and secondary polycythemia.

21) e.

The other causes of reactive thrombocytosis are systemic inflammatory disorders, vasculitides, trauma, and infections. Polycythemia rubra vera (PRV) may have a high platelet count but it is not a "reactive" process; it is part of the panmyelosis process seen in PRV. In treated PRV, the causes of death are thrombosis (29%), hematologic malignancies (i.e., AML or myelodysplastic syndromes; 23%), non-hematologic malignancies (16%), hemorrhage (7%), and myeloid metaplasia with myelofibrosis (3%).

22) e.

Myelodysplastic syndrome is mainly a disease of elderly people.

Cytopenias are very common and dysplastic changes in the bone marrow are seen, e.g., hypo -granular neutrophils, abnormal neutrophil nuclear lobulation, and vacuolated erythroblasts. The other causes of death are infection and bleeding. Secondary myelodysplastic syndrome (MDS) is seen after chemotherapy and radiotherapy. Refractory anemia with excess blasts in transformation is the commonest type.

Cutaneous manifestations of MDS are uncommon; 2 syndromes have appeared in examinations:
1. Sweet's syndrome (acute febrile neutrophilic dermatosis), when complicating the course of MDS, may herald transformation to acute leukemia. Paracrine and autocrine elaboration of the cytokines interleukin-6 and granulocyte colony-stimulating factor have been implicated in the pathogenesis of this condition.
2. Granulocytic sarcoma (chloroma) of the skin may herald disease transformation into acute leukemia.

Characteristics of 5q⁻ syndrome:
1. Female predominance (female to male ratio is 7:3) with a median age at diagnosis of 68 years.
2. Transfusion-dependent anemia (80%).
3. Low incidence of neutropenia, thrombocytopenia, infection, and bleeding.
4. Normal or increased platelet counts along with bone marrow hyperplasia of hypo-lobulated micromegakaryocytes.
5. Low incidence of transformation into acute leukemia (16%).

23) e.

In antiphospholipid syndrome (APS), the patient's aPTT is prolonged and does not correct with addition of normal plasma; this reflects the presence of inhibitors in the patient's plasma (unlike hemophilia, where the aPTT corrects itself upon mixing the patient's plasma with a normal one). Adrenal failure (with raised serum ACTH) may follow bilateral adrenal vein thrombosis. Thrombocytopenia and Coomb positive immune hemolytic anemia may occur. Chorea, migraine, transverse myelopathy, epilepsy, TIAs, stroke, and multi-infarct dementia are the usual CNS manifestations. Arterial and venous thromboses occur despite the prolonged aPTT; if hemorrhage occurs, it is either due to over-anticoagulation or to severe thrombocytopenia (which is uncommon). The aPTT is already prolonged; this would make the follow-up of heparin therapy difficult in APS patients.

24) e.

Von Willebrand disease is the commonest inherited bleeding tendency; many cases are mild and are discovered later in life. Because it induces further thrombocytopenia, DDAVP is contraindicated in type II B disease (the other disease subtypes have normal platelets count). In general, the only useful indication of bleeding time is to screen for von Willebrand disease. Some cases are associated with normal amount of von Willebrand factor antigen; however, vWF multimers are usually abnormal. Some cases are autosomal recessive; the dominant ones have a variable expression. The gene for vWF is located on the short arm of chromosome 12 and is composed of 178 kilobases (kb) and 52 exons; the vWF mRNA contains approximately 9 kb. A pseudogene is present on chromosome 22 that includes exons 23- 34 of the vWF gene; these exons correspond to regions of the authentic gene that encode domains A1, A2, and A3. Von Willebrand factor is synthesized in endothelial cells and megakaryocytes. vWF can bind factor VIII only when factor VIII has not been cleaved by thrombin. After thrombin cleavage, activated factor VIII (VIIIa) is released from vWF and becomes a fully functional, active cofactor in the ongoing thrombin generation that can be inactivated by activated protein C.

25) e.

About $1/3^{rd}$ of hemophilia A cases are kept in the community by fresh new mutations; therefore, no family history is obtained in such cases. The PT and the bleeding time are characteristically within their normal reference range, and the platelets level may be elevated after a bleeding episode, e.g., GIT hemorrhage, or due to recurrent bleedings causing iron deficiency anemia and reactive thrombocytosis. The prolonged aPTT does correct with the addition of normal plasma; no correction would indicate the development of factor VIII inhibitors. The normal factor VIII activity is between 50-150%. Hemarthrosis is a painful and physically debilitating manifestation of hemophilia. Bleeding originates from the synovial vessels, and hemorrhage occurs within the joint cavity. Distension of the synovial space and associated muscle spasm markedly increase intra-synovial pressure. In addition to these acute events, hemarthrosis is the initiating event in hemophilic arthropathy. Bleeding into muscles with hematoma formation most often affects the quadriceps, iliopsoas, and forearm. Iliopsoas muscle bleeding episodes may be large and may compromise neurovascular structures and produce a compartment syndrome. The hemorrhage may be localized with ultrasound. NB: patients with hemophilia and carriers of hemophilia appear to have a reduced risk of coronary heart disease mortality.

26) e.

The spleen in idiopathic thrombocytopenic purpura (ITP) is not enlarged in at least 90% of cases; the presence of splenomegaly would, however, point out towards a secondary cause (such as SLE or CLL). In contrast to children, adults with ITP usually have a chronic fluctuating course rather than an acute monophasic illness. About $2/3^{rd}$ of cases respond to glucocorticoid therapy initially; around $2/3^{rd}$ of patients respond to splenomegaly after failure of systemic steroids. The role of bone marrow studies as a diagnostic tool is limited; it is mainly used to exclude other diseases, such as ALL in children. Anti-platelets antibodies are not used routinely in the diagnosis. In classical cases, the bone marrow would show megakaryocytic hyperplasia (to compensate for the peripheral destruction). The disease is usually preceded by an upper respiratory tract infection in pediatric cases; such a history is usually absent in adults. There is no "gold standard" test that can establish the diagnosis of ITP. The diagnosis is in part one of exclusion, requiring that other causes of thrombocytopenia be ruled out. A presumptive diagnosis of ITP is made when the history (e.g., lack of ingestion of a drug that can cause thrombocytopenia), physical examination, complete blood count, and examination of the peripheral blood smear do not suggest other etiologies for the patient's isolated thrombocytopenia. The only recommended further tests in such patients are HIV testing (in patients with risk factors for HIV infection) and bone marrow aspiration in patients over 60 years of age (to rule out myelodysplastic syndromes).

27) e.

Treatment with steroids (together with melphalan) helps stabilize the disease and may produce improvement.

28) c.

High serum β_2 microglobulin level is a bad prognostic sign. Lab parameters of poor prognosis are: performance status 3 or 4; serum albumin <3 g/dl; serum creatinine >2 mg/dl; platelet count <150,000/mm^3; age >70 years, β_2 microglobulin >4 mg/L. plasma cell labeling index 1%; serum calcium >11 mg/dl; hemoglobin <10 g/dl; and bone marrow plasma cell >50%. Asymptomatic (smoldering) multiple myeloma should fulfill the following 2 criteria in order to be diagnosed: serum monoclonal protein >3 g/dl and/or bone marrow plasma cells >10% *and* no end organ damage related to plasma cell dyscrasia.

29) e.

Dehydration is common in myeloma; this is mainly the aftermath of hypercalcemia and polyuria. About 55% of myeloma cases are of IgG type; the IgA paraprotein is responsible for 21% of cases while the light chain paraprotein occurs in 22% of patients; others, including the so-called non-secretory myeloma, constitute about 1% of cases. Panhypogammaglobulinemia is common and is usually severe. Bone marrow failure reflects an advanced stage of the disease. Immune paresis is responsible for the infectious complications of myeloma. In multiple myeloma, the occurrence of headache, blurred vision and epistaxis usually point out towards hyperviscosity with paraproteinemia of IgA type; this usually calls for plasmapheresis. Raised serum alkaline phosphatase indicates a recent fracture. Myeloma does not raise serum alkaline phosphatase; therefore, the radio -isotope bone scanning is also normal (unless there is a recent fracture). About 1% of myeloma patients have no M-protein in the serum or urine on immunofixation at the time of diagnosis and are considered to have non-secretory myeloma. The condition remains non-secretory in (76% of cases, even with extended follow- up). Of interest, due to lack of light- chain excretion, renal failure due to multiple myeloma is uncommon in the non-secretory variant. Survival in non-secretory myeloma is no different, however.

30) e.

Stem "a" is one of the core features of hairy cell leukemia; hence, symptoms of infection are common. The disease relatively has a slow pace. The malignant B-cells characteristically carry CD25 and CD103 antigen; these can be detected through performing flow cytometry. The response to intravenous cladribine (and to deoxycoformycin) is favorable with long-term remission is the rule in many cases. Splenomegaly is seen in up to 90% of cases and may be massive, but lymph node enlargement is highly unusual (unlike CLL). The pathogenesis of hairy cell leukemia is unknown, although exposures to ionizing radiation, Epstein-Barr virus, organic chemicals, woodworking, and farming have been mentioned as possible causes.

31) b.

The treatment of MDS is mainly supportive; there is no curative one. Chromosomal analysis frequently reveals abnormalities in chromosomes 5 and 7.

The disease mainly targets the elderly population, but in theory, no age is immune. Secondary forms follow alkylating agents or etoposide administration. Macrocytosis is common, and dysplastic (not megaloblastic) changes are found in the bone marrow.

32) c.

The idiopathic form of immune hemolytic anemia constitutes up to 50% of cases; secondary causes include infections, tumors, and drugs. Smudge cells are seen in CLL; the presence of spherocytes may be a clue to immune hemolytic anemia, but not to its cause. The direct Coombs test detects IgG and complements, but not IgA or IgE antibodies; therefore, it may be falsely negative when immune hemolytic anemia is associated with IgA or IgE antibodies. A direct Coombs test requires at least 200 antibodies to be attached to the surface of RBCs to get a positive result; therefore, patients with low antibody tires may display false negative Coombs test. An association with HIV has been documented; ITP has a similar association.

33) d.

The commonest abnormalities are seen in beta spectrins and ankyrins of the red cell membrane; patients carrying these abnormalities are usually mildly affected. More severe cases either indicate a coexistent defect of a second different protein or coincidental polymorphism of α-spectrin. The development of pigment gall stones calls for splenectomy; the other indications for this form of surgery are growth retardation in children and death of a family member from the disease. About 25% of cases only arise from fresh new mutatio ns; therefore, family history is usually relevant. Coomb's tests must be negative.

34) e.

The prematurely released reticulocytes may have normal level of glucose-6-phosphate dehydrogenase; therefore, during the acute hemolytic episode, don't assess the enzyme level, as it may become transiently normal or even high. A useful clue in acute attacks is the demonstration of bite and blister cells on peripheral blood films. Heinz bodies can be seen only after staining the blood film with a supravital stain. The disease may come into light as neonatal jaundice, especially after receiving the water-soluble analogue of vitamin K.

The Caucasian and the Oriental types are the most severe when compared with the African variety. Medications that should be avoided in G6PD deficient patients are:

1. Analgesics: aspirin, phenacetin.
2. Anti-malarials: primaquine, quinine, chloroquine, pyrimethamine.
3. Antibiotics: sulfonamides, nitrofurantoin, ciprofloxacin.
4. Miscellaneous: quinidine, probenecid, vitamin K, dapsone.

35) d.

The blood film cannot differentiate between vitamin B_{12} and folic acid deficiencies; it just provides a clue to them, such as the presence of hypersegmented neutrophils. Neurological manifestations favor vitamin B_{12} over folic acid deficiency; however, peripheral neuropathy may be seen in folate deficiency. Anorexia, alcohol ingestion, and phenytoin can severely depress the serum level of folate in the absence of a true deficiency state, while a single meal may totally normalize serum folate in a severely deficient patient; therefore, serum folate is less useful than red cell folate in making the diagnosis of folate deficiency. Replacing the deficient vitamin may precipitate severe hypokalemia. Reticulocytes percentage may reach up to 50% within the first week of treatment (reflecting the rapid and favorable response to the hematinic replacement). Pregnancy (especially in the context of multiple pregnancy or multiparity) is the commonest cause of megaloblastosis world-wide.

36) c.

Hemosiderinuria, hemoglobinuria, and reduced (or absent) serum haptoglobin are the additional features of intravascular red cell destruction. Reticulocytosis is one of the core features of hemolysis (reflecting the bone marrow attempt to correct the reduction in the peripheral red cell population); reticulocytopenia in the background of hemolysis occurs in megaloblastic anemia (because of defective erythropoiesis) or during aplastic crisis of congenital hemolytic states.

37) d.

Serum ferritin levels down to 100 μg/L are still suggestive of iron deficiency anemia in the presence of underlying chronic inflammatory illnesses. Serum iron fluctuates by 30-50% on daily basis, and even on diurnal basis. It becomes low during acute phase responses (just like albumin, it is a negative phase reactant). Therefore, serum iron has a limited role in the diagnosis of iron deficiency anemia.

Serum transferrin level become low in liver diseases, nephrotic syndrome, malnutrition, during acute phase responses; it is raised in pregnancy and with the use of oral contraceptive pills. In the appropriate clinical setting, transferrin saturation of <16% is highly suggestive of iron deficiency anemia (the normal value is between 20-40%). Increased serum soluble transferring receptors level (which is measured by immunoassay) is very helpful in the diagnosis of iron deficiency anemia (and to differentiate this type of anemia from anemia of chronic diseases when the distinction is blurred).

38) e.

Bronchospasm, angioedema, and hypotension dominate the clinical picture of anaphylaxis. The following reactions are rare but can be fatal: acute hemolytic transfusion reaction; infusion of a bacterially contaminated blood or blood product; graft-versus- host disease; transfusion-associated lung injury; and severe allergic reaction or anaphylaxis.

39) e.

We should ensure a urine output of at least 100 ml/minute; frusemide should be given if the urine output falls.

40) a.

Stem "a" refers to a provoked episode of deep venous thrombosis following a high risk surgery; in addition, this is the first episode and therefore there is no need to screen for stem "e" patient. The occurrence of venous thrombosis at an unusual site, e.g. portal vein, should prompt a search for an underlying thrombophilia tendency.

41) b.

This patient presents with features of bone marrow failure, pancytopenia, and absence of oragnomegaly and lymphadenopathy. His bone marrow is difficult to be aspirated (a dry tap). Only aplastic anemia would fit this scenario. Acute myeloid leukemia may present as pancytopenia (and its consequences); however, the bone marrow is hypercellular and is easy to be aspirated.
Advanced Hodgkin's disease usually has multiple nodal as well as extra-nodal components (e.g., cervical lymph node enlargement, organomegaly). Early non-Hodgkin's lymphoma usually presents as a nodal or extra-nodal disease, not a dry tap.

ALL has a preference to involve lymphatic tissues; therefore, enlarged lymph nodes (and liver and spleen) would be usual; their absence in the background of profound leukemic bone marrow failure is unusual.

42) a.

The overall picture is that of a stage VI (or stage C) chronic lymphocytic leukemia (note the high white cell count and the mature-looking lymphocytes with smudged cells as well as thrombocytopenia).

43) b.

The combination of (muco-) cutaneous bleeding, thrombocytopenia, and absence of organomegaly and lymphadenopathy, together with normal bone marrow study (apart from megakaryocytic hyperplasia) would make idiopathic thrombocytopenia (ITP) the correct diagnosis. ITP may occur in SLE; the patient's scenario does not give extra-clues to fulfill the diagnostic criteria of SLE. Stem "e" presents with recurrent hemarthrosis and deep tissue hematomas (not a mucocutaneous type of bleeding). CML has a raised white cell count with full granulocytic series. Thrombocytopenic hemorrhages may be the initial manifestations of aplastic anemia, but the bone marrow study would show fatty hypocellular marrow.

44) a.

Medical treatment is used to control the cellular proliferation in CML and to reduce the peripheral white cell counts. Bone marrow transplantation as curative therapy would be unwise in this 75-year-old man (given the morbidity and mortality this operation incurs).

45) c.

Interferon alpha therapy has no place in the management of acute lymphoblastic or myeloblastic leukemia.

46) c.

The management of acute myeloblastic leukemia (AML) differs from that of its lymphoblastic counterpart.

After inducing a hematological remission in AML, the next best step is to do bone marrow transplantation (i.e., achieving cure). The other stems are part of acute lymphoblastic leukemia management protocol.

47) b.

This man has stage I chronic lymphocytic leukemia (CLL), and apart from fatigue, he has no symptoms. The best approach is to observe, as the long-term prognosis is excellent. The treatment of CLL depends upon the stage of the disease:
1. Clinical stage A; no specific treatment is required. Life expectancy can be entirely normal in those patients. Patients should be reassured.
2. Clinical stage B; chemotherapy with chlorambucil may suffice in symptomatic patients. Local radiotherapy to enlarged lymph nodes may be considered if these are causing discomfort.
3. Clinical stage C; anemia may require transfusions with packed red cells concentrate. Bone marrow failure, if present, should be treated initially with prednisolone. Some degree of bone marrow recovery is usually achieved.

48) c.

The clinical scenario is highly suggestive of hairy cell leukemia. Severe neutropenia, monocytopenia, and the characteristic hairy cells in the blood and bone marrow are typical. These cells are B lymphocytes, but characteristically express CD25 and CD103. A characteristic test is the demonstration that the acid phosphatase staining reaction in these cells is resistant to the action of tartrate. The neutrophil alkaline phosphatase score is almost always very high (it is low in CML).

49) c.

The overall clinical picture is typical of myelodysplastic syndrome. This syndrome consists of a group of clonal disorders which represent steps in the progression to the development of leukemia. It is characterized by macrocytosis, variable cytopenia, hypo-granular neutrophils with nuclear hyper-or hypo- segmentation, and a hyper-cellular marrow with dysplastic changes in all three cell lines (a differential diagnosis of megaloblastic anemia). Its exact incidence is uncertain, but it is thought to be more common than acute leukemia. Usually, the disease presents as a primary problem in elderly patients, although it may occur as a secondary complication of treatment for malignant disease in younger patients.

50) d.

Anemia of chronic diseases is a clinically common type of anemia. Characteristically:

1. The anemia occurs in the background of chronic infections, long-standing inflammatory disease, or malignancy.

2. The resulting anemia is not caused by underlying hemorrhage, red cell hemolysis, or bone marrow infiltration.

3. The anemia is usually mild (8.5-11.5 g/dl), and its MCV (and MCHC) is usually within its normal reference range (i.e., normocytic, normochromic); however, up to 25% of patients may have a reduced MCV value (i.e., microcytic), and thus, raising a differential diagnosis of iron deficiency anemia.

4. The serum iron is low, but the bone marrow iron stores are normal (or increased).

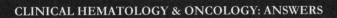
This page was intentionally left blank

Chapter 3
Clinical Pharmacology, Therapeutics, and Toxicology

Questions

1) A 67-year-old man presents with nausea, vomiting, headache, somnolence, and xanthopsia. He has atrial fibrillation for which he takes daily digoxin. One week ago, his physician prescribed some additional medications. Serum potassium is 5.7 mEq/L and serum digoxin level is high. There is normal renal function. Digoxin toxicity is enhanced by all of the following, *except?*
   a) Cardiac amyloidosis
   b) Hypomagnesaemia
   c) Hypothyroidism
   d) Concomitant treatment with amiodarone
   e) Hyperkalemia

2) A 65-year-old man had been receiving medical treatment for major depression over the past 3 weeks. Today, he was brought to the Acute and Emergency Department unconscious. He deliberately ingested 20 tablets of amitriptyline. 12-lead ECG reveals torsades de pointes ventricular tachycardia. With respect to tricyclic antidepressants overdose, all of the following are correct, *except?*
   a) Cardiac and CNS toxicities are responsible for most of the fatalities
   b) Skin blisters may be seen but they are rare
   c) Seizures are well-recognized features
   d) Rhabdomyolysis indicates severe poisoning
   e) Sodium bicarbonate infusion is indicated when the ECG shows short QT-interval

3) A 35-year-old plumber presents with right-sided wrist drop. Blood film shows basophilic stippling of red cells and there is hyperuricemia. Serum and urinary levels of lead are increased. All of the following statements regarding lead poisoning are correct, *except?*
   a) Red cell aminolevulinic acid dehydratase (ALA) activity is decreased
   b) Glycosuria and aminoaciduria may develop
   c) Blue lines on gums indicate that the poisoning is recent
   d) Urinary ALA and coproporphyrin are elevated
   e) Encephalopathy mainly occurs in children

4) A 25-year-old vagrant is brought to the Emergency Room in a deep coma. He has rapid and sigh breathing but you find no signs of trauma. Head CT scan is normal. Toxicology screen uncovers ethylene glycol poisoning. Regarding this type of poisoning, choose the *incorrect* statement?
   a) The anionic and osmolal gaps are increased
   b) The resulting renal failure is usually reversible
   c) Hypocalcemia develops

d) Hemodialysis is the best option in treating severe intoxications
e) Activated charcoal is helpful if poisoning occurs within 2 hours

5) A 19-year-old college student presents with extreme agitation and confusion. In addition, he has nausea, vomiting, and diarrhea. His friend states that the patient took some sort of amphetamine. Regarding ecstasy poisoning, which one of the following is the *wrong* statement?
a) Disseminated intravascular coagulation may occur
b) Serum creatinine phosphokinase is elevated
c) Hyponatremia is common
d) Hypotension may be due to myocardial infarction
e) The development of intracranial hemorrhage suggests an alternative diagnosis

6) A 24-year-old man is brought to the Emergency Room after developing clouded consciousness. He was found at the backyard of his house, behind his car. Neither signs of trauma nor intravenous drug injection sites are found. You suspect carbon monoxide poisoning. All of the following regarding this form of intoxication are correct, *except*?
a) The gold standard investigation is measuring the blood carboxyhemoglobin level
b) Despite severe intoxication, the pulse oximetry is usually normal
c) The earliest features are headache, nausea, and vomiting
d) When seizures occur, phenobarbitone should be avoided
e) Hyperbaric O_2 therapy is used when the patient's carboxyhemoglobin blood level is >5%

7) A 16-year-old male, who was diagnosed with idiopathic generalized tonic-clonic epilepsy few months ago, presents with tremor, diplopia, and ataxia for 2 days. He takes daily phenytoin, but recently he has increased the total daily doses by himself without consulting his physician. With respect to phenytoin toxicity, which one of the following is the *wrong* statement?
a) The occurrence of seizures is a rare event
b) Fatality is uncommon and is mainly the aftermath of cardiac toxicity
c) May be enhanced in renal and hepatic failures
d) Hemodialysis is a very helpful treatment method
e) Gingival hypertrophy indicates a chronic exposure, and hence reflects superadded toxicity rather than an accidental poisoning

8) A 43-year-old male has a bipolar disorder. Today he visits the physician's office because of tremor, diarrhea, and vomiting. He takes lithium carbonate and has been recently given a medication for essential hypertension. All of the following statements about lithium intoxication are correct, *except?*
   a) Enhanced by concomitant thiazide therapy
   b) There may be raised serum TSH with low serum T4 levels
   c) Seizures and hyperreflexia may be observed
   d) Fine tremor supports the diagnosis of overdose
   e) Hemodialysis is the treatment of choice in severe poisoning

9) A 27-year-old unemployed man was found unconscious on the street. The smell of alcohol is prominent and he has multiple bruises. Brain CT scan is unremarkable. With respect to ethanol intoxication, which one is the *wrong* statement?
   a) Ethanol intoxication produces nystagmus, dysarthria, and limb and gait ataxia
   b) In non-chronic alcoholics, the clinical manifestations decline over hours despite a stable blood ethanol level
   c) Plasma osmolality is useful in the assessment and should be normal in acute intoxication
   d) Should be differentiated from sedative-hypnotic drug intoxication
   e) May cause life-threatening hypoglycemia

10) A 23-year-old female ingested 30 tablets of diazepam 2 hours ago. Regarding sedative drugs intoxication, which one of the following is the *wrong* statement?
   a) Can present as confusional state or coma
   b) The pupils are usually reactive, and prominent pupillary abnormalities should prompt a search for an alternative diagnosis
   c) Nystagmus, gaze paresis, or decerebrate or decorticate posturing might be seen
   d) The mortality rate is very high and is due to the CNS depressant effect of these medications
   e) Forced alkaline diuresis is ineffective for short-acting barbiturates

11) A 53-year-old male presents with agitation, irritability, and insomnia. He takes lorazepam tablets each night to facilitate sleep. He has been out of lorazepam for 3 days. Regarding sedative drugs withdrawal, choose the *incorrect* statement?

a) The frequency and severity of these withdrawal syndromes depend on the duration of drug intake, total daily doses, and the half-life of the medication
b) The overall clinical picture can exactly resemble ethanol withdrawal syndromes
c) The diagnosis can be confirmed by phenobarbital challenge test
d) Like ethanol withdrawal seizures, seizures here should not be treated with anticonvulsants
e) A delirium tremens-like syndrome may be seen 3-8 days after abstinence

12) A 17-year-old male has ingested a large dose of morphine 9 hours ago in an attempt to end his life. Regarding opioid overdose, which one is the *incorrect* statement?
a) May present as an iatrogenic confusional state or coma in hospitalized patients
b) The cardinal features are pinpoint pupils and respiratory depression
c) Can simply be confirmed at the bedside by giving naloxone to the patient
d) The mortality rate is high
e) Because most opioids are long-acting, nalaxone should be given repeatedly as a treatment

13) A 4-year-old child ingested 18 tablets of hyoscine, which was used by her father who has irritable bowel syndrome. Regarding anti-cholinergic drugs intoxication, which one is the *wrong* statement?
a) Can be seen with antipsychotics overdose
b) Produces a characteristic picture of agitated delirium and fever, dry skin, fixed dilated pupils, blurring of vision, and hallucinations
c) May be confirmed by toxicology screen of blood and urine
d) The symptoms are usually progressive with a high mortality rate
e) Treated by physostigmine

14) A 16-year-old boy, in order to kill himself, ingested many tablets of ephedrine. He was brought to the Emergency Department by his father. All of the following statements about sympathomimetic intoxication are correct, *except*?
a) Results in hyperactivity, hallucinations, and schizophreniform paranoid psychosis
b) Cardiac dysrhythmia is the main cause of serious morbidity

c) Amphetamine and cocaine can cause thrombotic or hemorrhagic strokes

d) Beta blockers are very useful in cases complicated by severe hypertension

e) Haloperidol is very useful to offset the central dopaminergic effects

15) A 14-year-old teen is brought to the Emergency Room by his friend. The friend states that the patient bought a large amount LSD this morning and then ingested it 2 hours ago. Regarding LSD (lysergic acid diethylamide) intoxication, choose the *incorrect* statement?
   a) Produces nystagmus, ataxia, hypertonia, and hyperreflexia
   b) Visual and somatosensory illusions and hallucinations are the hallmark of this syndrome
   c) Seizures are very common and should be treated promptly
   d) There are dilated pupils and hyperthermia
   e) The treatment usually involves verbal calming and reassurance

16) You colleague asks about phencyclidine (PCP) and its implication as a drug of abuse. You tell him that this medication was originally developed as an anesthetic agent but the agitation that some people developed following such anesthesia quickly led to abandonment of PCP. Regarding PCP intoxication, which one of the following is the *wrong* statement?
   a) Considered to be a medical emergency with many fearful complications
   b) The patient can be drowsy or extremely agitated and may demonstrate amnesia, hallucinations, and violent behavior
   c) Phenothiazines should be given in the treatment of psychotic features
   d) In general, the signs and symptoms resolve with 24 hours
   e) May be complicated by severe hypertension, status epilepticus, malignant hyperthermia, coma, and death

17) Drug-induced confusional state is encountered in hospitalized patients. Regarding drug-induced confusional state, which one of the following is the *wrong* statement?
   a) This occurs when medications are prescribed in larger than their customary doses
   b) Many medications can induce prominent confusion in elderly people, even when prescribed in small recommended doses
   c) Especially seen when the metabolism of the medication is impaired by organ failure

d) Those with pre-existent cognitive impairment are more susceptible to drug-induced confusional states
e) Polypharmacy protects against the development of drug-induced confusional states

18) A 43-year-old male presented with repetitive seizures. His friend said that the patient tried to abstain from alcohol. Head CT scan is normal, as is the blood sugar level. With respect to ethanol withdrawal (rum) fits, which one is the *wrong* statement?
a) Usually develops within 19 days of abstinence
b) More than 90% of patients develop 1 to 6 seizures
c) Anti-epileptics are not usually required for the treatment
d) Focal seizures should always prompt a search for another pathology
e) All patients should be observed for any subsequent or concomitant complications of alcohol

19) A 19-year-old heavy alcoholic male tries to abstain from alcohol intake. He says that he just cannot do it, as he found in the internet that abstinence results in many problems, such as the delirium tremens thing. Regarding ethanol withdrawal subtype delirium tremens, which one is the *wrong* statement?
a) It is the most serious of all ethanol withdrawal syndromes
b) Characterized by confusion, agitation, fever, and hallucination
c) The mortality rate is low, around 1%
d) Should be treated with diazepam 10-20 mg intravenously and repeated every 5 minutes as needed until the patient is calm
e) Concomitant treatment with beta blockers is advisable

20) Alcoholic hallucinosis is one of the problems that are seen when alcoholics try to detoxify themselves from this type of substance abuse. Regarding ethanol withdrawal subtype tremulousness and hallucination, which one is the *wrong* statement?
a) In general, it is a benign self-limiting condition
b) Usually occurs within 2 days of drinking cessation
c) There is agitation, anorexia, hypertension, insomnia, and tachycardia
d) Prominent confusion dominates the clinical picture
e) Can be treated by chlordiazepoxide or diazepam

21) A 55-year-old man develops troublesome supraventricular tachycardia every now and then, for which he takes daily amiodarone. His body aches and he has weight gain and constipation. Serum TSH is high. Regarding amiodarone-associated hypothyroidism, which one is the *incorrect* statement?

a) Patients with positive anti-thyroid antibodies are more likely to develop persistent hypothyroidism
b) Thyroid function can be easily normalized by replacement with T4 while amiodarone is continued
c) Usually, larger than usual dose of thyroxin is required for the treatment
d) Amiodarone should be stopped
e) Overt hypothyroidism may manifest itself 2 weeks after starting amiodarone

22) A 5-year-old child is referred from a rural hospital for further treatment of iron intoxication. All of the following are toxic effects of iron on cells, *except*?
a) Mucosal cell necrosis
b) Inhibition of enzymatic processes in the Krebs cycle
c) Wide-spread vasoconstriction
d) Inhibition of plasma proteases
e) Uncoupling of oxidative phosphorylation

23) A 33-year-old man presents with rapid loss of vision. His final diagnosis turns out to be methanol poisoning. Which one of the following statements regarding this poisoning is *correct*?
a) Methanol is metabolized to oxalic acid
b) The blood's osmolal gap is reduced
c) The objective of using ethanol infusion as a treatment mode is to inhibit alcohol dehydrogenase
d) The main disadvantage of fomepizole is its bad side effect profile
e) Hemodialysis should be used only if metabolic acidosis is very severe

24) A 65-year-old man has been prescribed imatinib mesylate for his newly diagnosed Philadelphia chromosome positive chronic myeloid leukemia. Which one of the following is *incorrect* with respect to this medication?
a) Also used in the treatment of Philadelphia chromosome positive acute lymphoblastic leukemia
b) Promotes fluid retention
c) Results in dry eyes
d) Acute febrile neutropenic dermatosis is a rare adverse event
e) Decreases the intestinal absorption of digoxin

25) You plan to treat this 46-year-old woman with trastuzumab. She asks about this form of treatment. All of the following statements are correct with respect to trastuzumab therapy, *except*?

a) It is used in the treatment of HER2/neu over-expression metastatic breast cancer
b) Although it is given intravenously, it should not be given as a rapid bolus
c) Infusion reactions with fever and chills are common
d) Adult respiratory distress syndrome may occur during trastuzumab infusion or within 24 hours of administration
e) It reduces the incidence of cardiac dysfunction when combined with anthracyclines

26) A 32-old man who has type II diabetes visits your office. He says that he has read in the internet that there is an anti-diabetic medication called pramlintide and he asks whether he can use it or not to control his blood sugar. All of the following statements are incorrect with respect to this medication, except?
a) An amylin antagonist
b) Given orally as soft gel capsules
c) Gastroparesis is a contraindication for its use
d) Should be given simultaneously with insulin
e) Elimination half-life is about 24 hours

27) A 56-year-old man was prescribed exenatide for optimal control of his type II diabetes. He heard that this anti-diabetic medication is new and he declines its administration. Which one of the following statements about exenatide is correct?
a) Blocks the action of incretin
b) The majority of the drug is excreted unchanged in bile
c) Attenuation of its effect may occur due the development of anti-exenatide antibodies
d) Oral formulations should be given once daily
e) Useful in the management of diabetic ketoacidosis

28) A 67-year-old man receives many medications while he was admitted in the coronary care unit. One of his medications is ibutilide. Which one of the following is the correct statement with respect to this medication?
a) A class I anti-arrhythmic drug
b) Its oral formulation is not available because the drug is extremely bitter
c) Produces mild slowing of the sinus rate
d) Can result in severe hypotension resistant to intravenous fluids
e) Excreted unchanged in urine

29) A 61-year-old woman develops acute left ventricular failure. The Emergency Room physician gives her nesiritide in addition to other anti-failure measures. All of the following statements about nesiritide are wrong, *except*:

   a) Given as oral tablets of 250 µg every one hour for a maximum of 8 tablets

   b) Can increase serum creatinine

   c) Contraindicated in renal failure

   d) The drug of choice in cardiogenic shock

   e) Produces dose-dependent elevation in the pulmonary capillary wedge pressure

30) Your intern asks about the safety of various medications and drugs during pregnancy. You tell him that medications and drugs are categorized into different classes to guide physicians how to use them during pregnancy. All of the following statements about this categorization system are correct, *except?*

   a) Category A means that controlled studies show no risk to the fetus

   b) No evidence of risk in human fetuses refers to category B

   c) Category C indicates that fetal risk cannot be ruled out

   d) Positive evidence of fetal risk means category D

   e) Category X means that fetal risk is extremely low

Clinical Pharmacology, Therapeutics, and Toxicology

Answers

1) e.

Hypokalemia, hypomagnesaemia, hypoxemia, hypernatremia, and hypercalcemia increase the risk of digoxin intoxication; hyperkalemia occurs as a consequence in acute poisoning while chronic intoxication does not usually cause hyperkalemia. Hypothyroid patients are prone to digoxin toxicity secondary to decreased renal excretion and a smaller volume of distribution. Active cardiac ischemia, myocarditis, cardiomyopathy, cardiac amyloidosis, and cor pulmonale increase the sensitivity to digoxin intoxication.

2) e.

The toxicity of tricyclics results from inhibition of noradrenalin and serotonin reuptake at nerve terminals, their anti-cholinergic action, direct α-adrenergic blockade, and membrane stabilizing effect on the myocardium by blocking the cardiac myocyte fast sodium channels. Indications of bicarbonate infusion in tricyclic antidepressants toxicity are: long QT interval; severe hypotension; severe acidosis; life-threatening cardiac dysrhythmias; and signs of severer CNS toxicity (such as seizures). The objective is to raise the blood pH to a level of 7.45-7.55 with the serum potassium being in the upper part of its normal reference range.

3) c.

In lead poisoning, red cell aminolevulinic acid dehydratase (ALA) activity is decreased but the red cell free protoporphyrins are increased. Glycosuria and aminoaciduria may develop due to chronic interstitial nephritis and renal tubular acidosis. Blue lines on gums indicate chronic lead exposure; this staining occurs mainly due to deposition of sulfides and irritation of the nearby gum. Urinary ALA and coproporphyrin are elevated but urinary porphobilinogen is normal. Peripheral motor neuropathy is mainly seen in adults with chronic intoxication, while encephalopathy usually develops in children.

4) e.

Activated charcoal does not bind methanol and ethylene glycol. Initially, renal impairment is mild and reversible; it may become irreversible when there are delayed presentation and delayed treatment.

Ethylene glycol metabolites target the kidney and lead to reversible oliguric or anuric renal failure, which in turn slows elimination of ethylene glycol. Hypocalcemia in ethylene glycol overdose results from calcium oxalate formation.

5) e.

Amphetamine is one of the causes of unexplained intracranial hemorrhage in young people. Serum creatinine phosphokinase is elevated when rhabdomyolysis or prolonged seizures occur. Hyponatremia results from SIADH and drinking too much water (because of thirst). Hypotension may be due to myocardial infarction, shock, malignant ventricular dysrhythmias, and aortic dissection. However, hypertension is more common than hypotension.

6) e.

Carbon monoxide poisoning should always be looked for in the presence of reduced PaO_2 and normal level of O_2 saturation on pulse oximetry (note the discrepancy). The rose pink color of the skin is rare ante-mortem; cyanosis is much more common. Phenobarbitone should be avoided when treating seizures; this medication may further impair the release of oxygen to tissues. In smokers, the level of carboxyhemoglobin may reach up to 15%. Indications of hyperbaric O_2 therapy are: pregnancy; severe neurological impairment; coma; and carboxyhemoglobin blood level >40%.

7) d.

In phenytoin poisoning, diplopia, ataxia, and *coarse* tremor are common. Hemodialysis is not effective in clearing the blood from phenytoin molecules (because it is highly protein-bound).

8) d.

Lithium intoxication is enhanced by NSAIDs, renal failure, ACE inhibitors, and diarrhea. Lithium therapy can cause hypothyroidism and hyperparathyroidism (with high serum PTH). During poisoning, coma and up-going planter reflexes may occur. Fine tremor is commonly found at therapeutic levels; *coarse* irregular tremor indicates poisoning. Hemodialysis is the treatment of choice in severe poisoning (serum level >3-3.5 mmol/L).

9) c.

Ethanol intoxication produces acute confusional state, which is mainly seen in non-alcoholics. The severity and clinical features of the resulting encephalopathy correlate roughly with blood ethanol levels, generally speaking. Chronic heavy alcoholics might have a very high blood level although they do not appear to be intoxicated. The plasma osmolality is characteristically raised. The plasma osmolality roughly increases by 22 mOsm/Kg for every 100 mg/dl of ethanol present. Ethanol intoxication can generally be diagnosed by the presence of ethanol odor, increased plasma osmolality (in ethanol poisoning it is raised), and blood and urinary toxicology. This intoxication predisposes to head injury, lung aspiration, and seizures. Chronic alcoholism increases the risk of bacterial meningitis. The treatment is supportive only. All alcoholics should receive 100 mg of thiamin intravenously to prevent Wernicke's encephalopathy.

10) d.

Sedative drug poisoning results in respiratory depression, hypotension, and hypothermia. The patient's pupils are usually reactive; very large doses of phenobarbitone or glutethimide may result in large and fixed pupils. The mortality rate is low; this is mostly due to aspiration pneumonia (that results from impaired cough and swallowing) or due to iatrogenic fluid overload and pulmonary edema (especially when forced alkaline diuresis is being used). Despite severe intoxication, a patient who arrives at the hospital with adequate cardio-pulmonary function and support should survive without any sequelae. The treatment is mainly supportive while the drug is being eliminated. Stem "e" is mainly used to increase the urinary clearance of phenobarbital, but in general should be avoided, as it can lead to fluid overload. Hemodialysis is the best treatment option in severe resistant cases of barbiturate poisoning or when renal impairment impedes drug elimination.

11) d.

Withdrawal of intermediate- or short-acting agents is more likely to produce a withdrawal syndrome when these agents are stopped abruptly. This syndrome is usually seen within 1-3 days of stopping drug intake (for short-acting agents); however, up to 1 week or even more may be needed to result in a fully-fledged syndrome after suddenly withdrawing long -acting agents. Confusion, agitation, and seizures may dominate the clinical picture in some; these seizures, especially myoclonic ones, should be treated with anticonvulsants.

If stem "c" is positive, the patient should receive a long-acting phenobarbital to maintain a calm state without producing signs of intoxication. In most patients, it is possible to stop the drug gradually after progressive decrement in the daily doses within 2 weeks. Stem "e" is true; this is mainly seen in those who take very high and frequent doses.

12) d.

Morphine intoxication may occur as an iatrogenic overdose (in hospitalized patients or outpatients) as well as an accidental overdose in addicts (although needle tracks and marks of intravenous drug abuse might be detected, they are not diagnostic) and in suicidal attempts. Pontine hemorrhage is a differential diagnosis. This form of intoxication can simply be confirmed at the bedside by giving naloxone to the patient; the test is considered positive if the patient's pupils dilate and he/she regains full consciousness. However, when very large doses of opioids are taken or multiple drug ingestion is present, the pupils may slightly dilate (or even, no dilatation occurs at all). With appropriate treatment, patients should recover uneventfully; the mortality rate is low.

13) d.

Many classes of medications and drugs have a powerful anti -cholinergic effect; antidepressants, antihistamines, and antipsychotics have this property and their intoxication can be confirmed by doing urinary/blood toxicology screen (this is mainly used in antipsychotics or antidepressants overdose). There are flushing, urinary retention, and tachycardia as well. With proper management and support, symptoms usually resolve spontaneously and fatalities are rare (however, tricyclics have an increased mortality rates because of their cardiac toxicity). Physostigmine can produce severe bradycardia, seizures, and hyper-salivation; it is rarely indicated in clinical practice (mainly when there are life-threatening cardiac dysrhythmias).

14) d.

Sympathomimetic agents exert their effects via a variable combination of inhibiting the reuptake and/or increasing the release of noradrenalin and/or dopamine at receptor sites; the net result is a central stimulant and peripheral sympathomimetic effect. Cocaine can produce myocardial infarction via strong coronary vasospasm. Stroke may result from sudden severe hypertension, drug-induced vasculitis, or rupture of cerebral arteriovenous malformation.

Beta blockers should be avoided, especially in cocaine-induced myocardial infarctions, as this will allow the unopposed alpha receptor stimulation to further constrict the coronaries; therefore, alpha-blockers are useful to attack hypertension. Because amphetamines are longer acting than cocaine, amphetamine intoxication is more likely to require treatment.

15) c.

LSD intoxication results in prominent insomnia. Changes in the mental status are usually the most striking feature. Alterations in affect and mood may dominate the clinical picture. The presence of prominent seizure activity should prompt a search for another pathology or to revise the diagnosis. The pupils are dilated and there features of prominent sympathetic over-activity. Verbal calmness and reassurance are all that is required; when this fails, treatment with diazepam may be of benefit.

16) c.

Unlike other hallucinogens (e.g., LSD), PCP intoxication is a medical emergency. The other features of poisoning are *small* pupils, horizontal and vertical nystagmus, hypertonia, hyperreflexia, and myoclonus. There may be analgesia to a surprising degree. Phenothiazines reduce seizure threshold and may produce severe hypotension; haloperidol can be used safely in such cases. Diazepam can be used for sedation and treating muscle spasms. Most patients improve gradually within 1 day, although in some patients, this recovery may take days or even weeks. Severe hypertension, status epilepticus, malignant hyperthermia, coma, and eventual death are especially seen in poisoning with large doses.

17) e.

Accidental or intentional ingestion of larger than customary doses of medications can result in severe confusion in elderly. However, some patients become confused even when ingesting small recommended doses of medication; those patients may be wrongly diagnosed with a serious CNS illness. Renal and hepatic failures impair the metabolism/kinetics of many medications; acute confusional states can ensue rapidly. Dementia patients are very susceptible to side effects of medications. Polypharmacy is a very important cause of confusion in elderly.

18) a.

Rum fits of ethanol withdrawal usually develops within 48 hours of alcohol abstinence; however, in 70% of cases, they occur within 7-24 hours of abstinence. The interval between the first and last seizure is usually 6-12 hours (in up to 85% of cases). About 40% of patients develop 1 seizure only. The fits are generalized tonic-clonic ones. The occurrence of >6 seizures, prolonged duration of these seizures (>6-12 hours), development of status epilepticus, focal fits, or prolonged post-ictal phase should prompt a search for an alternative explanation for these seizures. These fits usually abate spontaneously; however, diazepam or chlordiazepoxide are given prophylactically, because $1/3^{rd}$ of patients develop delirium tremens.

19) c.

Delirium tremens is the most aggressive type of all ethanol withdrawal syndromes, which has a mortality rate of 15%; this is mostly due to concomitant infection, pancreatitis, cardiovascular collapse, or alcohol-associated trauma. It usually develops within 3-5 days of abstinence and may last for up to 3 days. Tachycardia and sweating also occur.

20) d.

Stem "d" is false because confusion, if present, is usually mild. Illusions and hallucinations, usually visual, occur in 25% of cases. It usually responds to diazepam 5-20 mg or chlordiazepoxide 20-25 mg orally every 4 hours.

21) d.

Patients with underlying Hashimoto thyroiditis or positive anti-thyroid antibodies are more likely to develop persistent hypothyroidism. This observation may explain the higher prevalence of amiodarone-induced hypothyroidism in women compared to men. Amiodarone is usually not discontinued unless it fails to control the underlying arrhythmia. However, if amiodarone is stopped, hypothyroidism in patients with no apparent preexisting thyroid disease often resolves. In contrast, hypothyroidism may persist after withdrawal of amiodarone in patients who have underlying chronic autoimmune thyroiditis with high titers of anti-thyroid antibodies and goiter, and they therefore require permanent T4 therapy.

22) c.

Ferric iron is toxic to a number of cellular processes. The primary mechanism for iron-induced tissue damage is free radical production and lipid peroxidation. Iron results in direct *vasodilatation* and impairment of capillary permeability. Normally, transferrin and ferritin protect cells by binding iron, limiting the amount that is circulating in the ferric state. However, these protective mechanisms become quickly overwhelmed during acute intoxications.

23) c.

Methanol is metabolized to formic acid which is responsible for the clinical picture of intoxication; methanol itself is relatively non-toxic and mainly results in mild CNS sedation. Ethylene glycol is metabolized to glycolate and oxalate. Methanol poisoning results in high anion-gap metabolic acidosis and raised serum osmolal gap. Fomepizole is easy to dose, easy to administer, and side effects are rare. Its main disadvantage is the high cost. It is thought be superior to ethanol as an antidote in methanol and ethylene glycol poisoning. Hemodialysis should be used as a treatment mode if metabolic acidosis is present (regardless of the degree).

24) c.

Imatinib inhibits tyrosine kinase. It has many clinical implications. It is used in the treatment of aggressive systemic mastocytosis, dermatofibrosarcoma protuberans, gastrointestinal stromal tumors, chronic eosinophilic leukemia, and myelodysplastic/myeloproliferative disease (that is associated with platelet-derived growth factor receptor gene rearrangements). The drug causes fluid retention and aggravates edemas and pleural, pericardial, and peritoneal effusions. Up to 20% of patients who receive this drug report increased lacrimation. Acute febrile neutropenic dermatosis, acute generalized exanthematous pustulosis, angioedema, aplastic anemia, avascular necrosis, breast enlargement, bullous eruption, cardiac failure, cardiac tamponade, and cerebral edema occur in <1% of recipients.

25) e.

The uses of trastuzumab are limited to the treatment of HER2/neu over-expression in metastatic breast cancer and as an adjuvant treatment of HER2/neu over-expression in node-positive breast cancer.

Its role in the treatment of ovarian, gastric, colorectal, endometrial, lung, bladder, prostate, and salivary gland tumors is still investigational. The drug should be given as a slow intravenous infusion. Infusion reactions occur in 20-40% of patients; fever and chills are the usual adverse events; these usually occur during infusion. Adult respiratory distress syndrome is very rare and may occur during infusion or up to 24 hours of administration. Anaphylaxis, anaphylactoid reactions, angioedema, apnea, and cardiac arrhythmia are fortunately rare. Congestive heart failure associated with trastuzumab may be severe and has been associated with disabling cardiac failure, death, mural thrombus, and stroke. Left ventricular function should be evaluated in all patients prior to and during treatment with trastuzumab; concurrent or prior administration of anthracyclines increases the risk of cardiac dysfunction.

26) c.

Pramlintide is a synthetic amylin analogue. Amylin is co-secreted with insulin by pancreatic beta cells. It reduces postprandial blood glucose levels through prolongation of gastric emptying time (hence, it is contraindicated in gastroparesis), reduction of postprandial glucagon secretion, and reduction of caloric intake through centrally-mediated appetite suppression. This anti-diabetic medication is given as mealtime subcutaneous injections and should not be given with insulin; dangerous hypoglycemia can rapidly ensue (both, in type I and II diabetes). Although pramlintide elimination half-like is about 48 minutes, its duration of action is about 3 hours. After a subcutaneous injection, the time to reach peak plasma level is 20 minutes. The primary mode of elimination is via urine.

27) c.

Exenatide is an incretin (glucagon-like peptide 1) analogue and is administered as subcutaneous injections, 60 minutes before morning and evening meals (i.e., twice daily). This medication increases insulin secretion, increases β-cell growth/replication, slows gastric emptying, and may decrease food intake. When it is combined with sulfonylureas and/or metformin, it results in additional lowering of hemoglobin A1c by approximately 0.5 to 1%. Its elimination half -life is 2.4 hours; the bulk of the drug is excreted in urine (it is contraindicated if the creatinine clearance is <30 ml/minute). Its mechanism of action requires the presence of insulin; therefore its use in type I diabetes or diabetic ketoacidosis is not recommended and is, actually, contraindicated. It is not a substitute for insulin in insulin requiring patients. It is mainly used as an adjunct (to metformin or sulfonylureas) in type II diabetes.

Anti-exenatide antibodies are common and because of their low titer, no decline in efficacy occurs; however, about 6% of patients develop high titers that may render the injection virtually useless.

28) c.

Ibutilide belongs to class III anti-arrhythmic drugs. It blocks IKr (the rapid component of the delayed rectifier potassium current that is responsible for phase III of cardiac depolarization). It used in the treatment of atrial fibrillation and flutter of *recent* onset and is administered as a rapid intravenous infusion. It has no oral preparations because it is subjected to extensive first pass metabolism; therefore, it cannot be used in the maintenance of sinus rhythm in the long-term. It results in slight slowing of the sinus rhythm and prolongation of the QT interval; however, the PR and QRS durations are unchanged (unlike the action of other class III anti-arrhythmics). The resulting hypotension (during or after infusion) is usually mild, transient, and responds well to fluids. Hepatic oxidation eliminates the drug.

29) b.

Nesiritide binds to guanylate cyclase receptors on vascular smooth muscle and endothelial cells. This would increase the intracellular cyclic GMP level resulting in smooth muscle cell relaxation. This produces dose-dependent reductions in pulmonary capillary wedge pressure and systemic arterial pressure. Although the drug is eliminated in urine, no dose modification is needed in renal failure. An increment in the baseline serum creatinine of greater than 0.5 mg/dl occurs in about 30% of patients upon receiving this medication. There are no oral preparations; it is given intravenously. Nesiritide is contraindicated in hypotension (systolic blood pressure <90 mmHg) and cardiogenic shock patients.

30) e.

Pregnant women, like any other human being, can become sick and may require drug therapy for a variety of causes. All physicians should be familiar with drug prescription in pregnancy, at least the commonly used ones. Category X means the drug is absolutely contraindicated because studies in animals or humans, or investigational or post-marketing reports have shown that fetal risk clearly outweighs any possible benefit to the patient.

Chapter 4
Clinical Sciences

Questions

1) A 65-year-old male has multiple myeloma. He presents with recurrent chest infections. You tell him that his blood levels of many immunoglobulins are low and this increases his susceptibility to infections. With respect to immunoglobulins, which one is the *incorrect* statement?
   a) IgA is mainly found in body fluids
   b) IgM is a high molecular weight immunoglobulin and is the largest of all immunoglobulins
   c) IgG is the only one that passes the placental barrier
   d) IgE is involved in type IV hypersensitivity reactions
   e) IgD is found on the surface of B-cells and may be involved in their maturation

2) A 62-year-old female, who was diagnosed with chronic lymphocytic leukemia 3 months ago, receives intravenous immunoglobulins because of recurrent infections. All of the following statements regarding immunoglobulins are correct, *except?*
   a) IgA may be responsible for hyperviscosity in multiple myeloma
   b) Serum IgE is very high in allergic bronchopulmonary aspergillosis
   c) IgG is the usual immunoglobulin detected as a rheumatoid factor by the slide agglutination test
   d) Selective IgA deficiency prevalence is 1/700 of the general population
   e) Anti-nuclear factor is usually an IgG autoantibody

3) A 34-year-old male has rapidly progressive glomerulonephritis and renal impairment. Serum complements are low. Regarding the complement system, choose the *wrong* statement?
   a) The classical pathway can be activated by IgG immune-complexes
   b) Properdin deficiency has an X-linked mode of inheritance
   c) Deficiency of late the complement components is associated with disseminated Neisseria infections
   d) SLE may be induced by congenital deficiency of the early complement components
   e) C1 esterase inhibitor deficiency is an autosomal dominant disease and its defect is qualitative in 90% of cases

4) A 21-year-old male complains of chronic cough that is productive of copious putrid sputum for 2 years. Examination shows bibasal coarse crackles and clubbing. Chest CT scan reveals wide-spread bronchiectatic changes. His serum immunoglobulins level is low. Hypogammaglobulinemia is associated with all of the following, *except?*

a) Thymoma
b) Phenytoin treatment
c) Myotonia dystrophica
d) Protein-losing enteropathy
e) Chronic active hepatitis

5) You attend a lecture about immune deficiency states. You have learned many things about common variable immune deficiency syndrome. Which one of the following you have *not* learned about this syndrome?
a) About 30% of cases have an associated lymphopenia and lymphadenopathy
b) Serum IgA is virtually absent but serum IgG is usually severely decreased
c) Although defective cell-mediated immunity is documented, opportunistic infections due to this defect are rare
d) The presentation may be at any age but usually in the 3rd decade
e) First-degree relatives have an increased risk of IgG subclass deficiency

6) A 32-year-old male, who was diagnosed with myotonia dystrophica, asks you about the consequences of having low serum immnunoglobulins, because he has this immune system abnormality. Long-term complications of hypogammaglobulinemia are all of the following, *except?*
a) Induction of varieties of autoimmune diseases
b) High risk of lymphoreticular malignancies
c) Bronchiectasis
d) Malabsorption
e) Chronic active hepatitis

7) A 9-month-old baby had developed recurrent chest infections. He was found to have a mutation in Bruton tyrosine kinase (BTK) gene. Regarding X-linked agammaglobulinemia, all of the following are true, *except?*
a) T-cell number and function are normal
b) There are no mature B-cells in the peripheral blood
c) Serum IgG is usually still detectable but is extremely low
d) The presentation is usually 3-6 months after birth
e) It is a common disease

8) A 4-year-old child presents with features of malabsorption. He was eventually diagnosed with celiac disease. The boy's serum IgA anti-endomysium antibodies are persistently negative, but serum IgG is positive.

You think he has co- existent selective IgA deficiency. Regarding selective IgA deficiency, which one is the *wrong* statement?
   a)   Incidence is 1/700 of the general population and is usually a sporadic disease
   b)   Ideally, checking serum IgA level should be done in all patients prior to truncal vagotomy
   c)   All patients should be tested for anti-IgA auto-antibodies
   d)   Accompanies 50% cases of celiac disease
   e)   Can be found in patients with rheumatoid arthritis

9) HIV infection is associated with impairment in many aspects of the immune system. Which one of the following statements regarding HIV infection is *wrong?*
   a)   There are M-tropic and T-tropic strains
   b)   b There is polyclonal B-cell activation
   c)   Natural killer (NK) cell number and function are usually impaired
   d)   There is no single mechanism to explain the progressive CD4+ cell attrition over time
   e)   The macrophages are CD4-negative cells and therefore are protected from infection

10) You are reading a journal article about major histocompatibility complex (MHC) and human leukocyte antigen (HLA). Which one of the following this article does *not* mention?
   a)   MHC class II antigen is restricted in distribution to certain cell types
   b)   Their inheritance is autosomal co-dominant
   c)   The loci encoding complement components are located nearby at chromosome number 6
   d)   Certain HLA haplotypes are associated with varieties of autoimmune diseases
   e)   HLA DR2 confers a high risk of developing type I diabetes mellitus

11) A 21-year-old male is referred to you for further evaluation because of primary infertility. His serum FSH and LH are high. He has Klinefelter habitus. With respect to Klinefelter syndrome, all of the following are correct, *except?*
   a)   Gynecomastia is observed in 30% of cases
   b)   The affected person may have 48 XXYY karyotype
   c)   Short stature is more common than tall stature
   d)   Cardiac anomalies are very rare
   e)   Some degree of low intelligence may be seen

12) A 17-year-old short girl has primary amenorrhea and her serum LH and FSH high. Abdominal ultrasonography reveals horseshoe kidneys. She has 45 XO karyotype. Regarding Turner syndrome, which one of the following is the *incorrect* statement?

a) Aortic coarctation is not the only cause of hypertension
b) Branch pulmonary artery stenosis suggests Noonan syndrome rather than Turner syndrome
c) Lymphedema of the hands and feet may be seen early in life
d) Large doses of growth hormone therapy can induce an increase in the patient's height
e) Hormonal replacement therapy can be used to induce fertility in the majority of patients

13) A father of newly diagnosed case of classical congenital adrenal hyperplasia asks you about the inheritance of this disease. You explain what autosomal recessive inheritance entails. The following are autosomal recessive diseases, *except?*

a) Phenylketonuria
b) Familial Mediterranean fever
c) Oculocutaneous albinism
d) Wilson disease
e) Familial hypercholesterolemia

14) A 28-year-old female has a history of repeated miscarriages due to fetal anencephaly. Now, she is 15 weeks pregnant and her amniotic fluid analysis reveals raised alpha-fetoprotein. She is due to test her serum alpha-fetoprotein. Maternal serum alpha-fetoprotein may be raised in all of the following, *except?*

a) Twin pregnancy
b) Threatened abortion
c) Spina bifida
d) Hepatocellular carcinoma
e) Post-date pregnancy

15) A 32-year-old woman gave birth to 2 children with neural tube defects. For the time being, she is 14 weeks pregnant. You plan to do amniotic fluid aspiration and analysis to see if the current baby is also affected. Regarding amniocentesis, all of the following are correct, *except?*

a) The risk of abortion incurred by this procedure is lower than that of chorionic villous sampling
b) Usually performed around the 14th-16th weeks of gestation
c) Cytological examination may be used to detect karyotyping

d) Is usually done blindly
e) Useful in the prenatal diagnosis of congenital adrenal hyperplasia

16) A 34-year-old male presents with frontal baldness, hyperglycemia, and repeated chest infections. His face is dull. You notice wasting of temporalis and masseters as well as bilateral ptosis. Grip myotonia is elicited. Myotonia dystrophica is caused by which trinucleotide (triplet) repeat expansion?
   a) GAA
   b) CTG
   c) CAG
   d) CGG
   e) GCC

17) Your intern asks you about genetic diseases due to chromosomal microdeletions. Which one of the following diseases is due to chromosomal microdeletion?
   a) Juvenile myoclonic epilepsy
   b) Glycogen storage disease type II
   c) Di-George syndrome
   d) Turner's syndrome
   e) Albinism

18) You tell a 24-year-old man why his vision has become poor after an orbital trauma. All of the following statements about the eyes are wrong, *except??*
   a) Rods are located in the central retina and are responsive to dim light
   b) The normal intraocular pressure is 30-40 mmHg
   c) The fovea has no overlying vascular network
   d) The total refractive power of the eye is 5.8 D
   e) The stromal layer forms 10% of the thickness of the cornea

19) A 66-year-old diabetic man visits the doctor's office because of orthostasis symptoms and gastroparesis. You tell him that these have resulted from his diabetic autonomic failure. Which one of the following about autonomic nervous system is *wrong*?
   a) The pre-ganglionic sympathetic fibers are non-myelinated
   b) Cell bodies of all pre-ganglionic parasympathetic neurons are located in the brainstem
   c) Anti-muscarinic medications accelerate the heart rate
   d) Acetylcholine is the neurotransmitter at all pre-ganglionic autonomic neurons

e) The intermediolateral cell column of the spinal cord contains sympathetic neurons

20) You are reading a journal article about adrenergic receptors and their implication in clinical medicine. Which one of the following is *not* present in this article?
a) α_1 adrenoceptors are found in the eye's dilator pupillae
b) Stimulation of β_1 receptors increases the force of cardiac contraction
c) Clonidine is a partial agonist of α_2 receptors
d) Blockage of the bronchial β_2 receptors results in dilatation of the bronchi
e) Adrenalin is a non-selective adrenergic agonist

21) A 22-year-old woman had aortic regurgitation. Her physician told her that she has increased cardiac output. All of the following increase the cardiac output, *except?*
a) Eating
b) Histamine
c) Pregnancy
d) Sleep
e) High environmental temperature

22) A 54-year-old man is brought to the Emergency Room after developing severe peptic ulcer-related hemorrhage. He is in shock and you resuscitate him. Which one of the following does *not* occur as a compensatory mechanism in hemorrhagic shock?
a) Venoconstriction
b) Increased plasma protein synthesis
c) Increased secretion of erythropoietin
d) Decreased secretion of vasopressin
e) Increased secretion of glucocorticoids

23) A 32-year-old man is brought to the Acute and Emergency Department. A&E. He has severe hypoxemia from severe chest infection. Hypoxia can result in all of the following, *expect?*
a) Impaired judgment
b) Loss of time sense
c) Dulled pain sensibility
d) Excitement
e) Increased appetite

24) You are reviewing a medical journal article about functions of the lung. You have learned that the lung is not only involved in respiration but it has many other additional tasks to do. Which one of the following substances is activated by the lungs to its active form?
a) Serotonin
b) Bradykinin
c) Acetylcholine
d) Adenine nucleotides
e) Angiotensin I

25) A 43-year-old man did renal biopsy because of persistent hematuria. Histopathological examination reveals expansion of the mesangium. Which one of the following agents causes relaxation of the glomerular mesangial cells?
a) Dopamine
b) Prostaglandin F_2
c) Leukotriene C
d) Thromboxane A_2
e) Platelet-derived growth factor

26) A 44-year-old woman has galls stones. Laparoscopic cholecystectomy has been scheduled. You tell her that her bile will no longer be stored in a reservoir and will flow directly into the duodenum. Which one of the following forms the largest constituent of normal human bile?
a) Bile salts
b) Bile pigments
c) Water
d) Alkaline phosphatase
e) Cholesterol

27) A 65-year-old woman complains of excessive mouth dryness. She has Sjögren syndrome. She desperately asks you if there is anything that can increase the flow of her saliva. Which one of the following statement about saliva is *correct*?
a) About 250 ml of saliva is secreted per day
b) Normal saliva has a pH of 5
c) Aldosterone increases the concentration of potassium in saliva
d) Salivary lipase is secreted by the sublingual glands
e) The parotids are responsible for about 70% of the daily secreted saliva

28) A 33-year-old man developed non-Hodgkin's lymphoma. You told him that his secondary lymphoid organs were also involved. Which one of the following is *not* a secondary lymphoid organ?
   a) Spleen
   b) Cervical lymph nodes
   c) Thymus
   d) Mucosa-associated lymphoid tissue
   e) Popliteal lymph nodes

29) A 21-year-old woman presents with fever and pallor. You do blood film examination. You believe that blood film should be repeated once again because there were many artifacts. Which one of the following statements regarding laboratory artifacts of blood film is *not* correct?
   a) When the blood film is too thin, red cell morphology cannot be assessed
   b) Pseudo-rouleaux formation is noted when the blood film is too thick
   c) Faulty drying of the film can result in the formation of pseudo-inclusions in RBCs
   d) In hyperlipidemia, leukocytes can burst even if the film is correctly prepared
   e) The nucleus of lymphocytes appears clove-like lobes if the patient's blood is stored in the refrigerator

30) A 22-year-old woman develops fine and fast postural tremor, palpitations, and heat intolerance. After a properly conducted work-up, you diagnose hyperthyroidism. The patient enquires about thyroid hormones. Which one of the following with respect to the physiology of thyroid gland is *correct*?
   a) Thyroid gland accumulates iodine from the circulation by an active energy-requiring process
   b) Inside the thyroid gland, iodine is organified by thyroid reductase
   c) The TSH receptors are intra-nuclear in location
   d) About 5% of the blood's T_4 is the active form
   e) Thyroxin-binding globulin binds 95% of the circulating T_4

31) A 28-year-old woman visits the doctor's office after getting pregnant. She in her 14th week of gestation. All of the following about endocrinal changes during pregnancy are incorrect, *except*?
   a) Post-prandial blood glucose level is lower than non-pregnancy levels
   b) Pituitary enlargement is not a normal occurrence during pregnancy
   c) Estrogen increases the half-life of thyroxin-binding globulin

d) The weight of the maternal adrenal glands increases gradually as pregnancy progresses

e) An increase in bone mineral content of the maternal skeleton is very common

32) A 6-year-old male has been found to have problems with lipid metabolism. All of the following about lipid degradation are incorrect, *except?*
a) Occurs after meals
b) The degradation site is the mitochondria
c) Mainly done in the adipose tissue
d) Insulin activates this process
e) Ends up in the formation of palmitate

33) A 43-year-old man presents with anemia and necrolytic migratory erythematous rash. He has overt hyperglycemia. His final diagnosis is proved to be glucagonoma. With respect to glucagon, which one is the *correct* statement?
a) Enhances glycogen synthesis in muscles and liver
b) Inhibits gluconeogenesis
c) Inhibits ketone body synthesis
d) Decreases the uptake of glucose by peripheral tissues
e) Stimulates the breakdown of protein

34) A 65-year-old woman develops cancer. Her cancer has developed because of prior history of ionizing irradiation exposure. Ionizing irradiation (e.g., X-ray) exposure increases the risk of developing the following types of cancer, *except?*
a) Thyroid
b) Stomach
c) Breast
d) Multiple myeloma
e) Liver

35) A 43-year-old man has been receiving radiotherapy as part of his cancer treatment. He is afraid of the adverse effects of this form of therapy. Which one of the following is *not* an adverse effect of radiotherapy?
a) Alopecia
b) Lassitude
c) Hypotension
d) Cardiac arrhythmia
e) Dental caries

36) A 67-year-old man has been told by his orthopedic surgeon that the right knee joint can be replaced totally by a prosthetic one to improve his osteoarthritis. Which one of the following about knee anatomy is *correct*?

a) It is composed of 5 sub-joints
b) The patellar tendon is the continuation of the tendon of iliopsoas muscle
c) The cruciate ligaments cover the surface of the joint
d) The menisci serve to deepen the surface of the head of tibia
e) In front of the joint, there is only 1 bursa

37) A 33-year-old man undergoes mediastinotomy in order to take biopsies from a suspicious-looking mass. All of the following are structures located within the middle mediastinum, *except*?

a) Lower half of the superior vena cava
b) Ascending aorta
c) Bifurcation of the trachea
d) Two major branches of the pulmonary veins
e) Azygos vein

38) The mother of this 16-year-old female asks you if there is anything that can accelerate her daughter's puberty because her periods have not started yet. You explain to her what puberty entails and its physiological changes. Which one of the following statements regarding female puberty is *correct*?

a) The earliest change is an increase in the fat content of thighs
b) The breasts develop gradually over 2 years before menses starts
c) The vagina shortens and becomes smooth
d) Ovulation usually occurs once menses begins
e) Pubic and axillary hairs stop developing gradually over few years

39) You are attending a workshop about epidemiological studies and their application in clinical medicine. Which one is the *incorrect* statement about epidemiology?

a) Case-control studies involve a group of persons with a disease and a matched group, which is similar in all respects except for disease
b) A cohort study provides the most direct evaluation of health and disease patterns in a population
c) Incidence is a number of new cases of a given disease at a specified time divided by the population at risk for that disease at that time
d) Prevalence is the number of existing cases of a given disease at a specified time divided by the population at risk for that disease at that time

e) Vital statistics are statistics enumerating blood pressure and heart rate of a given community

40) You tell your interns to attend a national symposium about evidence-base medicine (EBM) and to have an idea about it. Which one of the following about EMB is *wrong?*
   a) It is an approach to medicine that integrates the current best evidence to optimize clinical outcomes and quality of life
   b) The roots of EBM roots to the late 1970s
   c) The term evidence-based medicine is first used by Gordon Guyatt in 1990
   d) EMB recognizes that research evidence is never the sole determinant of clinical decision making
   e) Evidence-based journals and online services are poor resources for EBM

41) You are conducting a research about the use of trans-esophageal echocardiography in young patients with ischemic stroke. You intend to calculate the p-value while interpreting the results. Which one of the following is *not* true with respect to p-value?
   a) The p-value is not the probability that the null hypothesis is true
   b) The p-value does not indicate the size or importance of the observed effect
   c) The p-value is not the probability that a finding is "merely a fluke"
   d) The p-value is not the probability of falsely rejecting the null hypothesis
   e) The significance level of the test is not determined by the p-value

42) There is a presentation of results of one of your colleague's researches about the role of statins in preventing ischemic heart disease. He mentions "number needed to treat" (NNT) several times. Which one of the following is the *incorrect* statement about NNT?
   a) Is an epidemiological measure used in assessing the effectiveness of a health-care intervention
   b) Is the number of patients who need to be treated in order to prevent one additional bad outcome
   c) Is defined as the inverse of the absolute risk reduction
   d) Is an important measure in pharmacoeconomics
   e) If the number is negative, it means that the preventive intervention is very effective

43) You plan to conduct a test about the usefulness of aspirin in preventing stroke in atrial fibrillation patients. You have to calculate the relative risks and their reduction. Which one of the following is the *incorrect* statement about relative risks?
   a) Relative risk is a ratio of the probability of the event occurring in the exposed group versus a non- exposed group
   b) Relative risk is used frequently in the statistical analysis of binary outcomes where the outcome of interest has relatively low probability
   c) A relative risk of > 5 means the event is more likely to occur in the experimental group than in the control group
   d) Relative risk is used in randomized controlled trials and cohort studies
   e) It is used to compare the risk of developing a side effect in people receiving a drug as compared to the people who are not receiving the treatment

44) Your teaching hospital is involved in a large international randomized controlled trial about the effect of new medication called "drugin" to prevent lung cancer in smokers. With respect to randomized controlled trial, which one is the *wrong* statement?
   a) A type of scientific experiment most commonly used in testing the efficacy or effectiveness of healthcare services
   b) It involve the random allocation of different interventions (treatments or conditions) to subjects
   c) These trials could be open, blind, or double-blind
   d) Double-blind trials are preferred, as they tend to give the most accurate results
   e) In a single-blind trial, the researcher does not know the details of the treatment but the patient does

45) You intend to do a trial on a newly discovered anticancer medication called "ancerin". You surf the internet and know that you have to pass through many phases of this trial to approve the drug. Which one of the following about the phases of clinical trials of new medications is the *wrong* one?
   a) Phase 0 trials are also known as human microdosing studies and are designed to slow down the development of promising drugs
   b) Phase I trials are the first stage of testing in human subjects
   c) Phase IIA is specifically designed to assess dosing requirements
   d) Phase IIB is specifically designed to study efficacy
   e) Phase III studies are randomized controlled multicenter trials on large patient groups

This page was intentionally left blank

Clinical Sciences

Answers

1) d.

IgA is mainly found in body fluids; therefore, its deficiency may cause chronic or repeated sinopulmonary and gastrointestinal infections. Being the only immunoglobulin that crosses the placenta, it is responsible for the passive neonatal immunity and the false positive neonatal screening tests for many infections. IgE participates in type I hypersensitivity reactions. Serum IgD is very low (< 0.1 g/L) and its half-life is about 3 days. Along with IgM, IgD is predominant among the surface receptors of mature B cells.

2) c.

In Waldenström macroglobulinemia, the increased production of IgM is responsible for hyperviscosity, while IgA makes the hyperviscosity symptoms of multiple myeloma. IgE has the lowest concentration in normal human serum (250 µg/L is the upper normal limit); its half-life is 3 days. IgE plays a key role in immune responses to eukaryotic parasites and in allergic reactions. Very high serum levels of IgE (total and specific) are encountered in allergic bronchopulomonary aspergillosis. Slide agglutination tests for rheumatoid factor detect only IgM antibodies, although this factor could be of IgM, IgG, or IgA class.

3) e.

The classical complement pathway can be activated by IgG or IgM immune complexes; this pathway is activated when the C1q component of the C1 complex attaches to the Fc portion of an antibody. IgA activates the alternative pathway. The alternative pathway is an ancient pathway of innate immunity that preceded adaptive immunity; therefore, a prior contact with a microbe is not required for the alternative pathway to function. Properdin stabilizes C3bBb, a C3 convertase, by the formation of C3bBbP. Late complement components (C5-9) deficiency predisposes patients to recurrent Neisseria infections (including recurrent meningitis). Because of the inability to remove immune complexes, deficiency of early complement components (C1, C2, and C4) put patients at risk of developing SLE; however, isolated deficiency of C3 predisposes to recurrent infection with encapsulated bacteria. About 10% cases of C1 inhibitors have a qualitative defect in that enzyme; the rest (90%) are quantitative defects.

4) e.

In thymoma, hypogammaglobulinemia is considered to be a primary, not a secondary, phenomenon. Chlorpromazine, phenytoin, carbamazepine, sodium valproate, D-penicillamine, sulfasalazine, and hydroxychloroquine have been implicated in selective IgA deficiency. The other causes of hypogammaglobulinemia are nephrotic syndrome, severe ovarian hyperstimulation syndrome, chronic myeloid leukemia, and multiple myeloma. Chronic active hepatitis is associated with hypergammaglobulinemia, mainly of IgG type. Primary or congenital B-cell disorders that result in hypogammaglobulinemia are X-linked agammaglobulinemia (Bruton disease), non-X linked hyper-immunoglobulin M syndrome, isolated immunoglobulin deficiency (IgM, IgA), immunoglobulin G subclass deficiency, common variable immunodeficiency syndrome, transient hypogammaglobulinemia of infancy, and immunoglobulin E hypogammaglobulinemia. Immune deficiency states that are produced by a combined effect of T and B-cell disorders are X-linked severe combined immunodeficiency, Janus tyrosine kinase-3 deficiency, adenosine deaminase deficiency, recombinase-activating gene proteins 1 and 2 deficiency, X-linked immunodeficiency with hyper-immunoglobulin M, cartilage-hair hypoplasia, reticular dysgenesis, and Wiskott-Aldrich syndrome .

5) e.

In common variable immune deficiency, the spleen may be enlarged in 30% of cases. Serum IgM may be normal, low, or even elevated. The T-cells are usually functionally immature. The syndrome has no specific HLA association. First degree relatives have increased risk of selective IgA deficiency. Common variable immune deficiency is the most prevalent primary immunodeficiency syndrome. In those patients, 3 complications must be considered and tackled accordingly: recurrent infections, autoimmune phenomena, and malignancy.

6) e.

Chronic active hepatitis is associated with hypergammaglobulinemia. The presence of a broad-based peak or band, usually of gamma mobility, on serum protein electrophoresis suggests a polyclonal increase in immunoglobulins; this is most often due to an infectious, inflammatory, or reactive process. The commonest causes of this (in descending order) are chronic liver disease, connective tissue diseases, chronic infections, hematological disorders, and non-hematological malignancies.

7) e.

T-cell number and function are intact in this disease; therefore, there is no defective cell -medicated immunity. Because of defective maturation of B-cells in the bone marrow, no such cells can be found in the peripheral blood. Complete absence of serum IgG is rare; however, it is usually still detectable at a level below 50 mg/dl. The disease usually presents after 3-6 months of birth (usually with repeated chest infections) after disappearance of the passively transferred maternal antibodies. The disease is rare. Early detection and diagnosis is essential to prevent early morbidity and mortality from systemic and pulmonary infections. The disease can be diagnosed by finding an abnormally low or absent number of mature B lymphocytes as well as low or absent expression of the μ heavy chain on the surface of the lymphocyte. T-lymphocyte number is elevated. The gold standard diagnostic test is the detection of complete absence of BTK ribonucleic acid or protein. Rarely, the diagnosis is made in patients in their second decade of life; this is thought to be due to a mutation in the protein, rather than its complete absence.

8) d.

Relatives of IgA deficient patients may have common variable immune deficiency. Both serum and secretory IgA are lacking in most patients and, rarely, one or the other. Ideally, checking serum IgA level should be done in all patients prior to truncal vagotomy (as a treatment of peptic ulcer disease); this surgical operation would cause gut hypomotility (an additional risk factor for bacterial overgrowth). IgA deficient patients (especially those having undetectable levels of IgA) may develop anti-IgA antibodies; blood or blood product transfusion may cause severe anaphylactic reactions. Antibody reactions against cow's milk protein are also common. The disease accompanies 3-4% of celiac patients and may cause false-negative IgA anti-endomysium antibody testing. Selective IgA deficiency may occur in rheumatoid arthritis due to the disease itself or its treatment with gold and penicillamine.

9) e.

Depending on certain chemokine receptors and associated ligands, HIV is said to have 2 strains; M- tropic and T-tropic strains. Hypergammaglobulinemia is common and results from polyclonal activation of B cells. Natural killer cell number and function are usually impaired; this is usually overlooked in clinical practice.

Macrophages are CD4 positive cells and are infected early in the course of the disease; M (macrophage) strains of the virus are found in these cells. The T-strains are found in T-cells.

10) e.

MHC class II antigen is restricted in distribution to certain cell types (unlike the widely distributed class I). MHC class I antigens are HLA A, B, and C; this antigen is expressed on almost all cells of the body (except erythrocytes and trophoblasts) at a varying density. The class II antigens (HLA DP, DQ, and DR) are constitutively expressed on B cells, dendritic cells, and monocytes, and can be induced during inflammation on many other cell types (that normally have little or no expression). HLA DR2 may confer protection against the development of type I diabetes; unlike HLA B8/DR3 or DR4 which increase the risk of developing this autoimmune disease.

11) c.

Gynecomastia is observed in 30% of Klinefelter syndrome cases as a secondary phenomenon to hypogonadism. The classical karyotype is 47 XXY; however, many other karyotypes are also seen, e.g., 48 XXYY, 48 XXXY, and 46 XY/47 XXY mosaicism. Klinefelter patients are tall; the low levels of serum androgens would cause delayed epiphyseal closure. Those patients have a variety of behavioral abnormalities (which are unrelated to the hypogonadism) and this result in difficulty in social interactions throughout life. An interesting observation is that there is a predisposition to develop morbidities later in life that are unrelated to testosterone deficiency. The list includes pulmonary diseases (chronic bronchitis, emphysema, and bronchiectasis), cancers (including germ-cell tumors, particularly extra-gonadal tumors involving the mediastinum, breast cancer, and possibly non-Hodgkin lymphoma), varicose veins (leading to leg ulcers), and diabetes mellitus.

12) e.

Renal anomalies are common in Turner patients; hypertension secondary to renal disease may develop. Turner's syndrome confers an increased risk of left-sided cardiac lesions. Lymphedema of the hands and feet may be seen early in life (usually transient). Hearing loss, thyroid failure, and liver function abnormalities usually develop as these women get older. Liver enzymes are elevated in approximately 45% of adult Turner patients; this surprisingly improves with the use of hormonal replacement therapy.

Large doses of growth hormone may be required to induce growth and to increase patient's height; however, side effects of this therapy are common, e.g., gall stones development and neovascular retinal responses. Sex hormones are used to induce the appearance of secondary sexual characters; infertility is permanent and irreversible in Turner syndrome.

13) e.

The word "familial" indicates an autosomal dominant mode of inheritance; familial Mediterranean fever is an exception, however, by displaying an autosomal recessive form of inheritance. Inborn errors of metabolism are autosomal recessive diseases.

14) e.

Post-date pregnancy is not associated with elevated serum alpha fetoprotein. Alpha fetoprotein is a fetal specific globulin that is synthesized by the fetal yolk sac, gastrointestinal tract, and liver. Its function is still unknown; however, it may be involved in immune regulation during pregnancy and may act as an intravascular transport protein (because of its similarity to albumin). This fetal protein is excreted by the fetal kidneys into urine and then into the amniotic fluid. The concentration of amniotic fluid alpha fetoprotein is highest early in pregnancy, peaks between 12 and 14 weeks of gestation, then declines until it becomes undetectable at term.

15) d.

Amniotic fluid aspiration carries a risk of abortion; the other complications are membrane rupture, bleeding, infection, and fetal injury (direct and indirect). It is usually done when the size of the uterus is large enough (between 14-16 weeks of gestation), but still it is considered to be a late investigation; chorionic villous sampling can be performed earlier (so that an intervention can also be done early). The procedure is not blind; it should be guided by an ultrasound imaging. It is useful in the prenatal diagnosis of congenital adrenal hyperplasia by finding an elevated level of 17-hydroxy progesterone in the amniotic fluid.

16) b.

GAA trinucleotide repeats expansion occurs in the non -coding region of Frataxin in Friedreich's ataxia. Expanded CTG repeats are found in the non-coding area of DMPK gene of myotonia dystrophica. In Huntington's disease, Machado-Joseph disease, spinocerebellar ataxia (type 1, 2, 3, 6, and 7), dentatorubral-pallidoluysian atrophy, and spinobulbar muscular atrophy, CAG trinucleotide repeats expansion occurs in the "coding" region of their respective genes. Fragile X mental retardation has an abnormal CGG expansion of the non-coding region of FMR-1 gene, while expansions of GCC are found in fragile site mental retardation (non-coding area of the defective gene).

17) c.

The loss of a part (or deletion) of a chromosome may be microscopic and becomes visible only after using standard chromosome preparations. This deletion could also be submicroscopic where special techniques are required for its precise identification; the most helpful investigation is fluorescent in situ hybridization (FISH). The following are diseases due to chromosomal microdeletions: DiGeorge syndrome (the commonest one; there are facial dysmorphism, anomalies of the heart and palate, and absent parathyroid glands), William's syndrome (elfin face, hypercalcemia, mental retardation, and supravalvular aortic stenosis), WAGR (Wilms tumor, aniridia, genitourinary abnormalities, and mental retardation), and Angelman/Prader-Willi syndromes.

18) c.

The fovea is the central area of the macula and is located 3 mm lateral to the optic disc. The fovea has only cones receptors (which are present in very high density there) and no overlying vascular network; the overlying neural elements are displaced to allow maximum light access. The retinal cones are concentrated in the central part of the retina; they are less sensitive to light (when compared to rods) but they respond best to bright light. Their number is about 6 millions. The rods receptors are located mainly in the peripheral retina; they are very sensitive to light and they respond to dim light. The retina contains about 120 million rods. The total refractive power of the eye is 58.6 D (most of the power, and that is 42 D, is at the air-corneal interface). The cornea has 5 layers; outer epithelial layer, basement membrane, stroma (forms about 90% of the corneal thickness), basal lamina, and endothelial layer. The arrangement of the fibrils of the stromal layer gives the cornea its transparency.

19) b.

There are 2 clusters of parasympathetic neurons in the CNS; the brainstem neurons (which are parts of some cranial nerve nuclei) and sacral spinal cord neurons (at S_{2-4}). The fibers of the latter group exit the spinal cord with the ventral roots of spinal nerves. The intermediolateral column of the spinal cord extends from T_1 to L_2 segments of the cord; the cell bodies of 2nd order sympathetic neurons are located there. The M_2 muscarinic receptors are found in cardiac atrium and their stimulation results in slowing of the heart rate; anti-muscarinic medications, therefore, accelerates the heart rate. Acetylcholine is the neurotransmitter at all sympathetic and parasympathetic pre-ganglionic neurons; the post-ganglionic parasympathetic neurons also release acetylcholine. Most post-ganglionic sympathetic neurons release noradrenalin; only post-ganglionic sympathetic neurons of sweat glands release acetylcholine.

20) d.

Stimulation of $\alpha 1$ receptors results in contraction of smooth muscles in the blood vessels, bronchi, and dilator pupillae. Adrenoceptor $\alpha 2$ stimulation causes contraction of blood vessels smooth muscles, while reduction in central neurotransmitter release in the CNS occurs. Clonidine is a partial agonist of the central $\alpha 2$ receptors. The $\beta 1$ receptors are found in the heart; their stimulation causes an increase in the heart rate and force of contraction. Relaxation of smooth muscles in the blood vessels and bronchi results from stimulation of $\beta 2$ receptors; non-selective β-blockers (such as propranolol) may induce bronchospasm.

21) d.

The cardiac output is the result of stroke volume times the heart rate. About 70 ml of blood is ejected from the heart with every beat; this is the stroke volume. The cardiac output is about 5 L/minute. Factors that increase the stroke volume and/or heart rate will secondarily augment the cardiac output. Excitement and anxiety increases the cardiac output by 50-100%, while a 30% increase in the cardiac output occurs during and after eating. Exercise also augments the cardiac output by up to 700%. Histamine, adrenaline, pregnancy, and high environmental temperature increase the cardiac output. Sitting or standing from a supine position decreases the cardiac output by 20-30%; cardiac arrhythmia and myocardial dysfunction reduce the cardiac output as well. Sleep and moderate change in the environmental temperature incur no change in the cardiac output.

22) d.

During hemorrhagic shock, many mechanisms are activated to compensate for the collapse of the vascular volume. The secretion of the following hormones is increased: adrenalin and noradrenalin, vasopressin, renin and aldosterone, glucocorticoids, and erythropoietin. There is increased movement of interstitial fluid into capillaries. Plasma protein synthesis is augmented. There are increased intrathoracic pumping and skeletal muscle pumping of blood. Tachycardia, venoconstriction, and vasoconstriction happen.

23) e.

Many tissues are affected by hypoxia; the brain is the first victim. Generally, less severe forms of hypoxia result in a variety of mental aberrations not unlike those induced by alcohol, and these are drowsiness, impaired judgment, dulled pain sensation, disorientation, loss of time sense, excitement, and headache. The other features are anorexia, nausea, and vomiting. Tachycardia occurs and if hypoxia is severe, hypertension ensues.

24) e.

The lungs activate angiotensin I to its active form, angiotensin II. The following substances are partially removed by the lungs from the blood: serotonin, noradrenalin, acetylcholine, bradykinin, prostaglandins, and adenine nucleotides. Histamine, kallikrein, and prostaglandins are synthesized/stored in the lungs and then released into the blood stream.

25) a.

Dopamine, atrial natriuretic peptide (ANP), cyclic AMP, and prostaglandin E_2 relax the mesangial tissue. The following factors contract the glomerular mesangium: noradrenalin, vasopressin, angiotensin II, prostaglandin F_2, leukotrienes C and D, histamine, thromboxane A_2, platelet-derived growth factor, and platelet-activating factor.

26) c.

About 97% of the normal human bile is water while alkaline phosphatase forms zero percent of this bile. Bile salts constitute 0.7% of bile while 0.2% of bile is bile pigments. Cholesterol composes about 0.06% of bile.

27) c.

About 1500 ml of saliva is secreted daily; 70% of this comes from the submandibular glands, 20% is from the parotids, and 5% comes from the sublingual glands and the remaining 5% is from minor salivary glands. Saliva has 2 enzymes; lingual lipase (which is secreted by salivary glands on the tongue) and ptyalin (also called α-amylase; this is secreted by all salivary glands). Aldosterone increases the concentration potassium and decreases the concentration of sodium in saliva; an action that is analogous to its action on the kidneys. Normal saliva has a pH of 7; it helps neutralize gastric acid and relieve heartburn when gastric acid regurgitates into the esophagus.

28) c.

Lymphoid tissues are divided into primary and secondary lymphoid organs. Primary lymphoid organs are the sites where lymphocytes develop from their progenitor cells into mature and functional lymphocytes; these are the bone marrow and thymus. Secondary lymphoid tissues are the sites where lymphocytes interact with each other and with other non-lymphatic cells to generate immune reactions against antigens; spleen, lymph nodes, and mucosa-associated lymphoid tissue (MALT) are the secondary lymphoid organs. MALT is a diffusely organized aggregate of lymphocytes that protects the respiratory and gastrointestinal epithelium; these are the tonsils, adenoids, appendix, and Peyer's patches of the ileum.

29) e.

Artifacts are blood film abnormalities that occur when there is faulty or incorrect preparation of the film. The blood film is said to be too thin if too much pressure or too little blood was used during the preparation. In this situation, the RBCs have no central pallor and are usually polygonal in shape. As a result, RBCs morphology cannot be assessed. Too thick blood film occurs when little pressure or too much blood was used during the preparation of blood film. In this case, RBCs stack on each other and pseudo-rouleaux forms. Actually, this phenomenon is much more common than the rouleaux formation of myelomatosis. Leukocytes can crush easily upon doing the blood film. This happens if too much pressure is applied or if hyperlipidemia is present. The white cells could be fragile, as the smudge cells of chronic lymphocytic leukemia. During the drying process of the film, small dots develop and these can easily be confused with Pappenheimer bodies.

This can be solved by adjusting the microscope stage; these pseudo-inclusions become refractile. If the patient's blood is stored above room temperature after venipuncture for some time, the nuclei of lymphocytes appear as clove-like lobes.

30) a.

The thyroid gland traps iodine from the circulation by an active, energy-requiring process. In the thyroid gland, this iodine is organified by the action of the enzyme thyroid peroxidase. TSH acts on cell surface receptors to generate the formation of cAMP and phosphor-inositol as second messengers; the receptors for thyroid hormones are intra-nuclear. About 99.5% of the circulating T_4 is bound to certain plasma proteins; only 0.5% is the free form, which is the bioactive form. Thyroxin-binding globulin (TBG) binds 60-70% of the circulating T_4 at a high affinity but a relatively low capacity; transthyretin (pre-albumin) binds 20-30% of the blood's T_4 with a low affinity but a high capacity.

31) c.

In normal pregnancy, the blood glucose levels tend to have lower fasting values but higher post-prandial levels. The pituitary gland commonly increases in size during pregnancy because of the trophic effect of estrogen on the anterior pituitary cells; preexisting pituitary adenomas may also increase their volume. Estrogen increases the hepatic synthesis of TBG and lengthens its half-life; the net result in an increase in the total serum level of T_4 but the free bioactive fraction of this hormone is maintained. There is no change in the weight of the adrenal glands during pregnancy; however, there is a morphological increase in the width of zona reticularis. Loss of calcium from the maternal skeleton results in gradual decline in the bone mineral density which (if not compensated) may cause osteoporotic fractures.

32) b.

Lipid degradation occurs in response to fasting and prolonged exercise. Skeletal muscles and liver are the main sites involved in this process; it happens in their mitochondria. Lipid degradation is activated by adrenalin and glucagon while insulin inhibits this degradation. The process of lipid degradation ends up with the formation of acetyl CoA.

Lipogenesis, on the other hand, occurs after meals (in the fed state). Lipid synthesis happens in the cytosol of liver and fatty tissues. Insulin activates lipogenesis while adrenalin and glucagon have the opposite effect.

33) e.

The first 3 stems are the functions of insulin; glucagon has the opposite effect. Insulin increases the uptake of glucose by peripheral tissues (not liver) while glucagon has no effect on this process. Insulin decreases the rate of protein breakdown.

34) e.

Although liver cells are very sensitive to the damage incurred by ionizing irradiation, the latter does not seem to increase cancer development in the liver. Ionizing irradiation also increases the risk of developing cancers of the lung, urinary tract, and colon as well as acute leukemia.

35) c.

Exposure of the head/neck area to radiotherapy results in oral mucositis, accelerated dental carries, and oral candidiasis. The risk of thyroid cancer is increased, and intracranial meningioma tends to be multiple and occurs at a younger age. Mediastinal/chest radiotherapy may produce pericarditis, myocarditis and cardiac dysfunction, arrhythmia, acute adult respiratory distress syndrome, interstitial lung fibrosis, and esophagitis. Acute complications of pelvic irradiation are severe diarrhea and cramps. Hypertension gradually develops if the kidneys are damaged within the irradiation field.

36) d.

The knee joint is not a simple hinge joint. Actually, it is a complex one and is composed of 3 articulations in one area; 2 condyloid joints (one between each condyle of femur and the corresponding meniscus and condyle of the tibia) and a 3rd one between the patella and femur. The patellar tendon is the downward continuation of the quadriceps femoris tendon into the tibial tuberosity; it is about 8 cm in length. The cruciate ligaments are X-shaped and situated in the middle of the joint. There are 4 bursae in front of the knee joint; 4 bursae lie at the lateral surface of the knee while 5 bursae are on medial surface. The menisci are semilunar fibrocartilages; each one approximately covers the outer $2/3^{rd}$ of the corresponding articular surface of the tibia.

37) e.

The middle mediastinum is the broadest part of the inter-pleural space.

It houses the heart and its pericardium, major bronchi and peri-bronchial lymph nodes, and the phrenic nerves as well the structures in the first 4 stems in the question. The posterior mediastinum is an irregular triangular area that runs parallel to the vertebral column; this space is bounded by the pericardium above, diaphragm below, vertebrae behind (from T_4 to T_{12}), and on either side by the mediastinal pleura. The posterior mediastinum contains the thoracic duct and some lymph nodes, vagus and splanchnic nerves, descending thoracic aorta, azygos and 2 hemiazygos veins, and the esophagus.

38) b.

The earliest manifestation of female puberty is adrenarche. The rise in serum adrenal sex hormones usually has no clinical "signs"; therefore, the earliest clinically observed feature is the development of breasts buds. The low levels of serum estrogen gradually stimulate breast development; this usually progresses over 2 years before the start of menses. The vagina lengthens and becomes rugated and the labia majora and minora thicken and become rugated as well. Ovulation occurs within 6 months after the start of menses; therefore, the first few cycles are non-ovulatory. The development of pubic and axillary hairs as well as fat deposition (in the thighs and breast) continues for several years after the first menses.

39) e.

Vital statistics are statistics enumerating births, deaths, fetal deaths, marriages, and divorces. A case-control study is an epidemiological study in which one compares the exposure among cases (individuals with a disease) and controls (individuals without that disease). A retrospective evaluation is conducted to determine who was and who was not exposed. A cohort study is a study in which an exposed group and an unexposed group are followed over time to determine who develops the disease of interest.

40) e.

EBM integrates the current best evidence, clinical expertise, and patient values to optimize clinical outcomes and quality of life.

The roots of EBM date to the late 1970, when a group of clinical epidemiologists, led by David Sackett and his colleagues at McMaster University, began preparing a series of articles advising clinicians how to read clinical journals and apply evidence from the literature to direct patient care. However, the term evidence-based medicine was first used by Gordon Guyatt in 1990 while serving as a residency director of the internal medicine program at McMaster. In EBM, any empirical observation about the relationship between event and clinical outcome constitutes potential evidence. Nonetheless, all evidence should not be viewed as equal in making clinical decisions. Electronic evidence databases, evidence-based journals, and online services are the sources that provide significant current best evidence. These sources sharply contrast traditional medical textbooks, which are often not the most appropriate method of finding current best evidence. Although most medical textbooks often provide useful information on pathophysiology, they typically become quickly out-of -date with regard to information on cause, diagnosis, prognosis, prevention, and treatment for a given disorder

41) e.

The p-value is not the probability that the null hypothesis is true; this false conclusion is used to justify the "rule" of considering a result to be significant if its p- value is very small (near zero). The p-value is not the probability that a finding is "merely a fluke"; again, this conclusion arises from the "rule" that small p-values indicate significant differences. The significance level of the test is not determined by the p-value. The significance level of a test is a value that should be decided upon by the agent interpreting the data before the data are viewed, and is compared against the p-value or any other statistic calculated after the test has been performed. The p-value is not the probability of falsely rejecting the null hypothesis. This error is a version of the so-called prosecutor's fallacy.

42) e.

The number needed to treat (NNT) is an epidemiological measure used in assessing the effectiveness of a health-care intervention, typically a treatment with medication. The NNT is the number of patients who need to be treated in order to prevent one additional bad outcome (i.e., to reduce the expected number of cases of a defined endpoint by one). It is defined as the inverse of the absolute risk reduction. It was described in 1988. The NNT is an important measure in pharmacoeconomics.

If a clinical endpoint is devastating enough (e.g., death, heart attack), drugs with a high NNT may still be indicated in particular situations. If the endpoint is minor, health insurers may decline to reimburse drugs with a high NNT. A negative number would not be presented as a NNT, rather, as the intervention is harmful, it is expressed as a number needed to harm (NNH). The absolute risk reduction is the decrease in risk of a given activity or treatment in relation to a control activity or treatment. It is the inverse of the number needed to treat.

43) c.

In statistics and mathematical epidemiology, relative risk (RR) is the risk of an event (or of developing a disease) relative to exposure. Relative risk is a ratio of the probability of the event occurring in the exposed group versus a non-exposed group. Relative risk is used frequently in the statistical analysis of binary outcomes where the outcome of interest has relatively low probability. It is thus often suited to clinical trial data, where it is used to compare the risk of developing a disease, in people not receiving the new medical treatment (or receiving a placebo) versus people who are receiving an established (standard of care) treatment. Alternatively, it is used to compare the risk of developing a side effect in people receiving a drug as compared to the people who are not receiving the treatment (or receiving a placebo). A relative risk of 1 means there is no difference in risk between the two groups; a relative risk of < 1 means the event is less likely to occur in the experimental group than in the control group while a relative risk of > 1 means the event is more likely to occur in the experimental group than in the control group. In medical research, the odds ratio is favored for case-control studies and retrospective studies. Relative risk is used in randomized controlled trials and cohort studies.

44) e.

A randomized controlled trial (RCT) is a type of scientific experiment most commonly used in testing the efficacy or effectiveness of healthcare services (such as medicine or nursing) or health technologies (such as pharmaceuticals, medical devices or surgery). RCTs are also employed in other research areas, such as judicial, educational, and social research. As their name suggests, RCTs involve the random allocation of different interventions (treatments or conditions) to subjects. As long as numbers of subjects are sufficient, this ensures that both known and unknown confounding factors are evenly distributed between treatment groups.

In an open trial, also called an open-label trial, the researcher knows the full details of the treatment, and so does the patient. These trials are open to challenge for bias, and they do nothing to reduce the placebo effect. However, sometimes they are unavoidable, as placebo treatments are not always possible. Usually this kind of study design is used in bioequivalence studies. In a single-blind trial, the researcher knows the details of the treatment but the patient does not. Because the patient does not know which treatment is being administered (the new treatment or another treatment) there might be no placebo effect. In practice, since the researcher knows, it is possible for him to treat the patient differently or to subconsciously hint to the patient important treatment-related details, thus influencing the outcome of the study. In a double-blind trial, one researcher allocates a series of numbers to "new treatment" or "old treatment". The second researcher is told the numbers, but not what they have been allocated to. Since the second researcher does not know, he cannot possibly tell the patient, directly or otherwise, and cannot give in to patient pressure to give him the new treatment. In this system, there is also often a more realistic distribution of sexes and ages of patients. Therefore double-blind trials are preferred, as they tend to give the most accurate results.

45) a.

Phase 0 trials are also known as human microdosing studies and are designed to speed up the development of promising drugs or imaging agents by establishing very early on whether the drug or agent behaves in human subjects as was expected from preclinical studies. Distinctive features of Phase 0 trials include the administration of single sub- therapeutic doses of the study drug to a small number of subjects (10 to 15) to gather preliminary data on the agent's pharmacokinetics (how the body processes the drug) and pharmacodynamics (how the drug works in the body). A Phase 0 study gives no data on safety or efficacy, being by definition a dose too low to cause any therapeutic effect. Drug development companies carry out Phase 0 studies to rank drug candidates in order to decide which has the best pharmacokinetic parameters in humans to take forward into further development. They enable go/no-go decisions to be based on relevant human models instead of relying on sometimes inconsistent animal data. Questions have been raised by experts about whether Phase 0 trials are useful, ethically acceptable, feasible, speed up the drug development process or save money, and whether there is room for improvement. Phase IV trial is also known as Post Marketing Surveillance Trial. Phase IV trials involve the safety surveillance (pharmaco-vigilance) and ongoing technical support of a drug after it receives permission to be sold.

Phase IV studies may be required by regulatory authorities or may be undertaken by the sponsoring company for competitive (finding a new market for the drug) or other reasons (for example, the drug may not have been tested for interactions with other drugs, or on certain population groups such as pregnant women, who are unlikely to subject themselves to trials). The safety surveillance is designed to detect any rare or long- term adverse effects over a much larger patient population and longer time period than was possible during the Phase I-III clinical trials. Harmful effects discovered by Phase IV trials may result in a drug being no longer sold, or restricted to certain uses: recent examples involve cerivastatin, troglitazone, and rofecoxib.

This page was intentionally left blank

Chapter 5
Dermatology

Questions

1) A 32-year-old carpenter has developed oval red raised plaques with silvery white scales on the extensor surfaces of his elbows and knees. There are onycholysis with coarse pitting. With respect to psoriasis, all of the following statements are correct, *except?*
   a) The disease may be exacerbated by propranolol ingestion
   b) Psoriatic arthritis should not be treated by chloroquine
   c) Some areas of involvement may be hidden
   d) The guttate type is seen predominantly in middle-aged people
   e) The inheritance is polygenic

2) A 43-year-old man was recently diagnosed with chronic plaque psoriasis. Which one of the following statements with respect to psoriasis is *incorrect?*
   a) Dithranol can be very effective in treating psoriatic plaques of the face and genitals
   b) Topical vitamin D analogues are effective in treating psoriatic lesions
   c) Flexural psoriasis can be treated with topical steroids combined with topical antifungal preparations
   d) When withdrawing topical steroids gradually, it is advisable to combine them with topical tar preparations
   e) Methotrexate has an advantage of targeting both the skin and join disease

3) A 42-year-old male develops facial acnes few weeks after starting a medication. All but one of the following medications can induce the appearance of an acneform facial rash?
   a) Systemic steroids
   b) Androgenic steroids
   c) Lithium carbonate
   d) Phenytoin
   e) Isotretinoin

4) A 51-year-old female has an acneform facial rash. She denies chronic illnesses or being on medications or illicit drugs. Which one of the following statements regarding rosacea is *correct?*
   a) Usually develops in adolescence
   b) Prominent white and black comedons are present
   c) There is a poor response to systemic tetracycline
   d) Facial telangiectasia and erythema are rarely observed
   e) Conjunctivitis and keratitis occur

5) A 32-year-old male presents with polygonal purple flat-topped papules on the volar aspect of his wrists. The rash is intensely itchy. Regarding lichen planus, all of the following statements are true, *except?*
   a) Can result in scarring alopecia
   b) Although mouth mucosal involvement is common, but frank mucosal ulceration is rare
   c) Absence of itching should cast a doubt on the diagnosis
   d) The disease is disabling and persists for life with relapses and remissions
   e) A multitude of nail changes may develop

6) A 32-year-old female complained of generalized pruritis over the past 3 years. Many scratch marks were found and her nails are "polished". Liver function tests are abnormal and you find positive anti-mitochondrial antibodies. Eventually, primary biliary cirrhosis was diagnosed. Generalized pruritis may result from all of the following, *except?*
   a) Oral contraceptive pills
   b) Iron deficiency anemia
   c) Chronic renal failure
   d) Pregnancy
   e) Treatment with rifampicin

7) Your intern asks you about the causes of severe itching and its significance in making or refuting the diagnosis of diseases. You reply that intense disabling itching is characteristically observed in all of the following, except:
a. Scabies
b. Lichen planus
c. Dermatitis herpetiformis
d. Atopic eczema
e. Vitiligo

8) You are giving a lecture about the usefulness of examining the nails. Which one of the following you have *not* mentioned in your lecture?
   a) Blue nail lunulae are observed in Wilson's disease
   b) Beua's lines may indicate a serious disease in the past 2-3 months
   c) Koilonychia reflects severe long-standing iron deficiency anemia
   d) Diffuse symmetrical black nails point out towards severe chronic liver disease
   e) Half-and-half nails are a sign of chronic renal failure

9) A 43-year-old homosexual male presents with multiple skin lesions which are red-purple and resembles Kaposi sarcoma. The lesions are painful, easily bleed, and some are ulcerated. Histopathological examination of the lesions conforms the diagnosis of bacillary angiomatosis. All of following statements with respect to skin diseases in HIV/AIDS patients are correct, *except?*
   a) Seborrheic dermatitis is more common than in the general population
   b) Drug-induced skin rashes are very common and may limit many medications usage
   c) Wide-spread purple spots and plaques may reflect Kaposi sarcoma
   d) Impetigo is frequent and is one of the AIDS-defining illnesses
   e) Repeated unexplained oral candidiasis may be the first clue towards an underlying HIV infection

10) A 71-year-old male has pigmented basal cell carcinoma on his nose. With respect to basal cell carcinoma of the skin, choose the *incorrect* statements?
   a) The tumor mainly occurs in elderly people
   b) Metastases to the lungs develop early
   c) May appear as an ulcer with rolled-up edges with fine telangiectasia
   d) Best treated by surgery
   e) Has a predilection to the face

11) A 54-year-old farmer from Australia presents with a black nodule on his right shin. The nodule has a variegated appearance and irregular border with areas of ulceration. The right inguinal lymph nodes are enlarged. The liver is rocky-hard and nobly. Risk factors for the development of malignant melanomas include all of the following, except:
   a) Ultraviolet radiation exposure
   b) Fair skin
   c) Positive family history of malignant melanoma
   d) Atypical melanocytic nevi
   e) Diabetes mellitus

12) A 71-year-old retired architect presents with a bluish-black macule on his face. The cervical lymph nodes are not enlarged. You think that this is lentigo maligna melanoma. With respect to malignant melanoma of the skin, which one is the *wrong* statement?
   a) Two thirds of invasive cases are preceded by superficial and radial growth phases
   b) Lentigo maligna and lentigo maligna melanoma are mainly seen on the legs of young people
   c) Superficial spreading melanoma is the commonest subtype

d) About 30-50% of malignant melanomas appear to be derived from pre-existent melanocytic nevi

e) Some melanomas are non-pigmented

13) A 63-year-old female has stage II malignant melanoma. The tumor lies on the right hand. She is asking about her chances of survival, 5 years from now. Regarding the prognosis of malignant melanoma, which one is the *correct* statement?

a) Men generally fare better than women

b) Tumors of the legs are generally less aggressive than those of the face

c) The Breslow's tumor thickness is useful in predicting the prognosis in stage III of the disease

d) Those with stage I disease have a 5-year survival rate of 5%

e) Those with Breslow's tumor thickness of <1 mm have a very good prognosis

14) A 64-year-old retired outdoor manual worker develops a rapidly growing fungating mass on the dorsum of his right hand. Histopathological examination of this mass reveals squamous cell carcinoma. All of the following are risk factors for squamous cell carcinoma of the skin, *except?*

a) Long-term immune-suppressive therapy in solid organ transplantation

b) Dystrophic epidermolysis bullosa

c) Chronic cutaneous ulceration

d) Dark skin

e) Over-treatment with PUVA therapy for a long time

15) A 56-year-old female was referred for further evaluation. She had a small psoriasis plaque-like lesion on her left shin. Otherwise, she was well. She asked whether this lesion is a premalignant one. Regarding skin tumors and their precursors, all of the following statements are wrong, *except?*

a) Seborrheic warts are malignant tumors that mainly develop in young people

b) Actinic keratosis demonstrates full thickness dysplasia

c) Bowen's disease has partial thickness dysplastic changes in the hypodermis

d) On histological examination, keratoacanthoma may resemble squamous cell carcinoma

e) Seborrheic warts are tumors of the sebaceous skin glands

16) You advise your trainees to examine the nails of a 45-year-old female who has a systemic disease. All of the following statements about nail changes and signs are correct, *except*?

a) The presence of a ladder pattern of transverse ridges in the thumb may indicate a habit-tic dystrophy

b) In the absence of a history of nail trauma, a sub-ungual hematoma might be mistaken for a sub-ungual malignant melanoma

c) Splinter hemorrhages can be observed in psoriatic nails

d) The presence of shiny polished nails in chronic atopic dermatitis suggests an alternative diagnosis

e) Yellow brown discoloration with crumbling of the nail plate is seen in fungal nail infections

17) A 28-year-old African male presents with fever, cough, and arthralgia. Chest X-ray shows bilateral hilar lymphadenopathy. Further work-up confirms your clinical impression of sarcoidosis. All of the following are skin manifestations of sarcoidosis, *except*?

a) Dusky red painful and tender nodules on shins

b) Infiltration of long-standing skin scars

c) Purplish plaques on the nose

d) Hypo-pigmented papules

e) Erythroderma

18) A 32-year-old man presents with a strange-looking rash on his arms and legs. The lesions are wide-spread, non-itchy, and have an iris-like configuration. Some blisters are seen. The mucosal surfaces are normal-looking. Erythema multiforme may result from all of the following, *except*?

a) Mycoplasma infections

b) Barbiturate therapy

c) Treatment of internal malignancies with radiotherapy

d) Human Herpes virus type 8 infection

e) Orf

19) A 43-year-old male has inflammatory bowel disease and today he presents with a necrotic ulcer on his thigh. The ulcer has an indurated undermined purplish edge. With respect to pyoderma gangrenosum, which one is the *incorrect* statement?

a) In 50% of cases, no underlying cause can be identified

b) Skin biopsy is very helpful in securing the diagnosis

c) Can be a recurrent event

d) May respond to oral minocycline

e) Local infections should be treated aggressively

20) A 34-year-old man has alcoholic liver cirrhosis and porphyria cutanea tarda. Today, he presents with a variety of skin manifestations. He is on regular venesection and oral chloroquine. The following are skin manifestations of porphyria cutanea tarda, *except?*
a) Excessive skin fragility
b) Small skin erosions
c) Milias
d) Diffuse skin hypo-pigmentation
e) Facial hypertrichosis

21) A 61-year-old man has been referred to the physician's office with arthralgia and heart failure by his physician. Random blood sugar is 380 mg/dl and liver function testing is abnormal. His skin is diffusely hyper-pigmented. The final diagnosis turns out to be hereditary hemochromatosis. All of the following can result in generalized skin hyper-pigmentation, *except?*
a) Nelson syndrome
b) Primary biliary cirrhosis
c) Chronic renal failure
d) Phenylketonuria
e) Addison's disease

22) A 43-year-old male presents with a flare-up of his psoriasis. He denies new medication intake or infection. The following may trigger a flare-up of psoriasis, *except?*
a) A medication given to treat systemic hypertension
b) A medication given to a bipolar disorder patient
c) A prophylactic medication against malaria given to a male who is traveling to Africa
d) Skin surgery to remove a small benign nevus
e) A medication given to treat an acute attack of asthma

23) A 14-year-old girl has a papulovesicular rash on her wrists and ankles. She has severe itching and lichenification. There is a long history of atopic eczema. With respect to general principles in the treatment of eczemas, which one is the *incorrect* statement?
a) Explanation, reassurance, and encouragement of the patient
b) Avoidance of contacts with irritants
c) Regular use of greasy emollients
d) Appropriate use of topical steroids

e) Regular use of cyclosporine

24) A 19-year-old female has asthma. Today, she visits you and complains of excessive facial hair growth. She has been taking high doses of prednisolone over the past year. She is obviously hirsute. Drug-induced hirsutism in women may result from all of the following, *except?*
a) Phenytoin
b) Cyclosporine
c) Minoxidil
d) Androgens
e) Oral contraceptive pills

25) A 43-year-old male complains of a papulovesicular rash in the area of the wristwatch. He states that he is sensitive to nickel. You think this is a contact allergic eczema. All of the following investigations are helpful in eczema (in general), *except?*
a) Patch test
b) Serum total and specific IgE levels
c) Prick skin testing
d) Bacteriology swab from skin lesions
e) Mycobacterial PCR testing of skin biopsy specimens

26) A 14-year-old teen has a rash on his ankles and wrists. The rash is very itchy and interferes with sleeping. He has asthma and his older sister has allergic rhinitis. You tell him that this rash is atopic eczema. Complications of atopic eczema include all of the following, *except?*
a) Secondary super-infections may occur
b) Sleep disturbances are common
c) Loss of schooling is a well-recognized aftermath
d) Co-existent food allergy is rare
e) Topical irritant reactions should be anticipated

27) A 19-year-old female complains of excessive hair growth. She is obese and has secondary amenorrhea. You find no signs of virilization. She has polycystic ovarian syndrome and wants to get rid of this socially embarrassing hair. All of the following can be used in the treatment of hirsutism, *except?*
a) Depilatory creams
b) Waxing
c) Electrolysis
d) Simple shaving
e) Topical minoxidil

28) A 32-year-old female is referred to you for further evaluation of hirsutism. She is on no medications and she has no chronic illnesses. The dermatologist thinks that she has a systemic cause. Full endocrinology assessment should be done in the following cases of hirsutism, *except?*
   a)   If hirsutism occurs in early childhood
   b)   If hirsutism has a rapid onset
   c)   If hirsutism is associated with signs of virilization
   d)   If hirsutism is accompanied by prominent menstrual irregularity or cessation
   e)   If hirsutism is present in other family female members

29) A 32-year-old male has a small rounded patch of scalp hair loss. He says that this has resulted from a prior head trauma. All of the following may result in localized scarring alopecia, *except?*
   a)   Discoid lupus
   b)   Kerion
   c)   Pseudopelade
   d)   Idiopathic
   e)   Alopecia areata

30) A 45-year-old female presents with diffuse scarring hair loss and she is desperate to know why. Diffuse scarring alopecia is *not* caused by one of the following?
   a)   Scalp radiotherapy
   b)   Discoid lupus
   c)   Folliculitis decalvans
   d)   Lichen planopillaris
   e)   Androgenetic alopecia

31) A 45-year-old male presents with wide-spread appearance of flaccid roofed bullae and skin erosions. His month has many ulcers and raw red areas as well. Regarding immune-mediated skin blistering diseases, which one is the *incorrect* statement?
   a)   In pemphigus vuglaris, the target antigen is desmoglein-3
   b)   In bullous pemphigoid, the mucosa of the mouth is involved in 100% of cases
   c)   In dermatitis herpetiformis, there is coarse granular IgA deposition in the papillary dermis
   d)   Epidermolysis bullosa acquisita responds poorly to systemic corticosteroids

e) Pemphigoid gestationis usually attacks the peri-umbilical area and thighs

32) A 33-year-old woman has been referred to the dermatology clinic after developing skin blisters and mouth ulceration. All of the following can cause skin blistering associated with oral mucosal involvement, *except?*
a) Erythema multiforme
b) Eczema herpeticum
c) Steven-Johnson syndrome
d) Porphyria cutanea tarda
e) Toxic epidermal necrolysis

33) A 65-year-old male has been recently diagnosed with mycosis fungoides. Today, he presents with diffuse skin erythema. He has shivering, tachycardia, and hypotension. Serum albumin is low. You think that Sezary syndrome has complicated his illness. Erythroderma can be caused by all of the following, *except?*
a) Psoriasis
b) Drug reactions
c) Some types of icthyosis
d) Pityriasis rubra pillaris
e) Porphyria cutanea tarda

34) A 26-year-old primigravida woman, in her 3rd trimester, presents with generalized itching. She denies chronic illnesses or being on medications. Skin examination is unremarkable, apart from scattered excoriation marks. Pregnancy-associated pruritis may result from all of the following, *except?*
a) Prurigo gestationis
b) Pemphigoid gestationis
c) Pruritic folliculitis
d) Polymorphic eruption of pregnancy
e) Vitiligo

35) A 32-year-old female presents with diffuse and severe sunburn after going to the beach. She was diagnosed with SLE few months ago. All of the following are photosensitive dermatoses, *except?*
a) Solar urticaria
b) Polymorphic light eruption
c) Pellagra
d) Erythema multiforme
e) Pretibial myxedema

Dermatology

Answers

1) d.

Lithium, propranolol (and other beta blockers), and chloroquine may exacerbate psoriatic lesions; the latter is usually prescribed for joint symptoms. Methotrexate improves the skin and joint manifestations. The psoriatic lesions may be hidden in the genital and natal cleft areas; always examine these areas, especially in patients presenting with join disease without any apparent skin involvement. The guttate variety is seen in children; some may develop the chronic plaque type upon reaching adulthood. Psoriasis has many HLA associations, such as HLA DW6. The inheritance is polygenic.

2) a.

Dithranol is a very irritant agent and should always be avoided when treating facial and genital psoriatic lesions. Topical vitamin D and oral retinoids are useful in managing severe cases. To prevent relapse upon steroids withdrawal, topical tar should be combined with topical steroid preparations during withdrawal of the latter.

3) e.

Isotretinoin is used in the treatment of severe scaring acne. The other agents that might cause this acneform rash are chlorinated hydrocarbons, estrogenic steroids, oils, and tar.

4) e.

Rosacea mainly attacks middle -aged people. The presence of wide- spread comedons is suggestive of acne vulgaris, not rosacea. Oral tetracycline is the treatment of choice; sometimes prolonged courses are used for 1 year. Facial telangiectasia and fixed erythema are core parts of rosacea. Conjunctivitis and keratitis may complicate the disease and augment its morbidity.

5) d.

Lichen planopillaris can result in scarring alopecia. Oral involvement is usually seen as white lace-like lesions on the buccal mucosa, but the ulcerative variety is rare and is clinically challenging. Absence of itching should suggest an alternative diagnosis, as the skin lesions are very itchy.

The disease is self- limiting; usually ends within 1-2 years, leaving a prominent hyperpigmentation on the involved skin areas. Patients' nails may demonstrate trachyonychia; a fine roughness and white discoloration of the nail plate.

6) e.

Rifampicin has been shown to be effective against generalized pruritis in selected patients.

7) e.

The first four stems results in an intensely itchy skin rash. Although some vitiligo patients may itch, this itching is minor.

8) d.

Blue lunulae are non- neurological signs of Wilson disease. Beua's lines may not be seen in the fingers after 2-3 months because they disappear faster than those of the toes. The nails may also be short in iron deficient individuals; so-called brachynychia. Half- and-half nails reflect established uremia; a useful sign to differentiate it from acute renal failure.

9) d.

Seborrheic dermatitis is much more common in HIV patients than in the general population and is usually resistant to treatment. HIV patients are very sensitive to many drugs and medications and they may develop a multitude of adverse skin reactions; this would definitely limit medication choice when treating those patients. Although impetigo is common in AIDS patients, it is not an AIDS-defining illness.

10) b.

Basal cell carcinoma is a slowly-growing tumor that usually targets elderly people. Distant metastases are extremely rare; however, the tumor is locally invasive and destructive, sometimes to a considerable degree. Some types may be pigmented and may resemble malignant melanoma at a glance. The tumor is best treated by surgical removal; radiotherapy is an acceptable alternative in some patients. This tumor is the most common human cancer.

11) e.

The 1st 4 stems are correct; diabetes mellitus per se does not confer an increased risk for malignant melanoma development.

12) b.

The nodular variety of skin malignant melanoma is not preceded by a superficial growing phase. Malignant melanoma mainly develops on sun exposed areas (usually the face) of elderly people (mostly Caucasians). Any suspicious change in a melanocytic nevus should prompt the physician to think of melanoma development. The non-pigmented subtype is extremely rare; however, flecks of pigmentation may still be seen with an aid of a magnifying lens.

13) e.

Women generally fare better than males. Tumors at certain sites, such as the leg, generally are considered to be less aggressive than tumors of the head and neck. The Breslow's tumor thickness is a reliable predictor of the prognosis in stage I. Patients with stage I disease have a 5-year survival rate approaching 70%. Tumors with Breslow's tumor thickness of <1 mm have a very good prognosis; about 90% is the 5-year survival rate.

14) d.

About 50% of solid organ recipients (especially heart and kidney ones) may develop skin squamous cell. Dystrophic epidermolysis bullosa is a scaring genetic syndrome affecting the skin; there is a risk of developing cutaneous squamous cell carcinoma. Squamous cell skin cancer may develop in or around areas of chronic ulceration; chronic skin sinuses carry the same risk. Pale skins as well as exposure to UV radiation and X-ray irradiation are the additional risk factors.

15) d.

Seborrheic warts are benign skin tumors that mainly develop in elderly people. Partial thickness skin dysplasia is the histological appearance of actinic keratoses, while full thickness dysplastic changes are observed in Bowen's disease of the skin.

Although keratoacanthoma may exactly resemble cutaneous squamous cell carcinoma (clinically and histologically), their behavior is different and the former is a benign skin lesion. Seborrheic warts have nothing to do with sebum or sebaceous glands (a misnomer); they are basal cell papillomas.

16) d.

Habit tick dystrophy is a common disorder; there is repetitive picking or fiddling of the proximal nail fold of the skin. In the absence of a history of nail trauma, a sub-ungual hematoma might be mistaken for a sub-ungual malignant melanoma; follow-up is indicated to demonstrate movement of the underlying discoloration. Splinter hemorrhages are also observed in nail trauma (the commonest culprit) and infective endocarditis. Polishing of the nails indicates a chronic repetitive skin itching and rubbing with the nails. A yellow brown discoloration with crumbling of the nail plate is seen in fungal nail infections; usually few nails are infected.

17) e.

Stem "a" refers to erythema nodosum, while lupus pernio is stem "c". Infiltration of long-standing skin scars may occur with granulomas. Hypo-pigmented papules, nodules, or plaques are the additional skin manifestations. Erythroderma is not a manifestation of sarcoidosis.

18) d.

Human Herpes virus type 8 infections are linked to the development of Kaposi's sarcoma; Herpes simplex infection can cause erythema multiforme.

19) b.

Half of pyoderma gangrenosum patients have no identifiable cause for this skin lesion; however, it may be associated with many systemic illnesses, such as inflammatory bowel disease, rheumatoid arthritis, leukemia,...etc. There are no diagnostic changes on skin biopsy; the diagnosis is primarily a clinical one. The lesion itself is usually recurrent; after successful treatment of the associated disease, relapses become uncommon. It may respond to systemic steroids, sulfasalazine, dapsone, or cyclosporine.

20) d.

Diffuse skin hyperpigmentation is a striking feature of porphyria cutanea tarda. The 1st 3 stems are seen mainly on light-exposed areas. Acute intermittent porphyria lacks skin manifestations.

21) d.

Phenylketonuria is a cause of fair skin and fair hair. The other causes of pale skin (in the absence of anemia) are vitiligo, oculocutaneous albinism, and panhypopituitarism.

22) e.

Stem "a" refers to beta blockers, while lithium fits option b. Option c points towards chloroquine. Positive Köebner's phenomenon occurs in option d. None of the anti-asthma medications has been shown to exacerbate psoriasis; topical steroid preparations withdrawal can precipitate psoriasis flare-up, however.

23) e.

The first four stems are the mainstay in the management of eczema in general. There is no regular use of cyclosporine.

24) e.

Oral contraceptive pills are useful in the treatment of hirsutism. Only androgens causes true hirsutism; the other medications result in hypertrichosis.

25) e.

Skin patch testing is an investigation that can be applied for suspected contact allergic dermatitis. Serum total and specific IgE levels and skin prick tests can support the diagnosis of atopic eczema and to detect specific environmental allergens or aero-allergens (animal dander, house dust mites,...etc.). Stem "d" is useful in suspected secondary infections (which are common in eczema patients). Tuberculous skin lesions are infectious in nature, not eczematous and eczema patients have no increased risk of developing skin tuberculosis.

26) d.

Bacterial super-infections are very common in eczematous lesions and add more to the burden of the disease, particularly with Staph aureus, Herpes simplex (which may cause a severe diffuse skin rash, and that is eczema herpeticum); human papilloma virus and molloscum contagiosum infections are both common especially with the use of topical steroids. Loss of schooling, sleep abnormalities, and behavioral disturbances may complicate the picture. Patients with atopic eczema have an increased incidence of food allergy, particularly to eggs, cow's milk, soya, wheat, and fish. These types of food usually cause immediate urticarial lesions rather than exacerbation of the eczema. Because of the breaks in the skin barrier, topical irritant reactions are common.

27) e.

Minoxidil is a cause of hirsutism; minoxidil shampoos are used by bald people.

28) e.

Hirsutism is usually familial and racial; some degree of hirsutism is expected after menopause. The first four stems should prompt a search for an underlying systemic cause.

29) e.

Alopecia areata can result in localized or diffuse non-scarring alopecia.

30) e.

Androgenetic alopecia can cause localized or generalized non-scarring alopecia. Localized or diffuse areas of scarring alopecia may be observed in discoid lupus patients.

31) b.

In pemphigus vuglaris, the target antigen is desmoglein-3; involvement of the oral mucosa occurs in all patients and this may predate the skin manifestations. The oral mucosa is involved in 60% of cases of pemphigoid; the target antigen is BP-220 (which is part of hemidesmosomes).

Frank blisters are uncommonly detected in dermatitis herpetiformis, as itching is severe and ruptures these blisters; usually, only excoriations and denuded areas are noticed. Epidermolysis bullosa acquisita responds poorly to systemic corticosteroids; it may respond to cyclophosphamide or methotrexate. Pemphigoid gestationis usually attacks the peri-umbilical area and thighs; oral mucosa involvement is rare and the target antigen is type XVII collagen and BP-180.

32) d.

Porphyria cutanea tarda can cause skin blistering, increased skin fragility, ugly skin scars, milia, skin hyper-pigmentation, and hypertrichosis, but it does not involve the oral mucosa.

33) e.

Porphyria cutanea tarda causes generalized hyperpigmentation. The other causes of erythroderma are cutaneous T-cell lymphoma, eczema, and lichen planus.

34) e.

The so-called obstetric cholestasis and the first four options are specific causes of pruritus that are encountered only during pregnancy.

35) e.

The other photosensitive dermatoses are chronic actinic dermatitis, SLE, Herpes simplex, certain types of porphyria, and medication-induced (phototoxic and photo-allergic reactions).

Chapter 6
Endocrinology, Diabetes, and Metabolic Diseases

Questions

1) Because you are preparing a lecture about the endocrine system, you surf the internet in order to collect the required information. All of the following statements are correct, *except?*

a) The release of many hormones is pulsatile, and therefore, a random blood sample for hormonal assessment is usually useless

b) Many endocrine glands develop what is called incidentalomas

c) Several endocrine tumors are difficult to classify as malignant or benign during histopathological examination

d) If you suspect a hormonal excess, choose a suppression test to assess its serum level

e) Endocrinal diseases are rarely characterized by loss of normal regulation of hormonal secretion

2) A 32-year-old architect presents with impotence. After you thoroughly examine him, you conduct some tests. Your final diagnosis is a pituitary mass. Which one of the following about pituitary tumors is *incorrect?*

a) Rarely result in hydrocephalus

b) May be an incidental MRI finding

c) Downward extension of the mass presents as a nasal polyp

d) May produce a hypothalamic syndrome by an upward extension

e) Pituitary apoplexy, as a complication, is usually asymptomatic

3) A 44-year-old housewife is referred to you for further evaluation. She was diagnosed with secondary thyrotoxicosis. Her pituitary mass is thought to be a TSHoma. She asks about the pros and cons of surgical treatment of pituitary tumors. Surgical treatment is the first-line treatment option for the following pituitary/hypothalamic tumors, *except?*

a) Non-functioning pituitary macro-adenoma

b) Craniopharyngioma

c) Cushing's disease

d) Microprolactinoma

e) Acromegaly

4) A 23-year-old female has primary infertility because of hyperprolactinemia. She asks you about the possible causes behind her illness. This hormonal disturbance is *not* caused by which one of the following?

a) Stress

b) Primary hypothyroidism

c) Chronic renal failure

d) Chronic chest wall stimulation

e) Long-term treatment with pergolide

5) A 32-year-old woman visits the physician's office complaining of oligomenorrhea. Her serum FSH and LH are low while that of prolactin is elevated. Brain MRI reveals a pituitary mass. With respect to prolactinoma treatment, choose the *wrong* statement?

    a) Treatment with dopamine agonists is almost always effective in normalizing serum prolactin level and restoration of gonadal function

    b) After reaching menopause, treatment of microprolactinoma is only indicated if there is a troublesome galactorrhea

    c) Trans-sphenoidal surgery has a success rate approaching 80%

    d) Macroprolactinoma may enlarge rapidly during pregnancy

    e) External irradiation is a useful first-line treatment in the majority of patients

6) A 46-year-old primary school teacher complains of slowly progressive enlargement of his nose, hands, and lips over many years. He reports excessive sweating and headache. You think of acromegaly. Which one of the following is *not* true with respect to acromegaly?

    a) Although glucose intolerance occurs in 25% of cases, overt diabetes mellitus is seen only in 10% of cases

    b) There is a 2-3 folds increase in the relative risk of colonic cancer and coronary artery disease

    c) Trans-sphenoidal surgery as a treatment option has a high success rate

    d) Dopamine agonists may be used in those with co-existent hyperprolactinemia

    e) External irradiation has a rapid action against the tumor

7) A 15-year-old boy is about to do growth hormone assessment because you suspect growth hormone deficiency. All of the following are methods used to sample growth hormone, *except*?

    a) Sampling before exercise

    b) Frequent sampling during sleep

    c) Sampling 1 hour after going asleep

    d) Sampling during insulin-induced hypoglycemia testing

    e) Sampling after stimulation with arginine

8) A 43-year-old man is referred to the endocrinology outpatients' clinic as a suspected case of Addison's disease. He has developed tanning, lassitude, hypotension, and weight loss. You plan to do ACTH stimulation test. Regarding ACTH stimulation test, all of the following statements are correct, *except*?

a) Used in the diagnosis of primary and secondary adrenal insufficiency
b) The 250 microgram tetracosactrin should be given orally in the early morning
c) It relies upon ACTH-dependent adrenal atrophy in secondary adrenal failure
d) Normally, after 30 minutes, serum cortisol should be >550 nmol/L
e) It is usually useless in adrenal failure secondary to acute ACTH deficiency

9) A 27-year-old woman presents with undue fatigability. Her past notes show hypotension, postural hypotension, loss of pubic and axillary hair, and hypoglycemia. She has generalized hyperpigmentation and thin physique. Her plain abdominal X-ray reveals bilateral paravertebral calcifications. What is the likely diagnosis in this woman?
a) Wilson disease
b) Nelson syndrome
c) Addison disease
d) Adult-onset congenital adrenal hyperplasia
e) Chronic renal failure

10) A 65-year-old man developed post-traumatic panhypopituitarism. Which one of the following regarding panhypopituitarism is *incorrect*?
a) There is a striking pallor
b) Growth hormone is usually the earliest hormone to be lost
c) Coma is multi-factorial and may be due to water intoxication, hypoglycemia, or hypothermia
d) The skin is smooth and has a baby-like texture
e) Serum TSH should be measured frequently to assess the optimal T_4 replacement dose

11) Your medical ward nurse asks you about the details of a test called insulin tolerance test because they are planning to do this test. With respect to insulin tolerance test, which one is the *wrong* statement?
a) It is used in the assessment of hypothalamic-pituitary-adrenal axis
b) Contraindicated in ischemic heart disease and epilepsy
c) The aim of the test is to produce signs of hypoglycemia with serum glucose level < 2.2 mmol/L
d) Serial blood samples of glucose, growth hormone, and cortisol should be taken
e) The intermediate-acting NPH insulin, 0.15 u/Kg, is used in the test subcutaneously

12) A 54-year-old man presents with excessive thirst and frequent passing of large amount of dilute urine over the past month. Five weeks ago, he got a blunt head trauma. You consider cranial diabetes insipidus and you plan to do water deprivation test. With respect to this test, which one is the *wrong* statement?

a) It is used in the diagnosis of diabetes insipidus (DI) and to differentiate between cranial and nephrogenic DI

b) There should be no coffee, tea, or, smoking on the test day

c) The test should be stopped if the patient loses 3% of his total body weight

d) When trying to differentiate DI from compulsive water drinking, DDAVP is a very useful tool

e) If the initial urinary osmolality is 700 mOsm/kg, the test should be stopped and DI is excluded

13) A 41-year-old woman complained of irritability, sweating, palpitations, unintentional weight loss, and frequent defecation for the last 4 months. She is on no medications and she denies doing drugs. Her serum TSH is undetectable. Which one of the following regarding hyperthyroidism is *not* true?

a) The commonest cause is Graves' disease

b) If there is prominent anorexia, a malignant cause should be suspected

c) Vitiligo and lymphadenopathy favor Graves' disease over other etiologies

d) Apathy and osteoporosis are mainly encountered in elderly patients

e) Pruritus, palmer erythema, and spider nevi are more suggestive of an associated chronic active hepatitis

14) A 33-year-old woman is referred to you for further management of Graves' disease. She undergoes many blood tests. Some of her tests are abnormal. You tell her that her disease is the cause of these abnormal results. Non-specific biochemical abnormalities in thyrotoxicosis include all of the following, *except?*

a) Raised serum alkaline phosphatase

b) Raised serum ALT and AST

c) Hypocalcemia

d) Positive tests for urinary sugar

e) Raised serum gamma glutamyl transferase in the absence of enzyme-inducing drugs or alcoholism

15) A 28-year-old woman asks you about the various options available to treat her Graves' disease. With respect to the treatment of thyrotoxicosis, which is the *incorrect* statement?
   a) Following successful treatment with carbimazole, up 50% of patients will relapse following drug stoppage
   b) Subtotal thyroidectomy is contraindicated in patients with previous thyroid surgery
   c) Radioiodine is contraindicated in pregnancy
   d) Following subtotal thyroidectomy, up to 50% of patients will develop permanent hypocalcemia in the long-term
   e) Following treatment with radioiodine, up to 80% of patients will develop permanent hypothyroidism after 15 years

16) A 34-year-old man presents with a 1-week history of neck pain that is aggravated by swallowing. He has fever and fine postural tremor. The thyroid gland is slightly enlarged and tender. Regarding subacute thyroiditis, all of the following statements are correct, *except?*
   a) Usually viral
   b) The anterior neck pain is worsened by coughing
   c) The ESR is usually normal
   d) Usually responds well to treatment with non-steroidal anti-inflammatory drugs
   e) The resulting hyperthyroidism is usually mild and no treatment is needed

17) A 22-year-old nurse consults you about her hands' tremor. She states that she is trying to lose weight and she is using some sort of hormonal therapy. After thorough examination and investigations, you suspect factitious hyperthyroidism. Regarding this disease, which one of the following is *incorrect?*
   a) Uncommon condition that results from self-administration of T_4
   b) The radioiodine uptake scan is suppressed
   c) There is undetectable serum thyroglobulin
   d) The T_3:T_4 ratio is high
   e) Serum TSH is suppressed

18) A 28-year-old woman, who gave birth to healthy-looking baby 1 week ago, complains of irritability, sweating, and palpitation. There is fine and fast postural tremor. You find very low serum TSH and you consider post-partum thyroiditis. All of the following statements regarding post-partum thyroiditis are correct, *except?*

a) Occurs in 5-10% of women in the first 6 months following delivery
b) Thyroid biopsy shows lymphocytic thyroiditis
c) Tends to recur after subsequent pregnancies
d) There is an association between post-partum depression and post-partum thyroiditis
e) There is a negligible radioiodine thyroid scan

19) A 44-year-old man undergoes neck surgery for Graves' disease. On day 2 postoperatively, he develops high fever, sweating, tachycardia, and confusion. Your preliminary impression is thyroid storm. With respect to hyperthyroid crisis, choose the *wrong* statement?
a) The commonest precipitation factor is an infection in a previously undiagnosed or improperly treated hyperthyroid patient
b) It is a life-threatening increase in the activity of the thyroid gland
c) Rehydration and broad-spectrum antibiotics are parts of the management plan
d) If patients cannot swallow carbimazole, it should be given intravenously
e) After 10-14 days of treatment with various anti-thyroid measures, sodium iopodate and propranolol can be stopped

20) A 43-year-old woman presents with weight gain, apathy, thick dry skin, and constipation. Her serum TSH is raised and her thyroid gland is palpable. The list of causes of goitrous hypothyroidism includes all of the following, *except*?
a) Hashimoto thyroiditis
b) Dyshormonogenesis
c) Drug-induced hypothyroidism
d) Iodine deficiency
e) Post-ablative hypothyroidism

21) A 65-year-old woman is referred to your office because of severe depression. She is stocky, and you find non-pitting leg edema, dull face, husky voice, and thick coarse hair and skin. There is delayed relaxation of her ankle jerks. Serum TSH is elevated. She has hypothyroidism. All of the following are rare but well-recognized features of hypothyroidism, *except*?
a) Frank psychosis
b) Myotonia
c) Ascites
d) Ileus
e) Iron deficiency anemia

22) A 46-year-old woman presents with bilateral carpal tunnel syndrome. Her face is sallow and there are malar rash, slow voice, and bradycardia. You suspect hypothyroidism. All of the following biochemical abnormalities are encountered in hypothyroidism, *except?*
   a)   Raised serum LDH and creatine phosphokinase
   b)   Elevated serum cholesterol and triglycerides
   c)   Anemia with MCV of 110 fL
   d)   Low serum T_3 level
   e)   Raised serum parathyroid hormone

23) A 57-year-old man visits the physician's office for a scheduled follow-up. He has primary atrophic hypothyroidism, for which he takes 100 microgram of thyroxin each day. He feels well. Which one of the following statements regarding the treatment of hypothyroidism is *incorrect?*
   a)   Oral T_4 is the cornerstone of therapy
   b)   The correct dose of thyroxin should restore serum TSH to its normal reference range
   c)   Patients usually feel better 2-3 weeks after starting thyroxin therapy
   d)   Restoration of skin and hair texture is usually seen after 3 weeks of treatment
   e)   Elderly patients should receive low doses of thyroxin with gradual escalation

24) One of your patients is brought to the Emergency Room in a coma state. His past notes reveal that he is incompliant with his thyroxin tablets. You have not seen him over the past 2 years. Regarding myxedema coma, which one is the *wrong* statement?
   a)   The mortality rate is 50%
   b)   Convulsions are not uncommon, and the opening CSF pressure may be elevated
   c)   It is a medical emergency and the treatment must be started before biochemical confirmation
   d)   Unless there is an evidence of primary hypothyroidism, it should be considered secondary to pituitary or hypothalamic dysfunction
   e) T_4 should be given intravenously in all cases

25) A 43-year-old woman visits the endocrinology outpatients' clinic for a scheduled follow-up. She erratically ingests her thyroxin tablets and her physician is not satisfied with her blood tests. Which one of the following statements regarding the follow-up of hypothyroid patients is *wrong?*

a) It is important to ensure compliance, and thyroxin tablets should be taken infinitely
b) Once the dose of thyroxin is stabilized based on serum TSH, serum TSH and T4 should be measure every 1-2 years
c) The finding of elevated serum T4 and high serum TSH reflects good compliance
d) Some patients may take thyroxin erratically when they know that the half-life of the medication is long
e) Excessive sweating, tachycardia, and anxiety may indicate over-treatment

26) A 23-year-old male was found to have an isolated low serum TSH during a pre-employment assessment. He denies any symptoms, but he admits to taking some medications without prescription. All of the following medications can disturb thyroid hormones, *except*?
a) Amiodarone may cause hypothyroidism or hyperthyroidism
b) Lithium may result in hypothyroidism
c) Potassium iodide may induce hypothyroidism
d) Phenytoin may decrease the required daily dose of thyroxin
e) Oral contraceptive pills may increase the total, but not the free, thyroxin

27) A 36-year-old female complains of weight gain of 10 Kg over the last 3 months in spite of being on diet. She denies symptoms or being on any medications. Her menstrual cycle is regular and there is no change in her body hair. All of the following may result in obesity, *except*?
a) Cushing disease
b) Kallmann syndrome
c) Prader-Willi syndrome
d) Hypothalamic tumors
e) Addison disease

28) A 26-year-old man found a small nodule in his anterior neck while shaving. The nodule is in the area of the thyroid gland and is hard and fixed. There are palpable cervical lymph nodes. He is clinically euthyroid. Thyroid isotope scanning shows that the nodule is cold. Serum TSH is normal. You think of thyroid malignancy. Regarding thyroid carcinomas, choose the *incorrect* statement?
a) Each cancer type usually has a certain age group to affect
b) Papillary carcinoma is the commonest type
c) Some tumors are TSH-dependent

    d)   Follicular carcinoma can be diagnosed easily by thyroid FNA cytology
    e)   Very rarely, result in thyrotoxicosis

29) One of your friends has been recently diagnosed with thyroid gland cancer and he consults you about this tumor and the possibility of cure. All of the following statements about thyroid cancers are correct, *except?*
    a)   Papillary carcinoma usually targets people between 20-40 years of age
    b)   Medullary thyroid carcinoma in a 20-year-old man may reflect multiple endocrine neoplasia (MEN) type II
    c)   Anaplastic carcinomas usually develop in elderly people
    d)   Thyroid lymphomas may arise from pre-existent Hashimoto's thyroiditis
    e)   Thyroid gland is a favorable site for secondary tumors

30) A 50-year-old man has a rocky-hard thyroid mass. The final diagnosis turns out to be medullary thyroid carcinoma. All of the following statements regarding this tumor are correct, *except?*
    a)   The prognosis is generally poor when compared to differentiated thyroid carcinoma
    b)   Their occurrence young persons may be part of MEN type I
    c)   As a treatment option, total thyroidectomy is the preferred choice
    d)   High level of serum calcitonin secreted by the tumor rarely causes symptomatic hypocalcemia
    e)   The tumor cells do not respond to radioiodine treatment

31) A 58-year-old man developed recurrent renal stones. Abdominal X-ray reveals bilateral nephrocalcinosis. Serum calcium is 12.9 mg/dl. You find elevated serum parathyroid hormone level. The final diagnosis is primary hyperparathyroidism. Regarding primary hyperparathyroidism, which one of the following is the *wrong* statement?
    a)   The commonest cause is a single parathyroid adenoma
    b)   The commonest cause of asymptomatic hypercalcemia in outpatients
    c)   Together with systemic malignancy, they account for 90% of cases of hypercalcemia
    d)   Lithium-induced hyperparathyroidism may present as primary hyperparathyroidism-like picture
    e)   The parathyroid glands are usually palpable in the neck

32) A 43-year-old woman is referred to you because of hypercalcemia, hypercalciuria, and band-shaped keratopathy. Her blood pressure is 170/110 mmHg and serum calcium is 13.2 mg/dl.

There are elevated serum parathyroid hormone and serum alkaline phosphatase levels. All of the following can result in hypercalcemia with elevated serum parathyroid hormone, *except*?

a) Primary hyperparathyroidism
b) Tertiary hyperparathyroidism
c) Lithium-induced hyperparathyroidism
d) Familial hypocalciuric hypercalcemia
e) Bony metastases

33) A 63-year-old heavy smoker man presents with severe dehydration. He reports frequent urination and excessive thirst. Chest X-ray shows a right-sided hilar mass and you notice many bony osteolytic lesions. The serum calcium is high while that of the parathyroid hormone is suppressed. The treatment of malignancy-associated hypercalcemia entails all following, *except*?

a) Intravenous fluids have a very important role in the treatment
b) In the presence of very high calcium blood levels, intravenous pamidronate should be given initially
c) Forced diuresis may be useful to lower serum calcium
d) As in primary hyperparathyroidism, systemic glucocorticoids are very effective in reducing serum calcium level
e) In resistant cases, hemodialysis may be the only choice

34) You are evaluating a patient with long-standing hypocalcemia. Prolonged hypocalcaemia may result in all of following, *except*?

a) Hypertension
b) Cataract
c) Basal ganglia calcification
d) Papilledema
e) Parkinsonism

35) A 33-year-old man is being evaluated for low serum calcium. You also find high serum phosphate. This combination of electrolyte disturbance may result from all of the following, *except*?

a) Chronic renal failure
b) b- Hypoparathyroidism
c) c- Pseudohypoparathyroidism
d) d- Hypomagnesaemia
e) e- Pseudopseudohypoparathyroidism

36) A 43-year-old man was diagnosed with chronic renal failure. Today, he presents with a severe attack of carpopedal spasm.

You give intravenous calcium gluconate and an excellent symptomatic improvement occurs. With respect to hypocalcemia, all of the following statements are correct, *except?*

a) Carpopedal spasm is more common in children than adults
b) In adults, stridor is rare
c) Seizures are usually resistant to antiepileptic therapy
d) The cornerstone in the treatment of pseudohypoparathyroidism is high doses of oral phosphate
e) Regular follow-up should be encouraged with measurement of serum calcium

37) A 43-year-old woman has been taking oral prednisolone since 1 year. She has chronic asthma. You find round face, abdominal striae, and dorsal spine osteoporotic fractures. All of the following statements with respect to Cushing syndrome are correct, *except?*

a) Obesity is the commonest sign
b) Hypertension is absent in 25% of cases
c) Prominent skin hyper-pigmentation favors ectopic ACTH-secreting source over adrenal adenoma
d) Depression is the commonest psychiatric manifestation
e) Muscle biopsy examination shows type I muscle fiber atrophy

38) A 41-year-old female presents with progressive weight gain over several months. Examination reveals rounded plethoric face, central obesity, and hypertension. Serum potassium is 3.1 mEq/L. She denies drug ingestion and there is no family history of note. You suspect Cushing syndrome and you plan to do dynamic tests to confirm your clinical impression. Regarding the diagnosis and investigations of Cushing syndrome, which one of the following is the incorrect statement?

a) Acclimatization to hospitalization for 48 hours is very important beforehand
b) 24-hours urinary free cortisol or overnight low-dose dexamethasone suppression test are the preferred initial screening tools
c) Dexamethasone cross-reacts with the blood's cortisol immunoassay testing
d) Chronic alcoholism resembles Cushing's syndrome, clinically and biochemically
e) Suppressed serum ACTH level points out towards an adrenal source

39) 36-year-old female was diagnosed with Cushing disease. She asked about the various treatment modalities and their associated morbidities.

Regarding the treatment of Cushing syndrome, which one of the following statements is *wrong*?

a) In pituitary-dependent disease, trans-sphenoidal surgery is the preferred treatment option

b) If treated by bilateral adrenalectomy, the pituitary should be irradiated to prevent the future development of Nelson syndrome

c) Medical treatment is used as a bridge while preparing patients for surgery

d) Adrenal carcinoma should be removed surgically, the tumor bed is irradiated, and then the patient is given the drug opDDD (Mitotane)

e) Without any form of treatment, the 5-year survival rate is 90%

40) A 43-year-old female has resistant hypertension. She denies non-compliance with her medications. Plasma renin activity is low but you detect low serum aldosterone, low serum potassium, and elevated serum bicarbonate. All of the following may result in "apparent" hyperaldosteronism, *except*?

a) Treatment with carbenoxolone

b) 11-deoxycorticosterone-secreting tumors

c) Liddle syndrome

d) Ectopic ACTH syndrome

e) Glucocorticoid-suppressible hyperaldosteronism

41) A 43-year-old man presents with difficult-to-control hypertension. He has hypokalemic alkalosis. 12-lead ECG reveals left ventricular hypertrophy. The plasma renin activity is very low but there is elevated serum aldosterone. With respect to primary hyperaldosteronism, which one of the following statements is *incorrect*?

a) Hypertension is almost always found and is the commonest presenting feature

b) Serum potassium is normal in 70% of cases at the time of the diagnosis

c) Of all causes, only Conn's adenoma can be treated by surgery

d) Spironolactone is very effective in normalizing the blood pressure and biochemical abnormalities in the majority of cases

e) Bipedal pitting edema is very common

42) A 47-year-old male presents with repeated attacks of headache, palpitation, pallor, and elevated blood pressure. Serum TSH is normal, and you suspect pheochromocytoma as the underlying culprit. Which one of the following statements about pheochromocytoma is *wrong*?

a) May be part of specific syndromes
b) Predominantly elevated serum noradrenalin suggests either a large adrenal tumor or an extra-adrenal tumor as a cause
c) Weight loss always reflects an associated diabetes mellitus
d) Postural hypotension may occur in some patients
e) The blood pressure may rise during urination

43) A 34-year-old female presents with undue fatigability for several months. She is hyperpigmented. There is postural hypotension, as well as loss of axillary and pubic hairs. No abdominal scars are found. Regarding Addison disease, which one is the *incorrect* statement?
a) The commonest cause is autoimmune adrenalitis
b) Vitiligo develops in 20% of cases
c) Hyperglycemia points out towards an associated type I diabetes
d) Postural hypotension reflects glucocorticoids rather than mineralocorticoids deficiency
e) A common condition with an incidence of 8000 new case/million of population

44) Your colleague phones you to ask about congenital adrenal hyperplasia and its subtypes. All of the following statements with respect to congenital adrenal hyperplasia are correct, *except?*
a) The commonest cause is 21-α hydroxylase deficiency
b) All cases should demonstrate an autosomal recessive mode of inheritance
c) Results in ambiguous genitalia in females while precocious pseudopuberty is observed in males
d) 11-β and 17-α hydroxylase deficiencies are associated with hypotension
e) The condition can be prevented by appropriate prenatal diagnosis and giving dexamethasone to the pregnant mother

45) A 33-year-old man is referred from the surgical department. His breasts are somewhat enlarged. The referral letter states absence of surgical causes behind this gynecomastia. Which one of the following does *not* result in gynecomastia?
a) Digoxin
b) Cimetidine
c) Diethylstilbestrol
d) Spironolactone
e) Amiloride

46) A 28-year-old female has oligomenorrhea, obesity, and hirsutism. She asks you about her chances of getting pregnant in the foreseeable future. Abdominal ultrasound reveals ovarian hyperthecosis and there are multiple small ovarian cysts. You consider polycystic ovarian syndrome. Which one of the following statements with respect to this syndrome is *wrong*?
   a) Insulin resistance is thought to be the central key in the pathogenesis
   b) There is mild elevation in serum prolactin
   c) Serum androgens is mildly raised
   d) The blood level of estrogen is high
   e) The FSH:LH ratio is greater than 4.5:1

47) A 17-year-old high school girl from the Middle East complains of an excessive growth of her facial and arm hair. Her menstrual cycles are normal, and there are no signs of virilization. You tell her that her excess body hair is most likely racial. In the assessment of hirsutism, which one is the *incorrect* statement?
   a) Idiopathic hirsutism is the commonest cause
   b) High levels of serum androgens, which don't suppress with oral steroids or estrogens, suggest ovarian or adrenal tumors
   c) If the patient is a highly trained athlete, think of exogenous androgen intake
   d) Mooning of the face with central obesity and abdominal striae are clues to Cushing syndrome
   e) Family history of hirsutism is usually irrelevant

48) A 22-year-old man is being evaluated for suspected multiple endocrine neoplasia (MEN). With respect to MEN, choose the *incorrect* statement?
   a) Mutations in *MENIN* gene on chromosome 11 result in MEN type I
   b) Symptomatic hypercalcemia is the commonest presenting feature of MEN type II
   c) Family history of one relevant endocrine tumor may be positive
   d) Pheochromocytoma in MEN type II is bilateral in 70% of cases
   e) Carcinoid syndrome is an uncommon component of MEN type I

49) A 33-year-old woman was diagnosed with carcinoid tumor. Which one of the following regarding carcinoid tumor is *false*?
   a) The commonest site for carcinoid tumors is the terminal ileum
   b) Acute appendicitis is a well-recognized form of presentation
   c) The long-term prognosis is excellent in the majority of patients
   d) Carcinoid syndrome may present as right-sided heart failure

e) Attacks of cramping abdominal pain and constipation with facial pallor and stridor are the commonest presenting features of carcinoid tumors

50) Because of a newly discovered pancreatic endocrine tumor, a 55-year-old man visits the physician's office. He is desperate for help, and he thinks that he will lose his life soon. All of the following statements with respect to pancreatic endocrine tumors are correct, *except?*

a) Gallstones and diabetes mellitus may be the presenting manifestations of somatostatinomas
b) Glucagonomas may present with anemia and weight loss
c) Steatorrhea is a well-recognized feature of gastrinoma
d) VIPoma results in constipation and profound hyperkalemia
e) Insulinoma can cause dizzy spells

51) A 21-year-old woman visits the physician's office. She says that her older brother has hyperglycemia and she asks whether she may develop diabetes or not. Which one of the following diseases is *not* associated with hyperglycemia?

a) Myotonia dystrophica
b) Down syndrome
c) Lipodystrophy
d) Friedreich ataxia
e) Nesidiolastosis

52) A 15-year-old teen was diagnosed with type I diabetes mellitus and she asked about her disease, its treatment, and its possible complications. Regarding type I diabetes mellitus, which one is the *incorrect* statement?

a) It is a T-cell mediated immunological disease
b) There is mononuclear cell infiltration of the pancreas which results in insulinitis
c) There is an association with HLA B8/DR3 and DR4
d) The concordance rate in monozygotic twins is almost 100%
e) Immune suppressants can induce a long-lasting remission

53) A 62-year-old man undergoes some blood tests as part of his pre-medical insurance assessment. His random blood sugar turns out to be 310 mg/dl. He denies previous history of diabetes. Which one of the following with respect to type II diabetes is the *incorrect* statement?

a) About 70% of patients are older than age of 50 years at the time of diagnosis
b) Affects about 10% of the general population older than the age of 65 years

c) Around 80% of females who have a history of gestational diabetes will ultimately develop type II diabetes
d) Overeating, obesity, and under-activity are risk factors for the future development of type II diabetes
e) The concordance rate in monozygotic twins is around 50%

54) While doing your routine morning tour in the general medical ward, your intern asks about the genetic mutations which may result in diabetes mellitus. All of the following mutations (which cause diabetes mellitus) demonstrate an autosomal dominant mode of inheritance, *except*?
a) Hepatocyte nuclear factor 4α gene
b) Hepatocyte nuclear factor 1α gene
c) Glucokinase gene
d) Insulin promoter factor gene
e) Mitochondrial DNA

55) Because of preparing a poster about insulin, you have reviewed many books and journals and you find many useful information. Which one of the following you have *not* found with respect to the actions of insulin?
a) Insulin decreases lipolysis
b) The action of lipoprotein lipase is enhanced
c) It enhances potassium and amino acid entry into cells
d) Glycogenesis and glycolysis are both enhanced
e) It enhances fatty acid synthesis by the liver

56) A 20-year-old pregnant woman has been referred to the diabetes clinic for doing oral glucose tolerance test. Which one of the following statements regarding this investigation is the wrong one?
a) Is not used as a routine test for diagnosing diabetes mellitus
b) There should be unrestricted carbohydrate diet 3 days before the test
c) The patient may be allowed to smoke during the test
d) The patient should fast overnight
e) If the 2-hour plasma glucose level is between 7.8-11.1 mmol/L, the result is interpreted as an impaired glucose tolerance test

57) A 13-year-old schoolboy is brought by his father to your office. The boy has had excessive thirst and urination over the past 2 weeks. Urine dipstick testing turns out to be positive for glucose. With respect to the diagnosis of diabetes mellitus, choose the *incorrect* statement?

a) Glycated hemoglobin is not used for the diagnosis
b) The presence of glycosuria should warrant further investigations and should not be used as a diagnostic test alone
c) Ketonuria is not pathognomonic for diabetes and may be found in normal people after prolonged fasting or exercise
d) Fasting blood glucose is always preferred over random blood glucose testing for securing or refuting the diagnosis
e) A random blood glucose of >11.1 mmol/L on 3 or more occasions is required to diagnose diabetes mellitus with high certainty

58) You refer this 54-year-old man to the local dietician to advise him about the dietary manipulations for his newly diagnosed type II diabetes mellitus. All of the following statements about dietary management and life style adjustment for diabetes mellitus are correct, *except?*

a) About 50% of type II diabetes patients respond initially to diet alone
b) The total dietary fat should not exceed 10% of the total energy intake
c) All patients should be advised against ingesting alcohol
d) Salt restriction is advised
e) Encourage regular exercises

59) After securing the diagnosis of diabetes, a 40-year-old man asks about the anti-diabetic medications and their usefulness. Which one of the following statements with respect to anti-diabetes medications is *incorrect?*

a) Gliclazide has no favorable effect on lipid profile
b) Insulin does not reduce postprandial glycemia
c) Acarbose has no hypoglycemic effect
d) Pioglitazone does not raise serum insulin
e) Metformin does not increase body weight

60) A 65-year-old man takes daily repaglinide for his type II diabetes and he is compliant with it. However, he is afraid that this medication may result in serious side effects. Regarding the side effects of various anti-diabetic medications, which one is the *wrong* statement?

a) Metformin carries a risk of lactic acidosis
b) Chlorpropamide may cause SIADH
c) Prominent fluid retention may result from pioglitazone
d) Peripheral edema is a well-recognized side effect of insulin therapy
e) Acarbose produces severe constipation

61) A 32-year-old visually impaired male presents with seizures. His blood sugar is 21 mg/dl. The patient has long-standing type I diabetes.

His roommate states that the patient does many errors in insulin dosage and administration. Severe hypoglycemia in diabetics may result in all of the following, *except?*
a) Vitreous hemorrhage
b) Cardiac dysrhythmia
c) Hyperthermia
d) Risks of accidents
e) Stroke

62) A 61-year-old man has been experiencing repeated hypoglycemic attacks over the past 5 months. He denies mistakes in insulin dosage or administration. You intend to dig out for the underlying cause. Hypoglycemia in diabetics may result from all of the following, *except?*
a) Lack of exercises
b) Unrecognized co-existent endocrine disease
c) Missed, delayed, or inadequate meal
d) Gastroparesis
e) Deliberately-induced

63) A 15-year-old boy presents with drowsiness, severe dehydration, and fever. He has type I diabetes. His father states that the patient is not complaint with his insulin regimen. Urine is strongly positive for ketones. Which one of the following statements regarding diabetic ketoacidosis is the false one?
a) It results from severe and absolute insulin deficiency
b) The average fluid loss is 6 liters while that of potassium is 350 mEq
c) Sudden impairment in consciousness during treatment may indicate brain edema development
d) Sudden gastric dilatation may occur
e) Leukocytosis always indicates an underlying infection

64) A 32-year-old female is recovering from diabetic ketoacidosis (DKA). Today, she has developed right-sided bloody nasal discharge and peri-orbital pain. You suspect mucormycosis as a complication of this DKA. All of the following statements about the complications of diabetic ketoacidosis are correct, e*xcept?*
a) Disseminated intravascular coagulation is common
b) Confusion, drowsiness, and coma occurs in 10% of patients
c) Thromboembolism should always be anticipated
d) Acute circulatory failure is usually profound
e) Acute respiratory distress syndrome should always be suspected in breathless patients

65) A 65-year-old male, who has type II diabetes mellitus, is brought to the Acute and Emergency Department in a state of coma and severe dehydration. He lives alone. His past notes show a recent attack of pneumonia and that the patient declined hospital admission at that time. Plasma glucose is very high. You consider non-ketotic hyperosmolar coma. With respect to this metabolic complication of type II diabetes, which one is the *wrong* statement?
   a) Acidosis is not a prominent part of the picture
   b) Plasma hyperosmolality is a core feature
   c) Thromboembolic complications are common
   d) It confers a 1% mortality rate
   e) The condition usually targets elderly people with previously undiagnosed type II diabetes

66) A 54-year-old man presents with sudden right-sided visual loss. He has proliferative diabetic retinopathy and has declined LASER therapy recently. Examination reveals vitreous hemorrhage. Which one of the following is the *incorrect* statement regarding diabetic retinopathy?
   a) Micro-aneurysms are the earliest detected ophthalmoscopic sign in background retinopathy
   b) Hard exudates are characteristic
   c) Prominent soft exudates indicate an advanced retinopathy state or an associated hypertension
   d) IRMAs (intra-retinal micro-vascular abnormalities) signal the occurrence of the pre-proliferative stage
   e) Venous loops and beadings are seen mainly in the proliferative stage

67) A 65-year-old man complains of burning pain as well as pins and needles sensation in both feet. He takes daily metformin for long-standing type II diabetes. He reports similar but milder symptoms in his fingers. Your examination is consistent with peripheral polyneuropathy. All of the following statements about diabetic peripheral polyneuropathy are correct, *except?*
   a) There is a variable combination of axonopathy, demyelination, and thickening of Schwann cell basal lamina
   b) Overall, it occurs in 50% of diabetic patients and is usually not that symptomatic
   c) May result in Charcot joints
   d) Mainly motor and potentially reversible
   e) Trophic ulcerations in the feet may be detected

68) A 33-year-old man presents with progressive leg edema and you find overt proteinuria.

His type I diabetes was diagnosed at the age of 17 years. Abdominal ultrasound shows bilateral kidney enlargement. Past records disclose positive urine microalbuminuria tests, 5 years ago. Blood urea and serum creatinine are elevated. All of the following statements with respect to diabetic nephropathy are correct, *except*?

a) Occurs in 30-35% of type I diabetes cases
b) About 50% of diabetic nephropathy patients have type II diabetes
c) Hypertension is very common and may accelerate the course of nephropathy
d) The nodular glomerulosclerosis type is much more common than the diffuse one
e) ACE inhibitors are effective in reducing the rate of progression, even in the absence of hypertension

69) A 26-year-old woman visits the diabetes outpatients' clinic. She has type I diabetes and uses insulin. She asks about the possibility of getting pregnant and the treatment options of her high blood sugar during pregnancy. With respect to diabetes in pregnancy, choose the *incorrect* statement?

a) Ideally, all patients should have pre-pregnancy counseling
b) Diabetes should be optimally controlled before pregnancy
c) The objective of diabetic control is to decrease fetal congenital anomalies
d) Oral hypoglycemic agents are useful in the management
e) Insulin requirement varies throughout the course of pregnancy

70) A 54-year-old man uses daily repaglinide in order to control his type II diabetes. He will undergo elective laparoscopic surgery to remove gallstones. He is otherwise well. The surgeon calls you to know if there are any preparations or precautions that should be done beforehand. Regarding surgery in diabetics, which one of the following statements is *incorrect*?

a) Pre-operative cardiovascular and renal assessments are very important
b) Preoperatively, we should check signs of autonomic neuropathy
c) Assess the overall glycemic control by doing HbA1c
d) Stop metformin and long-acting sulfonylureas prior to the day of surgery
e) Prior hospitalization is never needed, as the assessment can be done on an outpatient basis

This page was intentionally left blank

Endocrinology

Answers

1) e.

The release of many hormones is pulsatile, and therefore, a random blood sample for hormonal assessment is usually useless. Accordingly, sequential or a dynamic testing should be used. Many endocrine organs have benign masses which have no endocrine consequences; therefore, a biochemical diagnosis (of hormonal excess) should be done before imaging these organs. Histological examination is usually fruitless in terms of differentiating benign from malignant endocrine tumors in general, e.g., benign from malignant pheochromocytoma. When there is a hormonal excess, choose a suppression test, while if you consider a hormonal deficiency state, choose a stimulation test (dynamic tests) to assess serum level of the hormone in question. One of the characteristic features of many endocrine diseases is the loss of normal regulation of hormonal secretion. For example, one of the screening tests for Cushing syndrome is to demonstrate loss of circadian release of serum cortisol.

2) e.

The presentation pf pituitary apoplexy is a catastrophic one. Lateral extension may involve the third, fourth, and sixth cranial nerves resulting their palsies (and double vision). TSHoma is responsible for 1% of all functioning pituitary tumors; it is a very rare cause of secondary thyrotoxicosis.

3) d.

Radiotherapy is a second-line treatment option for non-functioning pituitary macro-adenoma, acromegaly, and Cushing disease. In Cushing disease, radiotherapy is more effective in children. Radiotherapy is also used to irradiate the pituitary gland to prevent the development of Nelson syndrome. Dopamine agonists are the first-line agents for pituitary prolactinoma, unless it is cystic and/or very large. Unlike the situation in prolactinoma, octreotide does not cause tumor shrinkage in acromegaly.

4) e.

Pergolide is a dopamine agonist that is used in the treatment of hyperprolactinemia. The other causes of this endocrine hyper-functioning are pregnancy, lactation, and medications (oral contraceptive pills, metoclopramide, reserpine, methyldopa, antipsychotics, antidepressants, opiates, and anti-androgens).

However, the commonest causes in clinical practice (apart from pregnancy and lactation) are:
1. Disconnection hyperprolactinemia: as in non-functioning pituitary macro-adenomas.
2. Prolactinoma, usually microprolactinoma.
3. Primary hypothyroidism.
4. Polycystic ovarian syndrome.

5) e.

External irradiation is used for some macro-adenomas to prevent re-growth if dopamine agonists are stopped. The other stems are true.

6) e.

Although external irradiation results in tumor shrinkage, but the growth hormone level takes long time to return back to its normal reference range and there is a high risk of developing panhypopituitarism in the long-term.

7) a.

Blood sampling for growth hormone assay can be taken after exercise, not pre-exercise. About 76% of cases of adult growth hormone deficiency results from pituitary tumors (or the consequences of treatment of these tumor, including surgery and/or radiation therapy), while juxta-pituitary masses (such as craniopharyngioma) comprise about 13% of cases; about 8% of adult growth hormone deficiency have no detectable underlying cause.

8) b.

The 250 microgram tetracosactrin *injection* can be given at any time during the day. Because it relies on ACTH-dependent adrenal atrophy in secondary adrenal failure, ACTH stimulation test may not diagnose acute adrenal failure secondary to acute ACTH deficiency. Serum cortisol levels fail to increase in response to the exogenously administered ACTH in patients with primary or secondary adrenal insufficiency. These 2 can be distinguished by measurement of random sample of serum ACTH (which is low in ACTH deficiency and high in Addison disease). If the ACTH assay is unavailable, long ACTH stimulation test can be applied using a 1 mg depot. ACTH is given intramuscularly, daily for 3 days.

In secondary adrenal insufficiency, there is a progressive increase in plasma cortisol with repeated ACTH administrations, whereas in Addison disease, serum cortisol remains <700 nmol/l at 8 hours after the last injection. In a patient who is already receiving glucocorticoids, short ACTH stimulation test can be performed first thing in the morning, >12 hours after the last dose of glucocorticoid, or the treatment can be changed to a synthetic steroid such as dexamethasone (0.75 mg daily), which does not cross-react with the plasma cortisol radioimmunoassay.

9) c.

Addison disease patients can have a multitude of biochemical abnormalities. There are raised blood urea, eosinophilia, hyperkalemia, hyponatremia, and hypoglycemia. Adrenal calcification is seen in 50% of cases. Autoimmunity is the commonest cause in the Western countries, while tuberculosis is the leading cause in the African and Asian populations.

10) e.

In secondary thyroid failure, measuring serum TSH (for the diagnosis or the follow-up of patients who receive thyroxin therapy) is not helpful. This is because the pituitary may secrete certain glycoproteins, which are usually detected by the TSH assay; these are not bioactive, however. Therefore, the aim of replacing thyroid hormones is to keep the serum T_4 in the upper part of reference range.

11) e.

Insulin tolerance test uses intravenous soluble insulin, 0.15 u/Kg. The objective is to produce hypoglycemia (plasma glucose <2.2 mmol/L) and signs of neuroglycopenia. This would incite secretion of growth hormone and cortisol into the blood stream. Blood samples are taken at 0, 30, 45, 60, 90, and 120 minutes post-administration. The tests should not be done in patients with seizure disorders, coronary artery disease, and in severe hypopituitarism. It is a useful test in the assessment of hypothalamic-pituitary-adrenal axis and growth hormone deficiency, when other tests are equivocal. At the end of the test, serum growth hormone level should be >20 mu/L and that of cortisol >550 nmol/L.

12) d.

DDAVP should not be given if the patient is a compulsive water drinker; those patients are at risk of developing severe water excretion if they are given this injection (this is more likely to occur in those who continue their excessive water drinking post-testing).

13) e.

Pruritus (with increased sweating), palmer erythema, spider nevi, skin pigmentation (actually, vitiligo is much more common), and clubbing are features of thyrotoxicosis.

14) c.

Thyrotoxicosis results in many non-specific biochemical abnormalities that should be kept in mind when investigating hyperthyroid patients. Liver function tests are usually deranged; there are slightly raised serum levels of bilirubin, AST, ALT, and alkaline phosphatase. Mild hypercalcemia occurs in 5% of cases. Hyperglycemia and positive urinary sugar testing may reflect diabetes mellitus or lag storage disease. A well-recognized finding is the slightly elevated serum gamma glutamyl transferase enzymatic activity (in the absence of alcohol ingestion or enzyme-inducing medication use).

15) d.

About 0.2% patients who receive the anti-thyroid medication carbimazole develop severe, but reversible, agranulocytosis. In general, the surgical option for treating hyperthyroid patients should be avoided in the presence of previous thyroid surgery and those who are dependent on their voice, e.g., opera singers. Radioiodine therapy is absolutely contraindicated in pregnancy and planned pregnancy within 6 months. About 10% of patients develop transient hypocalcemia following thyroid surgery, but 1% only will develop long-term permanent hypocalcemia that necessitates life-long calcium replacement therapy. Clinical improvement is observed 2 weeks after starting carbimazole while the biochemical normalization takes at least 4 weeks to return to their normal reference range.

16) c.

Subacute thyroiditis is usually virally-induced (especially by coxsackie, mumps, and adeno viruses). The resulting anterior neck pain is usually aggravated by swallowing, coughing, and movement of the neck. The pain may radiate to the jaw and ears. The ESR is usually elevated, and a low titer of thyroid antibodies (non-specifically) can be detected transiently. Most patients respond symptomatically to non-steroidal anti-inflammatory drugs; however, systemic steroids are occasionally used for severe cases. Those who have features of hyperthyroidism needs nothing apart from re -assurance that the course is self-limited and benign; oral propranolol may be used in certain cases to alleviate tachycardia, otherwise, there is no need for anti-thyroid medications.

17) d.

The combination of suppressed serum TSH, negligible radioiodine uptake, low serum thyroglobulin, and greatly raised $T_4:T_3$ ratio (usually around 70:1) is diagnostic of factitious hyperthyroidism.

18) d.

There is no association between post-partum depression and postpartum thyroiditis. The latter is supposed to be an autoimmune in origin.

19) d.

Thyroid storm can be precipitated by surgery, radioiodine treatment in a previously inappropriately treated patient, and by systemic infections. All hyperthyroid patients should be rendered euthyroid before doing thyroid operation to prevent the development of thyrotoxic crisis. The usual anti-thyroid measures applied are oral or intravenous propranolol, oral sodium iopodate (a radio-contrast medium which restores serum thyroid hormone levels to normal within 2-3 days), oral potassium iodate or Lugol's solution, and oral or per rectal carbimazole. The latter medication has no intravenous preparation; if it cannot be given orally, then give it rectally or by a nasogastric tube. After 10-14 days of treatment with various anti-thyroid measures, sodium iopodate and propranolol can be stopped and the patient should continue on carbimazole.

20) e.

Post-ablative hypothyroid cases and primary atrophic hypothyroidisms are not goitrous.

21) e.

Iron deficiency anemia is a common finding in pre-menopausal middle-aged hypothyroid women because of menorrhagia; amenorrhea is more common in the younger age group. There are 8 well-recognized features of hypothyroidism but they are rare in clinical practice. These are myotonia, cerebellar ataxia (as an isolated syndrome), frank psychosis (myxedema madness), pleural and pericardial effusions, heart failure, ileus, impotence (while in thyrotoxicosis it is common), galactorrhea, and ascites.

22) d.

T_3 alone is not a reliable test to differentiate between euthyroid and hypothyroid states. Anemia in hypothyroid patients could be normochromic normocytic (as part of anemia of chronic diseases), hypochromic microcytic (from iron deficiency anemia because of menorrhagia), or macrocytic (from an associated vitamin B_{12} deficiency/pernicious anemia). Low voltage ECG, bradycardia, and non-specific ST-T changes reflect prolonged and severe hypothyroidism.

23) d.

After starting thyroxin therapy for hypothyroidism, subjective improvement is reported by patients after 2-3 weeks, and the reduction in body weight and peri-orbital puffiness occurs relatively rapidly, while restoration of skin and hair texture and resolution of any effusion may take 3-6 months.

24) e.

Myxedema coma is a rare presentation of hypothyroidism rather than being encountered as a complication in an already diagnosed patient. The CSF profile may be abnormal; the opening pressure may be elevated and the protein content may be raised as well; those patients may be missed as a primary CNS disease. Intravenous hydrocortisone should be started before the biochemical diagnosis is achieved, and this should be followed by intravenous T_3.

Thyroxin is not usually available for parenteral use and tri -iodothyronine (T3) is given as an intravenous bolus of 20 μg followed by 20 μg 8-hourly until there is a sustained clinical improvement. In survivors, there is a rise in body temperature within 24 hours and, after 48 -72 hours, it is usually possible to substitute oral thyroxin in a dose of 50 μg per day. The mortality rate is 50% and survival depends on early recognition and treatment of hypothyroidism and other factors contributing to the altered consciousness level, e.g., drugs, cardiac failure, pneumonia, dilutional hyponatremia, and hypoxemia and hypercapnia due to hypoventilation.

25) c.

Stem "c" is the classical finding in hypothyroid patients who are non - compliant with their thyroid replacement therapy; patients usually ingest large doses few days before the follow-up visit! Non-compliance with medications is one of the major issues in treating chronically ill patients, and thyroid failure is no exception.

26) d.

Phenytoin is an enzyme inducer medication; the daily dose of thyroxin may need to be increased in hypothyroidism and concomitant use of phenytoin.

27) e.

Obesity is rarely due to a single gene mutation. Endocrine diseases associated with obesity are Cushing syndrome, hypothyroidism, and polycystic ovarian syndrome.

28) d.

Follicular thyroid carcinoma and adenoma have a similar appearance on FNA cytological examination; therefore, thyroid biopsy is needed to demonstrate vascular invasion by the malignant cells in cases of follicular carcinoma.

29) e.

Secondary thyroid tumors are rare. Papillary carcinoma targets young people and is the commonest type of thyroid malignancy; it constitutes up to 70% of all thyroid cancers.

30) b.

Medullary thyroid cancer is an aggressive malignancy with a gloomy outcome. It is usually found as part of MEN type II; however, many cases could be sporadic and have no underlying genetic syndromes. There is no curative treatment. Serum calcitonin in those patients is measured as a tumor marker; it rarely, if ever, results in symptomatic hypocalcemia. Because the para-follicular cells do not trap iodine, radioiodine treatment is not an option.

31) e.

Parathyroid adenoma can rarely be palpated in the neck (usually, the tumor is located during surgical exploration of the anterior neck).

32) e.

Secondary bony metastases results in hypercalcemia but serum parathyroid hormone level is undetectable. High serum calcium *and* low serum parathyroid hormone combination can also be encountered in vitamin D intoxication, milk-alkali syndrome, sarcoidosis, and Addison disease. The 1st 4 stems in the question are the cause of hypercalcemia *and* normal or high serum parathyroid hormone level.

33) d.

The hypercalcemia of primary hyperparathyroidism does not respond to systemic steroids. High doses of glucocorticoids are very effective in the treatment of malignancy-associated hypercalcemia. The following measures are used to treat malignancy-associated hypercalcemia:
1. Rehydration: hypercalcemia induces nephrogenic diabetes insipidus and most patients are profoundly dehydrated. A range of 4-6 liters is usually required to re-expand the blood volume.
2. Bisphosphonates: this class of medications is given intravenously initially. All patients should be maintained on oral preparations, unless the underlying systemic cancer is well-controlled.
3. Forced diuresis with saline and frusemide, systemic glucocorticoids, calcitonin, and hemodialysis are reserved for severely ill patients and very high serum calcium level.
4. Try to treat the cause (if possible).

34) a.

Hypercalcemia can result in hypertension. Prolonged hypocalcemia may cause seizures and Parkinsonian-like features. Pseudopseudohypoparathyroidism resembles pseudohypoparathyroidism phenotypically, but serum calcium is normal in the former.

35) e.

Pseudopseudohypoparathyroidism displays a phenotype of pseudohypoparathyroidism (i.e., short stature, moon face, mental retardation, and short 4th metacarpals) but demonstrates normal biochemical profile. Prolonged hypomagnesemia can cause functional hypoparathyroidism; it inhibits PTH release and causes PTH receptor defects.

36) d.

Pseudohypoparathyroidism is treated by vitamin D metabolites (e.g., alpha calcidol tablets) and their follow-up entails regular measurement of serum PTH and serum calcium. For chronic control of hypocalcaemia, commercial preparations of PTH are unsatisfactory, because they have to be given by frequent injections and these injections soon become ineffective due to serum antibody formation against them.

37) e.

In Cushing syndrome, weight gain is the commonest symptom and obesity is the commonest sign. Obesity as a sign is observed in 97% of cases, which could be central or generalized; 3% of patients are not obese. Hypertension with hypokalemic metabolic alkalosis occurs in 75% of cases. Depression and low mood are the commonest psychiatric manifestations while psychosis is rare. Proximal muscle weakness with type II fiber atrophy is common. Striae and skin bruising develop in 50% of patients.

38) c.

Cushing's syndrome encompasses many screening tests. All require some sort of acclimatization to the hospital environment; those patients should be admitted 2 days before conducting these investigations.

The stress of hospitalization may interfere with many tests, and some normal individuals will even fail to suppress upon doing overnight dexamethasone suppression test (i.e., they behave like Cushing patients), because the hypothalamic-pituitary-adrenal axis escapes such suppression (due to powerful endogenous stress mechanisms). Dexamethasone does not cross-react with the blood cortisol radio-immune assay; that is why it can replace prednisolone in those who ingest the latter and who are due to undergo suppression tests. Alcoholic individuals can resemble Cushing patients, both clinically and biochemically; hence the designation pseudo -Cushing syndrome. The finding of very low serum ACTH in patients who have gross cushingoid features reflects an adrenal source of excess blood cortisol, while very high serum ACTH values indicate an ectopic source.

39) e.

Without treatment, the 5-year survival rate of Cushing syndrome patients is 50%. Nelson syndrome encompasses progressive growth of the pituitary tumor, skin hyperpigmentation, and visual field defects; it develops in Cushing disease patients who have undergone bilateral adrenalectomy but without pituitary irradiation; with the loss of negative feedback control of adrenal cortisol on the pituitary, the pituitary adenoma enlarges to a considerable size. Benign tumors causing ectopic ACTH syndrome (e.g. bronchial carcinoid) should be removed. During treatment or palliation of other malignancies, it is important to reduce the severity of Cushing syndrome using medical therapy; a number of drugs are used to inhibit corticosteroid biosynthesis, such as metyrapone, aminoglutethimide, and ketoconazole. The dose of these agents is best titrated against the 24-hour urinary free cortisol.

40) e.

"Apparent" hyperaldosteronism, and that is low serum aldosterone *and* reduced plasma renin activity, hypokalemia, hypertension, occurs with carbenoxolone and licorice abuse, congenital adrenal hyperplasia due to 11-β hydroxylase and 17-α hydroxylase deficiency, Liddle syndrome, 11-β HSD deficiency, and 11-deoxycorticosterone -secreting adrenal tumors. Glucocorticoid-suppressible hyperaldosteronism, Conn's adenoma, and idiopathic bilateral adrenal hyperplasia are responsible for primary hyperaldosteronism (i.e., high serum aldosterone and reduced plasma rennin activity).

41) e.

Although there is an avid sodium retention in hyperaldosteronism, leg edema is uncommon and suggests secondary aldosteronism. Serum potassium is normal in up to 70% of cases at the time of diagnosis, because many patients are already treated by salt restriction; this would make less sodium available to be exchanged for potassium at the distal renal tubules. In patients with Conn's adenoma, aldosterone antagonists, such as spironolactone, are usually given for a few weeks to normalize body electrolytes before doing unilateral adrenalectomy. Laparoscopic surgery cures the biochemical abnormality but hypertension persists in 70% of cases, probably because of irreversible damage to the systemic microcirculation.

42) c.

Pheochromocytoma may be part of syndromes; MEN type II, neurofibromatosis, and von Hippel -Lindau are the best examples. Predominantly elevated serum adrenalin indicates an adrenal tumor that is not large enough to outgrow its blood supply. Weight loss is common and occurs in the absence of diabetes; pheochromocytoma may impair glucose tolerance test, however. Postural hypotension in pheochromocytoma patients reflects high blood levels of dopamine (in dopamine-secreting tumors). With urinary bladder tumors, patients may develop hypertensive episodes upon urination. Although pallor is commonly seen during the episode, occasionally flushing occurs. Constipation, not diarrhea, is a disease manifestation.

43) e.

The commonest cause of Addison disease in the Western World is autoimmune adrenalitits, while tuberculosis is the cause of most African and Asian cases; therefore, chest plain X-ray films should be obtained in all cases. Addison disease causes fasting hypoglycemia. It is a rare disease with an incidence of eight new cases per million of population. Male to female ratio is 1:2.

44) d.

The 21-α hydroxylase deficiency is the commonest type of congenital adrenal hyperplasia (CAH) and is responsible for up to 90% of cases; a late-onset form with isolated hirsutism is a well-recognized entity. CAH may result in salt-losing nephropathy and crises in neonates (usually in males).

Hypertension in CAH occurs due to increased serum levels of 11-deoxycorticosterone (which is a powerful mineralocorticoid) in 11-β and 17-α hydroxylase deficiencies. High levels of plasma 17OH-progesterone are found in 21-hydroxylase deficiency. In late-onset cases, this may only be demonstrated after ACTH administration. To avoid salt -wasting crisis in infancy, 17OH-progesterone is routinely measured in heel prick blood spot samples taken from all infants in the first week of life.

45) e.

The other gynecomastia-inducing drugs are estrogens and GnRH analogues (given for prostatic carcinomas). Unilateral enlarged breast in males should always be taken seriously; it may indicate an underlying breast cancer.

46) e.

The LH:FSH ratio in polycystic ovarian syndrome is more than 2.5:1 (usually >3.0). In addition, there are hypertension, hyperglycemia, hyperlipidemia, hirsutism, oligomenorrhea or secondary amenorrhea, and infertility. The management depends on the "presenting" clinical problem. Infertility may be treated under specialist supervision with clomifene or exogenous gonadotrophins. Although patients with polycystic ovarian syndrome may have amenorrhea, hormone replacement therapy is not required to prevent osteoporosis, since they have elevated, rather than low, circulating levels of estrogens and androgens.

47) e.

In idiopathic hirsutism, family history is very important (especially in Asians or Mediterranean people). The severity of hirsutism is subjective. Some women suffer profound embarrassment from a degree of hair growth which others would not consider remarkable. Important observations are a drug and menstrual history, calculation of body mass index, measurement of blood pressure, examination for virilization (clitoromegaly, deep voice, balding, and breast atrophy), and associated features including acne vulgaris or Cushing syndrome. Hirsutism of recent onset associated with virilization is suggestive of an androgen-secreting tumor, but this is rare.

48) b.

Mutations in the *RET* proto-oncogene on chromosome 10 are responsible for MEN type II. Hypercalcemia is the commonest presenting feature of MEN type I, while in type IIa, it is uncommon; in type IIb, hypercalcemia is absent. Pheochromocytoma in MEN type II is bilateral in 70% of cases, unlike the sporadic ones (which are 10% bilateral). Cushing syndrome may occur as part of MEN type I.

49) e.

The commonest site for carcinoid tumors is the terminal ileum; next is the cecum and appendix (obstruction of the mouth of the latter organ may result in acute appendicitis). The long- term prognosis is excellent in the majority of patients, because such tumors have a low malignant potential. Pulmonic stenosis, tricuspid regurgitation, and right-sided endocardial fibrosis occur in carcinoid syndrome (in general) or bronchial carcinoids (the metabolite of which bypass the hepatic circulation and directly affect the right cardiac side). Cramping abdominal pain and diarrhea with flushing and wheeze are the commonest presenting features of carcinoid *syndrome* (not tumors)

50) d.

Somatostatinoma may also result in steatorrhea and achlorhydria. Skin rash (necrolytic migratory erythema) and diabetes mellitus are found in glucagonomas. Steatorrhea is a well-recognized feature of gastrinoma, as are watery diarrhea and multiple severe peptic ulcerations. Pancreatic VIPoma results in severe watery diarrhea and hypokalemia; hence the designation "pancreatic cholera." Insulinomas could cause dizzy spells due to hypoglycemia.

51) e.

Nesidioblastosis is a form of diffuse beta cells hyperplasia of the pancreas and is a cause of neonatal and infantile recurrent hypoglycemia.

52) d.

The concordance rate in monozygotic twins is around 40-50%; it is almost 100% in type II diabetes.

53) e.

The concordance rate in monozygotic twins is almost 100%, implying strong, yet unidentified genes involved in the pathogenesis of type II diabetes mellitus. Its incidence is 1/1000 population per year and its prevalence is 1-2/100 population.

54) e.

Hepatocyte nuclear factor 4α genetic mutations are the cause of MODY (maturity onset diabetes of the young) type I; a rare progressive early onset form. MODY type III results from mutations involving hepatocyte nuclear factor 1α gene; a progressive early onset. MODY type II is mild and relatively stable early onset type, which is caused by genetic mutations in glucokinase gene. Mitochondrial cytopathic diseases have a *maternal* mode of inheritance; diabetes is a cardinal feature of these syndromes.

55) c.

It inhibits the action of lipoprotein lipase, and hence it enhances lipogenesis (compare it with the weight loss incurred by insulin deficiency!).

56) c.

No smoking is allowed and the patient should sit during glucose tolerance testing. In some people, in whom an oral glucose tolerance test is normal, an abnormal result is observed under conditions which impose a burden on the pancreatic beta cells, e.g., during pregnancy, systemic infections, myocardial infarction or other severe stress, or during treatment with diabetogenic medications, such as corticosteroids. This "stress hyperglycemia" usually disappears after the acute illness has resolved, but blood glucose should be re-measured.

57) e.

Fasting blood glucose of >7 mmol/L on 2 or more occasions is diagnostic for diabetes mellitus; a single random blood glucose level greater than 11.1 mmol/L *with* symptoms of diabetes is diagnostic. The diagnostic criteria for diabetes in pregnancy are more stringent than those recommended for non-pregnant subjects.

Pregnant women with abnormal glucose tolerance should be referred urgently to a specialist unit for full evaluation. Glycated hemoglobin provides an accurate and objective measure of glycemic control over a period of weeks to months. This can be utilized as an assessment of glycemic control in a patient with known diabetes, but is not sufficiently sensitive to make a diagnosis of diabetes and is usually normal in patients with impaired glucose tolerance. Glycated serum proteins (fructosamine) can be measured and, because of their shorter half-life, give an indication of glycemic control over the preceding 2 weeks. Other than diabetic follow-up in pregnancy, this is generally too short a period to make clinical decisions on therapeutic management.

58) b.

The recommended composition of the daily diet for diabetics is as follows: the total fat 30-35%, saturated fat <10%, monounsaturated fat 10-15%, polyunsaturated fat should be <10%, carbohydrates 50-55%, and total protein 10-15%. The ideal management for diabetes would allow the patient to lead a completely normal life, to remain not only symptom- free, but in good health, to achieve a normal metabolic state and to escape the long-term complications of diabetes.

59) b.

Insulin does reduce post-prandial glycemia and has a favorable effect on lipid profile. All anti-diabetic medications improve the postprandial blood glucose level. Hypoglycemia can be caused by insulin, sulfonylureas, and meglitinides and amino acid derivatives. In addition, metformin and acarbose can improve serum lipid profile slightly (but less efficiently than insulin).

60) e.

Metformin causes anorexia and weight loss in a substantial number of patients, as well as carrying a risk of lactic acidosis; metformin should avoided when serum creatinine is >1.5 mg/dl. Lactic acidosis is a rare, but a potentially severe consequence of therapy with metformin. Lactic acidosis should be suspected in any diabetic patient receiving metformin who has evidence of acidosis but when evidence of ketoacidosis is lacking. We should discontinue metformin in clinical situations predisposing to hypoxemia (as in cardiovascular collapse, respiratory failure, acute myocardial infarction, acute congestive heart failure, and septicemia). Chlorpropamide may cause SIADH and cholestatic jaundice.

Prominent fluid retention may result from pioglitazone; therefore, it is contraindicated in heart failure. Due to its salt and water retention in the short-term, insulin can cause transient leg edema. Acarbose produces diarrhea and flatulence.

61) c.

Hypoglycemia causes hypothermia, myocardial infarction, TIAs, brain damage, convulsions, intellectual decline, and coma. Hypoglycemia (blood glucose <3.5 mmol/l) occurs often in diabetic patients treated with insulin, but it is relatively infrequent in those taking sulfonylureas. The risk of hypoglycemia is the most important single factor limiting the attainment of the therapeutic goal, namely near-normal glycemia. Severe hypoglycemia, which is defined as "hypoglycemia requiring the assistance of another person for recovery", can result in serious morbidity and has a recognized mortality of 2-4% in insulin-treated patients.

62) a.

Unaccustomed or unusual exercise may cause hypoglycemia if the original insulin dose is not reduced. The other causes are associated malabsorption, errors in anti-diabetic agents dosing and schedules and administration, impaired hypoglycemia awareness, alcohol, poorly designed insulin regimen, lipohypertrophy and dumping syndromes.

63) e.

Leukocytosis occurs in diabetic ketoacidosis (DKA), even in the absence of infection. Confusion, drowsiness and coma occur in 10% of DKA cases. Every diabetic ketoacidosis patient is potassium-depleted, but the plasma concentration of potassium gives very little indication of the total body deficit. Plasma potassium may even be raised initially due to disproportionate loss of water and catabolism of protein and glycogen. However, soon after insulin treatment is started, there is likely to be a precipitous fall in the plasma potassium due to dilution of extracellular potassium by administration of intravenous fluids, the movement of potassium into cells as a result of treatment with insulin, and the continuing renal loss of potassium.

64) a.

Disseminated intravascular coagulation is rare in DKA. Cerebral edema is caused by a rapid reduction in blood glucose as well as use of hypotonic fluids and/or bicarbonate; it has a high mortality, and should be treated by intravenous mannitol with high concentration, high flow oxygen.

65) d.

The mortality rate of non-ketotic hyperosmolar coma usually exceeds 40%, even at the best centers. Thromboembolic complications are common, and prophylactic subcutaneous heparin is recommended. These patients are usually relatively sensitive to insulin and approximately half the dose of insulin recommended for the treatment of ketoacidosis should be employed. The patient should be given 0.45% saline until the osmolality approaches normal, when 0.9% saline should be substituted. The rate of fluid replacement should be regulated based on the central venous pressure, and plasma sodium concentration should be checked frequently.

66) e.

Venous loops and beadings are seen mainly in the *pre*-proliferative stage. Intra-retinal microvascular abnormalities (IRMAs) are dilated and tortuous capillaries which represent the remaining patent capillaries in an area where most have been occluded. Venous changes include venous dilatation (an early feature probably representing increased blood flow), "beading" (sausage-like changes in caliber) and increased tortuosity including "oxbow lakes" or loops. These latter changes indicate widespread capillary non-perfusion and are a feature of advanced pre-proliferative retinopathy. Microaneurysms are the earliest clinical abnormality detected on ophthalmoscopy. They appear as tiny, discrete, circular, and dark red spots near to, but apparently separate from, the retinal vessels.

67) d.

The peripheral symmetric diabetic polyneuropathy is mainly sensory and is frequently asymptomatic. The most common signs found on physical examination are diminished perception of vibration sensation distally, "glove-and-stocking" impairment of all other modalities of sensation, and loss of tendon reflexes in the lower limbs. Sensory abnormalities dominate the clinical presentation.

Symptoms include paresthesia in the feet and, rarely, in the hands, pain in the lower limbs (dull, aching and/or lancinating, worse at night, and mainly felt on the anterior aspect of the legs), burning sensations in the soles of the feet, cutaneous hyperesthesia and an abnormal gait (commonly wide-based), often associated with a sense of numbness in the feet. Muscle weakness and wasting develop only in advanced cases, but subclinical motor nerve dysfunction is common. The toes may be clawed with wasting of the interosseous muscles, which results in increased pressure on the plantar aspects of the metatarsal heads with the development of callus skin at these and other pressure points. Electro-physiological tests demonstrate slowing of both motor and sensory conduction, and tests of vibration sensitivity and thermal thresholds are abnormal. A diffuse small-fiber neuropathy causes altered perception of pain and temperature and is associated with symptomatic autonomic neuropathy; characteristic features include foot ulcers and Charcot neuroarthropathy.

68) d.

The nodular glomerulosclerosis type is less common than the diffuse type in diabetic nephropathy. Risk factors for developing diabetic nephropathy are long duration of diabetes and poor control of blood sugar, presence of other microvascular complications, ethnicity (Asians, Pima Indians), pre-existent hypertension, family history of diabetic nephropathy, and family history of hypertension. Microalbuminuria is an important indicator of risk of developing overt diabetic nephropathy, although it is also found in other conditions. It is therefore most reliable as an indicator of diabetic nephropathy within the first 10 years of type I diabetes (the majority will progress to overt nephropathy within a further period of 10 years), and less reliable in older patients with type II diabetes, in whom it may be accounted for by other diseases. Progressively increasing albuminuria, or albuminuria accompanied by hypertension, is much more likely to be due to early diabetic nephropathy.

69) d.

All diabetic female patients should have pre-pregnancy counseling to assess the risks and benefits of treatment and complications of diabetes. Diabetes should be controlled before pregnancy rather than during its course, so that it is controlled at the time of conception. The objective of diabetic control is to decrease fetal congenital anomalies (and maternal and fetal morbidity and mortality). All oral hypoglycemic agents are contraindicated; insulin is used. Due to the physiological and stress changes during pregnancy, insulin requirement varies throughout the course of pregnancy.

Hyperglycemia in early pregnancy can cause fetal malformations, and will promote increased somatic growth later in gestation. Pregnancy in diabetic women is associated with an increased perinatal mortality rate (that is, stillbirths and neonatal deaths within the first week of life). The main causes are intrauterine death in the third trimester of pregnancy, prematurity (due to a high incidence of spontaneous premature labor and of elective premature delivery in an attempt to avoid late intrauterine death), low birth weight and congenital malformation. Birth trauma is also more common due to a high incidence of excessively large, macrosomic babies.

70) e.

Much of the pre -operative assessment can be done on an outpatient basis but, if cardiovascular or renal function is impaired, there are signs of neuropathy (particularly autonomic), diabetic control is poor or alterations need to be made to the patient's usual treatment, then admission to hospital some days before operation will be required. Emergency surgery in a patient with well-controlled insulin -treated diabetes depends on when the last subcutaneous injection of insulin was given. If this was recent, an infusion of glucose alone may be sufficient, but frequent monitoring is essential.

Chapter 7
Gastroenterology and Hepatology

Questions

1) You are reviewing a physiology textbook about gut hormones and their various modes of actions and functions. Regarding gut hormones, all of the following are correct, *except?*
   a) Gastrin is mainly secreted by the gastric antral G-cells
   b) Secretin inhibits gastric acid secretion
   c) Somatostatin stimulates insulin secretion
   d) Motilin secretion is stimulated by dietary fat
   e) Gastric inhibitory polypeptide is secreted from duodenum and jejunum to inhibit gastric acid secretion and stimulate insulin secretion

2) You refer a 65-year-old man with a history of lower GIT bleeding and unintentional weight loss to the radiology department in order to do barium enema. You suspect colorectal cancer. All of the following statements about gastrointestinal contrast radiology are correct, *except?*
   a) The main limitations of barium swallow are risk of aspiration and poor mucosal details
   b) The main limitations of barium meal are low sensitivity to detect early cancer and inability to assess H. pylori status
   c) The main limitations for barium-follow-through are time-consuming and greater risk of radiation exposure
   d) The main limitations for barium enema are difficult in elderly or those with incontinence and being somewhat uncomfortable
   e) Barium enemas usually miss polyps less than 10 cm in size

3) A 62-year-old man, who has a history of alcoholic liver cirrhosis and ascites, presents with severe upper gastrointestinal bleeding and shock. After an appropriate resuscitation, you consult the surgical unit for further management plan. They decide that he should undergo esophagogastroduodenoscopy (OGD). Regarding upper gastrointestinal endoscopy, all of the following statements are contraindications, *except?*
   a) Severe shock
   b) Recent myocardial infarction
   c) Severe respiratory distress
   d) Possible visceral perforation
   e) Anemia

4) A 32-year-old woman, who has a history of recurrent peptic ulcers, presents with coffee-ground vomitus. She is not compliant with her omeprazole, which was prescribed by her physician. She will undergo an upper gastrointestinal endoscopy. She declines this investigation because of its complications. Which one of the following is *not* a complication of upper gastrointestinal endoscope?

a) Aspiration pneumonia
b) Visceral perforation
c) gastrointestinal bleeding
d) Cardiopulmonary depression due to over-sedation
e) Negligible risk of infective endocarditis in all patients

5) One of your junior house officers asks you about gastrointestinal dynamic tests. He wonders if such investigations are helpful to assess their celiac disease patient. Which one of the following statements about these dynamic tests is *incorrect?*

a) Lactose hydrogen breath test is used in lactose intolerance and although non-invasive and accurate, but it may provoke abdominal pain and diarrhea in sufferers
b) ^{14}C-triolein breath test is used in fat malabsorption and although fast and non-invasive, it is non-quantitative
c) ^{75}SeHCAT test is used in fat malabsorption and is accurate and specific but requires 2 visits and involves radiation
d) Pancreolauryl test is used as a test for pancreatic exocrine function and is accurate and avoids duodenal intubation but needs accurate collection of urine
e) ^{14}C-glycocholate breath test is a useful screening test for bacterial overgrowth syndromes

6) A 45-year-old man presents with massive lower gastrointestinal bleeding. Colonoscopy is done by an experienced gastroenterologist but it fails to detect any abnormality. You consider isotope scanning to detect the bleeding site. Which one of the following statements is *incorrect* with respect to radioisotope tests?

a) ^{13}C and ^{14}C urea breath tests are used for H. pylori detection
b) ^{99m}Tc pertechnetate is used for the detection of Meckel's diverticulum
c) ^{99m}Tc HMPAO-labeled leukocytes is used for the detection of visceral abscesses
d) ^{51}Cr-albumin is used as a test for gastrointestinal epithelial permeability
e) ^{99}Tc-sulphur is used as a test for protein-losing enteropathy

7) A 54-year-old man is referred to you for further evaluation. He has long-standing history of heartburn and regurgitation, for which he takes chewable anti-acids. Upper gastrointestinal endoscopy detects Barrette's esophagus. Regarding Barrette's esophagus, which one is the *incorrect* statement?

a) Pre-malignant condition that increases the risk of esophageal adenocarcinoma by 90-150 folds
b) Always symptomatic with heartburn and regurgitation
c) Results from long-standing gastric acid reflux
d) The risk of malignancy is particularly very high when the metaplastic tissue is composed of intestinal origin with goblet cells
e) Long-term acid suppression therapy is not useful in reversing the histological abnormality

8) A 49-year-old woman presents with a 3-year history of heartburn, water-brash, and regurgitation. She surfed the internet and found that she might develop many complications because of her acid reflux, and she is desperate for help. Which one of the following is *not* a well-recognized complications of long-standing gastroesophageal reflux disease?
a) Iron deficiency anemia
b) Benign stricture formation
c) Barrette esophagus
d) Esophagitis
e) Esophageal squamous cell carcinoma

9) A 62-year-old male presents with halitosis. You find a mass in the upper neck. The mass is reducible with a gurgling sound. He reports regurgitation of food particles. What is the likely diagnosis?
a) Thyroid cyst
b) Branchial cyst
c) Pharyngeal pouch
d) Cystic hygroma
e) Nasopharyngeal carcinoma

10) A 42-year-old female presents with a 9-month history of dysphagia. At the beginning, it was for solids only, and now she has difficulty swallowing fluids. She reports two episodes of aspiration pneumonia. Chest X-ray reveals patchy pneumonia with enlarged mediastinum. Barium swallow shows mega esophagus. Which one of the following statements regarding achalasia of the cardia is *incorrect?*
a) A similar picture may be seen in Chagas disease
b) May result in esophageal squamous cell carcinoma, even after successful treatment
c) Heartburn is prominent
d) Usually seen in middle-aged women, although no age is exempt
e) Treatment with botulinum toxin is effective, but the effect is transient

11) A 63-year-old Japanese man presents with pallor, global weight loss, and dysphagia for solids and liquids. Esophagoscopy reveals a large fungating mass just above the cardia. He has esophageal cancer. Which one of the following is *not* a well-recognized etiological factor in the development of esophageal carcinoma?
    a) Celiac disease
    b) Tylosis
    c) Cigarette smoking and alcohol ingestion
    d) Chewing tobacco
    e) Ulcerative colitis

12) A 65-year-old man takes daily aspirin after developing hemispheric transient attack. Today, he visits the physician's office because of upper abdominal discomfort and a single attack of hematemesis. Upper gastrointestinal endoscopy reveals acute erosive gastritis. All of the following may result in acute gastritis, *except?*
    a) Iron medications
    b) Bile reflux following gastric surgery
    c) CMV infection
    d) Acute infection with H. pylori
    e) Pernicious anemia

13) Your junior house officer read a microbiology book. He asked you about the clinical consequences of H. pylori bacterial infections. Which one of the following diseases is *not* associated with H. pylori infection?
    a) Gastro-esophageal reflux disease
    b) Duodenal ulcer
    c) Gastric MALToma
    d) Gastric adenocarcinoma
    e) Gastric ulcer

14) A 36-year-old man presents with dyspepsia. He reports recurrent epigastric pain that is relieved by food and anti-acids. You order upper gastrointestinal endoscopy and this reveals kissing ulcers at the duodenal bulb. He tests positive for H. pylori. With respect to infections with H. pylori diagnosis, which one of the following is the *incorrect* statement?
    a) H. pylori serology although being rapid and useful for population studies, but it cannot differentiate between acute and past infections and it lacks sensitivity and specificity
    b) Although urea breath test has a high sensitivity and specificity but ^{14}C requires radioactivity and ^{13}C requires mass spectrometer

c) Rapid urease test on an antral biopsy specimen has a high sensitivity and is cheap and quick

d) Microbiological culture of an antral biopsy specimen is the gold standard method and defines antibiotic sensitivity, but it is slow and lacks sensitivity

e) Histopathological examination of an antral biopsy specimen although sensitive and specific, but false-negative results occur and it takes several days to process

15) A 76-year-old man presents with a transient ischemic attack. You intend to prescribe aspirin but you consider aspirin-induced peptic ulceration. Risk factors for NSAID (non-steroidal anti-inflammatory drugs) induced peptic ulceration are all of the following, *except?*

a) History of peptic ulcer
b) High dose NSAIDs or multiple NSAIDs ingestion
c) Concomitant steroid therapy
d) Age <50 years
e) Treatment with azapropazone

16) A 67-year-old man receives anti-H. pylori eradication therapy after developing duodenal ulceration. He is non-compliant with his medications because he has been experiencing many side effects. Which one of the following is *not* a well-recognized side effect of anti-H. pylori therapy?

a) Diarrhea is uncommon
b) Metallic taste in the mouth is commonly seen with metronidazole
c) Headache
d) Skin rash
e) Abdominal cramps and vomiting

17) A 43-year-old woman has been recently diagnosed with benign gastric ulcer and is due to receive anti-ulcer medications, including anti-H. pylori ones. Side effects of individual anti-H. pylori medications include all of the following, *except?*

a) Cimetidine can cause confusion
b) Omeprazole may result in hypergastrinemia
c) Sucralfate enhances the effect of digoxin and warfarin
d) Misoprostol produces diarrhea in up to 20% of cases
e) Colloidal bismuth blackens teeth, tongue, and stool

18) A 34-year-old male presents with diarrhea, dyspepsia, and epigastric pain that is relieved by food ingestion.

Upper gastrointestinal endoscopy reveals multiple ulcerations at the lower esophagus, gastric antrum, and the first 2 parts of the duodenum. His fasting serum gastrin is very high. Regarding Zollinger-Ellison syndrome, which one of the following is the *incorrect* statement?

a) Accounts for 0.1% of all cases of peptic ulceration
b) Commonly seen between 30-50 years of age
c) About 90% of cases result from gastrin-secreting pancreatic tumors
d) Malignant in 50-70% of cases
e) Around 20-60% of cases are part of MEN type IIb

19) A 46-year-old man is referred to you for further evaluation. He has been developing recurrent peptic ulcers overt the past several months, despite full compliance with his anti-H. pylori medications. You run a battery of investigations and his final diagnosis turns out to be Zollinger-Ellison (ZE) syndrome. With respect to ZE syndrome, which one is the *incorrect* statement?

a) Should be suspected whenever there are severe and multiple peptic ulcerations
b) Diarrhea is common and may the presenting feature in 30% of cases
c) Ulcer bleeding and perforations are common
d) Barium meal may reveal thin gastric mucosal folds
e) The ulcers may occur at atypical sites such as jejunum and esophagus

20) A 45-year-old female was diagnosed with Zollinger-Ellison syndrome (ZE) and she asked about her treatment options and complications. Regarding the management of Zollinger-Ellison syndrome, all of the following statements are correct, *except?*

a) Some patients present with metastatic disease and therefore surgery is inappropriate
b) Proton pump inhibitors should be given in large doses in order to be effective
c) Octreotide may result in symptomatic improvement
d) The overall 5 years survival is 1-2%
e) All patients should be monitored for the future development or other features of MEN type I

21) A 68-year-old retired man complains of with unintentional weight loss of 15 Kg, lassitude, early satiety, and progressive pallor over the past 6 months. Examination reveals a mass in the epigastrium and a hard knobbly liver. The left supra-clavicular lymph nodes are palpable and are rocky hard.

Gastroscopy shows a large antral mass, and histopathological examination of a biopsy specimen is suggestive of gastric adenocarcinoma. Predisposing factors for the development of gastric cancer are all of the following, *except?*
a) Previous partial gastrectomy
b) Autoimmune gastritis
c) Adenomatous gastric polyps
d) Familial adenomatous polyposis
e) Tylosis

22) A 71-year-old man from the Far East presents with a 5-month history of anorexia, anemia and asthenia. Barium meal reveals a large somewhat rounded filling defect within the gastric antrum. This finally turns out to be gastric adenocarcinoma. With respect to gastric cancer, which one of the following is the *incorrect* statement?
a) The incidence of gastric cancer is rising in the Western World
b) The overall prognosis remains very poor
c) Almost all tumors are adenocarcinomas
d) Clinical examination may reveal nothing
e) Dysphagia occurs with tumors near the cardia

a) **23)** A 29-year-old man presents with a 6-month history of unintentional weight loss, leg swelling, and pallor. Blood film shows dimorphic blood picture. Barium follow-through reveals small bowel mucosal flocculation and clumping. Upper gastrointestinal endoscopy is normal. Histopathological examination of a duodenal biopsy uncovers subtotal villous atrophy. You consider celiac disease. With respect to celiac disease, which one is the *incorrect* statement?
   a. The disease may occur at any age
   b. The presentation is highly variable, depending on the severity and extent of the small bowel involvement
   c. An association with HLA B8, DR17, and DQ2 is present
   d. The earliest histological finding is an increase in the intraepithelial lymphocytes
   e. The disease is seen worldwide but is very rare in northern Europe

24) A 16-year-old female is brought by her parents to consult you. She has oligomenorrhea and short stature. You find pallor, angular stomatitis, and spoon-shaped nails. She denies loss of appetite and she eats a well -balanced diet. The final diagnosis is celiac disease. Regarding the treatment of celiac disease, choose the *wrong* statement?

a) Gluten-free diet should be advised indefinitely
b) Dietary restriction of wheat, rye, barely, and possibly oats should be encouraged
c) Rice, maize, and potatoes should be eliminated from diet
d) Minerals and vitamins may be given, but in practice this is uncommon
e) Poor compliance with dietary advices remains the commonest cause of treatment failure

25) A 36-year-old celiac patient visits the physician's office for a scheduled follow-up. He admits to being non-compliant with his gluten-free diet and he has failed to achieve any weight gain. You educate him about the possible disease course and complications. Which one of the following is *not* a complication of celiac disease?
a) Ulcerative jejunitis
b) Esophageal adenocarcinoma
c) Small bowel T-cell lymphoma
d) Intensely itchy skin rash
e) Small bowel carcinoma

26) A 28-year-old housekeeper presents with weight loss and secondary amenorrhea over a period of 7 months. She denies other symptoms. She does not report diarrhea, anorexia, abdominal pain, or vomiting. After thorough investigations, the final diagnosis turns out to be celiac disease. Regarding the diagnosis of celiac disease, which one of the following statement is *not* true?
a) Of all detected antibodies, IgA anti-endomysium has the highest sensitivity and specificity
b) Small bowel biopsy is the gold standard investigation
c) Barium-follow-through reveals non-specific findings of dilated loops with coarse folds and contrast clumping
d) Dimorphic blood picture may be seen
e) Co-existent IgA deficiency would not affect the serological tests

27) Your colleague asks you about the possible autoimmune diseases that might be associated with celiac disease. The following diseases are associated with celiac disease, *except*?
a) Type II diabetes
b) Splenic atrophy
c) Hypothyroidism
d) Primary biliary cirrhosis
e) Inflammatory bowel disease

28) A 45-year-old businessman visits the physician's office with a 5-month history of steatorrhea after returning from Jamaica. He has weight loss, pallor, and cheilitis. Serum vitamin B_{12} is very low. You think of tropical sprue. With respect to tropical sprue, all of the following statements are true, *except?*
   a) The disease occasionally occurs in epidemics and an infective etiology has been suggested
   b) Partial villous atrophy is much more common than total villous atrophy on small bowel biopsy examinations
   c) The most important differential diagnosis in endemic areas is infective diarrhea
   d) Tetracycline for 28 days results in cure or long-term remission and is the treatment of choice
   e) The disease is rare in Malaysia and Indonesia

29) A 34-year-old male presents with a 6-month history of gastrointestinal complaints. He did some form of gastric surgery because of perforated peptic ulcer 3 years ago. Culture of small bowel juice reveals overgrowth of E. coli. Regarding bacterial overgrowth syndrome, all of the following statements correct, *except?*
   a) May be caused by pernicious anemia
   b) ^{14}C-glycocholate breath test is a good screening tool
   c) Serum levels of vitamin B_{12} and folate are low
   d) The treatment of choice in most patients is tetracycline for 1 week
   e) Patients usually present with watery diarrhea and/or steatorrhea

30) Because of a 5-month history of steatorrhea and seizures, a 45-year-old man visits the physician's office. You find weight loss, leg edema, and skin hyper-pigmentation. You consider Whipple disease. With respect to this disease, choose the *incorrect* statement?
   a) Caused by a small Gram-negative bacilli
   b) Usually seen in middle-aged men with clubbing and low-grade fever
   c) Almost any organ can be involved
   d) Usually fatal if not treated
   e) Following successful therapy, follow-up is very important because 30% of cases will relapse

31) A 67-year-old retired mechanic underwent extensive small bowel resection because of small bowel ischemia and gangrene. These resulted from atrial fibrillation-associated emboli. Before hospital discharge, you educate him about the possibility of developing short bowel syndrome. Which one of the following statements regarding short bowel syndrome is the *wrong* one?

a) May be caused by necrotizing enterocolitis in children and Crohn disease in adults
b) Dehydration and weight loss are common
c) Some patients need treatment with octreotide to lessen bowel secretions and diarrhea volume
d) Anti-diarrhea agents are contraindicated
e) Some patients need total parenteral nutrition for survival

32) A 46-year-old female presents with steatorrhea. She was diagnosed with a pelvic tumor few months ago. She received radiotherapy. You interview the patient and examine her properly. At the end, you consider radiation enteritis as the cause of this steatorrhea. With respect to radiation enteritis, which one is the *incorrect* statement?
a) Usually occurs in the context of radiotherapy for abdominal or pelvic malignancy
b) The terminal ileum, sigmoid, and rectum are the usual victims
c) May produce malabsorption through different mechanisms
d) Small bowel adhesions and fistulae may occur
e) Cholestyramine is useless in the treatment

33) You are discussing the possible causes of protein-losing enteropathy with your colleagues because one of your medical ward patients has this condition. With respect to protein-losing enteropathy, which one is the *incorrect* statement?
a) Patients may present with peripheral edema, hypoalbuminemia, normal liver function tests, and normal 24-hour urinary protein excretion
b) The treatment is that of the underlying disorder
c) Nutritional support is important
d) The diagnosis is confirmed by measuring the rate of fecal clearance of intravenous radio-labeled albumin
e) Can be caused by very few gastrointestinal diseases

34) A 45-year-old male presents with a 4-month history of progressive leg edema that pits on pressure. Cardiac, liver, and renal functions are intact. Serum albumin is very low, and 24-hour urinary protein excretions is 100 mg. You consider protein-losing enteropathy and one of your colleagues suggests intestinal lymphangiectasia as the culprit. Which one of the following statements about intestinal lymphangiectasia is *not* true?
a) May be caused by Whipple disease, filariasis, lymphoma, and constrictive pericarditis
b) There is prominent peripheral blood lymphocytosis
c) Results in severe hypogammaglobulinemia

d) Small bowel biopsy reveals dilated lacteals
e) Medium-chain triglyceride supplements are useful in the treatment

35) Because of progressive abdominal distension over the past few weeks, a 33-year-old man consults you. You investigate him and find that he has ascites and there are many eosinophils in the ascitic fluid. One of your interns says something about eosinophilic gastroenteritis and this drags your attention. Regarding eosinophilic gastroenteritis, choose the *wrong* statement?
a) Any part of the gastrointestinal tract may be involved
b) The inflammatory process may involve the mucosa, muscularis, and/or serosal layer
c) Peripheral blood eosinophilia is present in 80% of cases and 50% of patients have some form of allergy
d) Full-thickness intestinal biopsy is usually used for the diagnosis, although multiple endoscopic biopsies are done in clinical practice
e) The prognosis is poor in the majority

36) A 34-year-old soldier presents with abdominal pain, bloody diarrhea, and anemia over 2 months. Sigmoidoscopy reveals friable congested sigmoidorectal mucosa, which bleeds easily when touched. Histopathological examination of this mucosa shows intense cryptitis. You diagnose ulcerative colitis. With respect to inflammatory bowel disease, all of the following statements are correct, *except*?
a) More common in Jewish people
b) HLA DR103 is found in those with severe ulcerative colitis
c) Associated with low residue and high-refined sugar diet
d) There is a possible association with measles and atypical mycobacterial infections
e) Crohn disease is more common in ex-smokers

37) A 43-year-old man presents with severe bloody diarrhea and shock. He has a history of ulcerative colitis. He receives daily sulfasalazine. The following signs reflect a severe attack of ulcerative colitis, *except*?
a) Stool volume >400 gram/day
b) Hemoglobin <10 g/dl
c) Daily bowel motions >twice per day
d) ESR >30 mm/hour
e) Serum albumin >30 g/L

38) A 33-year-old woman presents with an 8-month history of abdominal pain, diarrhea, weight loss, and mouth ulcers. She is pale and has lip fissuring and multiple peri-anal fistulae. What does the woman have?
a) Cecal tuberculosis
b) Small bowel lymphoma
c) Crohn disease
d) Ulcerative colitis
e) Celiac disease

39) A 31-year-old female develops painful and tender bluish-red nodules over both shins over 4 days. She was diagnosed with Crohn disease 3 years ago. You tell her that her new skin complaint represents one of the extra-intestinal complications of Crohn disease. All of the following systemic complications of inflammatory bowel disease tend to occur during active relapses, *except*?
a) Mouth ulceration
b) Episcleritis
c) Pyoderma gangrenosum
d) Deep venous thrombosis
e) Sacroiliitis

40) A 41-year-old man is referred to you for further evaluation and management of his Crohn disease. He has scleritis and pyoderma gangrenosum. He reports colicky abdominal pain every now and then as well as discharging anterior abdominal wall fecal fistulae. Which one of the following statements about the management of inflammatory bowel disease is the incorrect one?
a) Sulfasalazine side effects are usually dose-dependent and reversible
b) Specific nutritional therapy in Crohn disease can be very effective but it is expensive and poorly tolerated by most patients
c) The most important indication for surgery is impairment of the quality of life, including schooling, occupation, and social and family life.
d) Colectomy should be done when the diameter of the transverse colon in an acute ulcerative colitis attack exceeds 6 cm, as this means impending perforation
e) Panproctocolectomy is rarely curative in ulcerative colitis

41) A 28-year-old male presents with chronic diarrhea. He denies weight loss, appetite change, or vomiting. There is no blood or pus in stool. He is on no medications. Thorough investigations fail to detect any organic cause and you consider irritable bowel syndrome as his final diagnosis. Regarding irritable bowel syndrome, all of the following are correct, *except*?

a) Diarrhea-predominant type should be differentiated from microscopic colitis, lactose intolerance, and bile salt diarrhea
b) The commonest presentation is abdominal pain
c) 1% of cases would meet the criteria of psychiatric diseases
d) Reassurance of the patient has a very important role in the management
e) Amitriptyline may be used in selected patient

42) A 54-year-old woman presents with watery diarrhea and hypokalemia over the past 4 months. Colonoscopy reveals a large rectal adenoma that turns out to have a villous architecture. Indicators of potential malignancy in GIT adenomatous polyps include all of the following, except:
a) The presence of dysplastic changes on histology
b) The presence of multiple polyps
c) Large size polyps, of more than 2 cm
d) The presence of a villous architecture on histology
e) The presence of metaplastic polyps

43) A 21-year-old male consults you about the possibility of developing colonic polyps. He states that two of his older brothers had a disease called familial adenomatous polyposis (FAP) and died later on because of colonic cancer. Regarding FAP, all of the following statements are correct, *except?*
a) It is due to germ-line mutations in the APC gene on chromosome 5
b) About 50% of cases have adenomatous polyps in the stomach and 90% of patients have polyps in the duodenum
c) Congenital hypertrophy of the retinal pigment epithelium, when found in an at risk patient, is 100% predictive for the presence of FAP
d) Gardner syndrome, Turcot Syndrome, and an attenuated form called attenuated FAP are variants of the disease
e) Desmoid tumors are highly malignant tumors that are usually found in the chest wall and occur in up to 10% of cases

44) You are preparing a lecture about gastrointestinal tract polyposis syndromes and you intend to give it to postgraduate medical students and interns. All of the following statements regarding gastrointestinal polyposis are correct, *except?*
a) Although hamartomatous polyps of Peutz-Jeghers syndrome have no malignant potential, but still there is a risk of small bowel carcinoma
b) Juvenile polyposis has no risk of colorectal cancer
c) In Cowden disease, there is a risk of thyroid cancer and polyp formation throughout the gastrointestinal tract

d) In Cronkhite-Canada syndrome, there are hair loss, skin pigmentation, and nail dystrophy with polyps throughout the gastrointestinal tract
e) Of all gastrointestinal polyposis syndromes, esophageal polyps are found only in Cronkhite-Canada syndrome and Cowden disease

45) A 61-year-old man presents with a 3-month history of constipation. He reports unintentional weight loss of 8 Kg, lassitude, and bloody stool. You detect a hard mass in the left lower abdomen and a hard knobbly liver. There are hypochromic microcytic anemia and positive occult blood on stool examination. Sigmoidoscopy reveals a large rectosigmoidal mass. Risk factors for the development of colorectal cancer include all of the following, *except?*
a) Acromegaly
b) Prior pelvic irradiation
c) Alcohol and smoking
d) Diet rich in meat and fats
e) Diet rich in fibers and fruits

46) A 45-year-old man presents with severe epigastric pain radiating to the mid-back, which is relieved by leaning forward. Abdominal ultrasound reveals a swollen pancreas, and serum amylase is very high. The commonest causes of acute pancreatitis are all of the following, *except?*
a) Alcohol
b) Post-ERCP
c) Idiopathic
d) Gall stones
e) Viral infections

47) A 61-year-old man is referred to your tertiary hospital for further management and care. He has severe attack of hemorrhagic pancreatitis. Adverse prognostic factors in acute pancreatitis (Glasgow's criteria) are all of the following, *except?*
a) PaO_2 <8 kPa
b) Blood sugar >10 mmol/L
c) Serum calcium (corrected) <2.00 mmol/L
d) Very high serum amylase
e) Serum albumin <30 g/L

48) Pancreatitis is one of the causes of acute abdomen, management of which is shared by surgeons and physicians. The following statements about acute pancreatitis are correct, *except?*

a) Serum amylase is only useful in the first 24-48 hours of the illness, after that urinary amylase:creatine ratio becomes a diagnostic aid
b) Necrotizing pancreatitis is better assessed by CT scan
c) Visible gases in the pancreatic tissue suggest the development of abscess formation
d) Persistently elevated serum amylase suggests the development of pancreatic pseudocyst
e) C-reactive protein has a very limited role in the follow-up

49) A 52-year-old woman presents with recurrent upper abdominal pain, weight loss, and steatorrhea. Plain X-ray of the abdomen reveals pancreatic calcification. Pancreatic functional assessment uncovers failure in the exocrine pancreas. With respect to chronic pancreatitis, which one is the *wrong* statement?
a) The commonest cause is chronic alcohol ingestion
b) About 20% of patients becomes opiate-dependent
c) Around 15% of cases presents with steatorrhea but no abdominal pain
d) Pancreatic ascites is an indication for terminal pancreatectomy
e) Most patients stop drinking alcohol after knowing their diagnosis

50) A 61-year-old alcoholic man has been referred to your office for further management because of severe post-prandial abdominal pain. He has chronic pancreatitis and he is not compliant with his current medications. With respect to chronic pancreatitis, choose the *incorrect* statement?
a) The overall incidence of diabetes is 30% but this figure rises to 70% in calcific pancreatitis
b) Steatorrhea occurs when at least 90% of the exocrine function has been destroyed
c) Pain may continue despite total pancreatectomy
d) Chronic pancreatitis is a risk factor for pancreatic carcinoma development
e) Pancreatic enzymes supplements are useful to lessen malabsorption but have no effect on abdominal pain

51) You are reviewing the liver and its physiology because you intend to give a lecture about this topic. Regarding the normal liver, all of the following statements are correct, *except*?
a) About 15% of the liver is composed of cells other than hepatocytes
b) Clearance of bacteria, viruses, and erythrocytes is done by Kupffer cells
c) Ito cells have a role in the uptake and storage of vitamin A

d) Vitamin K and folic acid are stored in a huge amount
e) Hepatic synthesis of urea, endogenous proteins, and amino acid release by the liver are suppressed during fasting

52) One of your colleagues is in a dilemma over a case he has referred to you. The patient is 15 years old and has liver dysfunction. The physician is unable to elucidate the cause of this abnormality. The following statements are correct with respect to liver function, *except?*
a) Low blood urea is seen in many acute and chronic liver diseases
b) High blood urea in the context of severe liver damage may indicate gastrointestinal hemorrhage or hepatorenal syndrome
c) Hyponatremia is very common in severe liver disease and is usually multi-factorial
d) Raised γ-GT enzyme level may occur during treatment with carbamazepine
e) Large increase in serum aminotransferases with a small rise in alkaline phosphatase favors biliary obstruction over acute hepatitis

53) A 56-year-old epileptic man presents with embolic stroke. He takes daily warfarin for atrial fibrillation. Previous INR was 3.1, but the current one is 1.2. His sister says that the patient is compliant with warfarin and phenytoin. You consider drug-drug interaction as a cause of this low INR. All of the following can induce hepatic microsomal enzymes activity, *except?*
a) Chronic ethanol ingestion
b) Glucocorticoids
c) Griseofulvin
d) Carbamazepine
e) Cimetidine

54) A 63-year-old man, who has sigmoid cancer, presents with left hypochondrial pain. The liver is enlarged, rocky-hard, and knobbly. Abdominal ultrasound examination shows many liver target lesions, highly suggestive of malignant secondary tumors. Regarding imaging in liver diseases, all of the following statements are correct, *except?*
a) Ultrasound of the liver is a rapid, cheap, and easy method and is usually the first imaging to be done
b) Color Doppler studies are very useful and used to investigate hepatic veins, portal vein, and hepatic artery diseases
c) MRI is usually used for pancreaticobiliary diseases rather than parenchymal liver diseases

d) Outlining the biliary tree can be done by injecting a contrast medium into the biliary tree through the skin or by an endoscopic approach
e) Plain abdominal radiographs are very helpful in focal liver diseases

55) A 14-year-old male presents with jaundice and dystonia. His older brother has Wilson disease and you consider the same diagnosis. You plan to perform liver biopsy. With respect to liver biopsy, which one of the following statements is the *incorrect* one?
a) The patient should be cooperative
b) PT prolongation, if present, should be less than 4 seconds above the upper normal control value
c) Severe COPD is a contraindication
d) Marked ascites will make the procedure easier
e) Local skin infection should not be present

56) A medical student asks you about the significance of bilirubin in urine and why does the stool become clay-colored in obstructive jaundice. Regarding the metabolism of bilirubins, all of the following statements are correct, *except?*
a) Every day, about 300 mg of indirect bilirubin is produced
b) Jaundice is obvious clinically if serum total bilirubin exceeds 50 μmole/L
c) Roughly, 100-200 mg of stercobilinogen is lost in stool
d) approximately 400 mg of urobilinogen is excreted in urine
e) Indirect bilirubin is conjugated in the endoplasmic reticulum of hepatocytes to become water-soluble

57) A 32-year-old baker presents with pallor and jaundice. Her serum indirect bilirubin is 3.2 mg/dl and there are normal serum ASL, ALT, and alkaline phosphatase. Hemoglobin is 7.3 g/dl. Coombs test is positive and the final diagnosis is immune-hemolytic anemia. Which one of the following does *not* result in indirect hyperbilirubinemia?
a) Vitamin B_{12} deficiency
b) Wilson disease
c) Gilbert syndrome
d) Rotor syndrome
e) Major ABO incompatibility reaction

58) When you examine a patient with serum direct bilirubin of 30 μmol/L, the presence of all of the following are useful in guessing the underlying cause of this jaundice, *except?*

a) Palpable gall bladder
b) Upper abdominal paramedian scar
c) Irregular hard knobbly liver
d) Upper midline abdominal mass
e) Wide-spread skin scratching marks

59) A 54-year-old man presents with massive upper gastrointestinal bleeding. He has alcoholic liver cirrhosis and ascites. Upper gastrointestinal endoscopy is done after proper resuscitation in the Emergency Room. This reveals grade IV bleeding esophageal varices. Which one of the following is *not* a local measure to arrest upper gastrointestinal variceal hemorrhage?
a) Banding
b) Sclerotherapy
c) Esophageal trans-section
d) Balloon tamponade
e) Terlipressin infusion

60) A 45-year-old man is recovering from upper gastrointestinal variceal bleeding and is about to leave the hospital. You tell him that there are many ways to prevent further bleeding episodes. Which one of the following is not a prophylactic measure to prevent recurrent upper gastrointestinal variceal hemorrhage?
a) Oral propranolol
b) Sclerotherapy/banding
c) Trans-jugular intrahepatic portosystemic shunt (TIPSS)
d) Esophageal trans-section
e) Selective or non-selective portosystemic shunt surgery

61) A 33-year-old male, who is a known case of chronic liver disease and ascites, is brought to the Emergency Room. He has developed upper gastrointestinal hemorrhage. He denies recent alcohol or aspirin ingestion. With respect to hematemesis in cirrhosis patients, which one is the *incorrect* statement?
a) Upper gastrointestinal endoscopy should be done in all patients, as the bleeding is non-variceal in 20% of cases
b) Despite advances in the management, the mortality rate is still high
c) Porto-systemic shunting surgery may have a mortality of 50%
d) Vasopressin is contraindicated in ischemic heart disease
e) Esophageal trans-section is commonly used as the first-line treatment

62) A 43-year-old liver cirrhosis patient is referred to you from the gastrointestinal department. He asks you about the indications of trans-jugular intrahepatic portosystemic shunt (TIPSS) and its efficacy in preventing future variceal bleedings, its complications, and the possibility of its removal. Regarding TIPSS, choose the *wrong* statement?

a) It is done by placing a stent between the hepatic vein and the portal vein in the liver under radiological control
b) The objective is to produce porto-systemic shunting to reduce the portal venous pressure
c) Prior patency of the portal vein should checked beforehand by angiography
d) May precipitate or worsen hepatic encephalopathy
e) When re-bleeding occurs, the shunt should be removed

63) A 43-year-old man, who was diagnosed with cryptogenic cirrhosis and ascites, visits the Emergency Room. He developed fever and confusion before days. After a proper examination, you decide to do abdominal paracentesis, looking for spontaneous bacterial peritonitis (SBP). Regarding SBP, all of the following statements are correct, *except?*

a) In $1/3^{rd}$ of cases, abdominal signs are mild or absent
b) Almost always a mono-microbial infection state
c) Recurrence is common but there is no way to prevent it
d) The commonest organisms are enteric Gram-negative ones, but the source of infection is usually unknown
e) The ascitic fluid is cloudy and has >250 neutrophils/mm^3

64) A 52-year-old pub owner presents with confusional state. He has alcoholic liver cirrhosis. The disease was well-compensated and properly managed. His wife states that he developed coffee-ground vomitus yesterday. You think that his decompensation is due to upper gastrointestinal hemorrhage. Which one of the following is not a precipitating factors for hepatic encephalopathy in cirrhosis patients?

a) Occult infection
b) Aggressive diuresis
c) Diarrhea or constipation
d) Over-treatment with oral neomycin
e) Excess dietary proteins

65) A 59-year-old man is brought to the Emergency Room. He developed progressive confusion over the past 2 days. His wife denies head trauma or illicit drug ingestion, but she says that he has liver disease.

You order many investigations to discover the cause of this presentation. Which one of the following is not a differential diagnosis of hepatic encephalopathy?

a) Primary psychiatric disease
b) Hypoglycemia
c) Wernicke's encephalopathy
d) Subdural hematoma
e) Dissociative fugue

66) A 22-year-old female is referred to you. She intentionally ingested 20 tablets of paracetamol in an attempt to end her life. She is confused. Liver function tests are abnormal and there are hypoglycemia and prolonged PT. With respect to acute fulminant hepatic failure, all of the following statements are correct, *except?*

a) The commonest causes are viral hepatitides and medications
b) The hallmark is the presence of acute hepatic encephalopathy
c) Absence of jaundice is against the diagnosis
d) There are long-listed consequences and these usually complicate the presentation further
e) The patient should be managed in an intensive care unit or a high dependency unit once the PT is prolonged

67) A 41-year-old man presents with a 3-day history of oliguria. He was diagnosed with liver cirrhosis before 1 year. After examining him and doing some tests, you consider hepatorenal syndrome. Regarding hepatorenal syndrome, all of the following statements are correct, *except?*

a) Carries a very bad prognosis unless hepatic transplantation is carried out
b) One of the causes of fractional Na^+ excretion of >2
c) Characteristically presents as rapidly evolving uremia with bland urinary sediment
d) Encountered in advanced cirrhosis
e) Renal dose dopamine has a minor role in the management

68) You are discussing the subtypes and various causes of hepatic steatosis with a pathologist. Which one of the following does *not* result in hepatic micro-vesicular steatosis?

a) Fatty liver of pregnancy
b) Rye syndrome
c) Treatment with didanosine
d) Wolman disease

e) Treatment with amiodarone

69) A 19-year-old college student presents with anorexia, nausea, vomiting, and right upper abdominal pain for 1 week. A tinge of jaundice is present and you detect tender hepatomegaly. Serum ALT and AST are raised. Serum IgM antibodies against hepatitis A virus are positive. Which one of the following does *not* result in acute hepatitis?
a) Halothane
b) Wilson disease
c) Autoimmune hepatitis
d) Cytomegalovirus
e) Hemochromatosis

70) A 43-year-old woman presents with a 1-year history of anorexia, easy fatigability, and a tinge of jaundice. Anti-HBs and IgG anti-HBc antibodies are positive. Liver biopsy reveals changes that are consistent with liver cirrhosis. Which one of the following is *not* a recognized cause of chronic liver disease and cirrhosis?
a) α1-antitrypsin deficiency
b) Chronic hepatitis C infection
c) Hemochromatosis
d) Autoimmune hepatitis
e) Epstein-Barr virus infection

71) A 53-year-old man who has liver cirrhosis is referred for liver transplantation because of severe impairment in the quality of life. Regarding liver cirrhosis, all of the following statements are correct, *except*?
a) Skin hyperpigmentation is found in hemochromatosis and prolonged biliary obstruction
b) Spider telangiectasia occurs early in the course of cirrhosis
c) Parotid gland enlargement suggests alcoholic etiology
d) Ascites is an early sign of cirrhosis
e) Finger clubbing is a non-specific sign

72) A 54-year-old Swedish man is about to travel to the Middle East. He asks hepatitis vaccines. With respect to hepatitides viruses, which one is the *wrong* statement?
a) Hepatitis A is an RNA enterovirus which does not lead to chronic carrier states
b) Hepatitis B is a DNA virus that is 42 nm in diameter and leads to chronic infection in to 10% of adults

c) Hepatitis C is an RNA flavivirus that is the commonest cause of chronic liver disease in the Western World

d) Hepatitis D is a defective RNA virus that can be prevented by preventing hepatitis B infection in high risk groups by using hepatitis B vaccine and immunoglobulin

e) Hepatitis E is an RNA calicivirus that carries a mortality of 2% if the infection occurs in pregnancy

73) A 23-year-old refugee from China was recently diagnosed with hepatitis B associated chronic liver disease. He asks about the treatment modalities and their effectiveness. Which one of the following does *not* predict poor response to α-interferon therapy in chronic hepatitis B viral infection?
  a) Male
  b) Pre-core mutant strains of the virus
  c) Asian
  d) Very high pretreatment serum hepatitis B DNA level
  e) Absence of cirrhosis on histopathological examination

74) A 34-year-old woman developed autoimmune hepatitis. With respect to autoimmune hepatitis, choose the *incorrect* statement?
  a) Type I is ANA and anti-smooth muscle antibodies positive
  b) Amenorrhea is the rule and cushingoid faces may be seen
  c) 25% of patients present as acute hepatitis-like picture
  d) Corticosteroids are effective in the treatment of flare-ups
  e) Hepatocellular carcinoma, as a complication, is the usual cause of death

75) A 51-year-old man has bilateral parotid enlargement and moderate ascites. He admits to drinking heavily every day for the past 30 years. You consider alcoholic cirrhosis. Which one of the following is *not* a histological change of alcoholic liver disease?
  a) Mitochondrial swelling
  b) Siderosis
  c) Lipogranulomas
  d) Autoimmune interface hepatitis
  e) Depletion of endoplasmic reticula

76) A 46-year-old female presents with generalized pruritus for the last 2 years. No dermatological disease has ever been found and her physician thinks of psychogenic pruritus. However, you find that serum alkaline phosphatase is very high and there is little increment in serum ASL and ALT.

She is hyperpigmented and has many scratch marks especially on her shins and forearms. You consider primary biliary cirrhosis. With respect to this disease, which one is the *wrong* statement?

a) Serum anti-mitochondrial antibodies are positive in 96% of cases
b) Early in the course, there is proliferation of small bile ductules
c) Hypercholesterolemia is common and substantially increases the risk of coronary artery disease
d) Finger clubbing occurs
e) Ursodeoxycholic acid may improve liver function tests

77) A 65-year-old man was referred to the gastrointestinal department because of unexplained obstructive jaundice. He displays non-dilated biliary tree on hepatic ultrasound examination. Further investigations confirm primary sclerosing cholangitis. Regarding primary sclerosing cholangitis, all of the following statements are correct, *except?*

a) About 80% of cases occur in the context of ulcerative colitis
b) Spontaneous ascending cholangitis is uncommon but usually occurs after biliary instrumentation
c) Risk factor for cholangiocarcinoma
d) There is an association with HIV infection and retroperitoneal fibrosis
e) Corticosteroids and immune-suppressants are useful in the treatment

78) A 51-year-old man has hemochromatosis, which is managed by regular venesections. Today, he visits the physician's office with progressive abdominal swelling. Abdominal ultrasound reveals ascites and a mass in the right hepatic lobe. You biopsy the mass and finally diagnose hepatocellular carcinoma. Regarding hepatocellular carcinoma (HCC), all of the following statements are correct, *except?*

a) Occurs in the background of cirrhosis in 80% of cases
b) Chronic hepatitis B infection is the commonest cause worldwide
c) May be treated by liver transplantation
d) Cirrhosis patients should be screened for the development of HCC by serial serum α-fetoprotein and liver ultrasound
e) The fibrolammellar variant has a very poor prognosis

79) A 56-year-old female, who has with long-standing type II diabetes and hypertension, visits you. She has been experiencing recurrent colicky abdominal pain and dyspepsia over the past several months. You do abdominal ultrasonography and find gallstones. She declines surgical treatment. The criteria for using ursodeoxycholic acid, as a medical treatment for gallstones solubilization, are all of the following, *except?*

a) The stone should be radiolucent
b) The stone diameter does not exceed 15 mm
c) The gall bladder should be functioning
d) The patient should be moderately obese
e) Prominent symptoms ascribed to the stone

80) A 23-year-old woman visits the physician's office because of gall stones. Her general practitioner told her that these stones are of the "pigment type." Which one of the following is *not* a risk factor for the development of pigment gallstones?
a) Liver cirrhosis
b) Biliary parasites
c) Chronic long-term hemolysis
d) Ileal resection/disease
e) Pregnancy

This page was intentionally left blank

Gastroenterology and Hepatology

Answers

1) c.

Somatostatin decreases gastrin secretion, inhibits gastric acid secretion, inhibits insulin secretion (risk of diabetes), and inhibits gall bladder contraction (risk of gallstone formation) . It is secreted from D-cells (present throughout the gastrointestinal tract) in response to fat ingestion.

2) e.

Barium enemas usually miss polyps less than 1 cm in size. Gastrointestinal contrast studies are very useful in many gastrointestinal diseases, including functional assessment (such as gastric emptying studies).

3) e.

The contraindications for doing esophagogastroduodenoscopy are:
1- Severe shock.
2- Recent myocardial infarction, unstable angina, and cardiac arrhythmias.
3- Severe respiratory disease.
4- Atlanto-axial subluxation.
5- Possible visceral perforation.

Items 2, 3, and 4 are *relative* contraindications and can safely be performed by a highly experienced endoscopist; careful patient selection is very important.

4) e.

Certain high- risk groups (e.g., cardiac prosthetic valves and those with a history of infective endocarditis) should be given antibiotic prophylaxis when undergoing upper gastrointestinal endoscopy.

5) c.

Bile salt malabsorption most often occurs when more than 100 cm of terminal ileum has been removed but can also result from disease of the terminal ileum (usually Crohn's disease), HIV infection, or primary abnormalities in bile salt absorption. Presentation of the excess bile salts to the colon can lead to diarrhea (cholerheic enteropathy).[75]SeHCAT test is used in bile salt malabsorption; however, the SeHCAT test is not used commonly.

Quantitative measurement of bile acids in stool in patients who did not respond to cholestyramine may be the method of choice to diagnose cholerheic enteropathy.

6) e.

^{13}C and ^{14}C urea breath tests are used for H. pylori detection in peptic ulcer disease, but both tests differ in certain aspects. ^{13}C uses radioactivity and ^{14}C requires an expensive mass spectrometer. ^{99m}Tc pertechnetate is used for the detection of Meckel's diverticulum (e.g., in obscure lower gastrointestinal bleeding in children). ^{99m}Tc HMPAO- labeled leukocytes is used for the detection of visceral abscesses and in inflammatory bowel disease. ^{51}Cr-albumin is used as a test for gastrointestinal epithelial permeability in protein-losing enteropathy. ^{99}Tc-sulphur is used as a test for gastric emptying studies, e.g., in gastroparesis.

7) b.

Barrett esophagus is asymptomatic and may be discovered during upper gastrointestinal endoscopy done for some reason or another. Proton pump inhibitors are not effective in reversing the histological abnormalities. Until now, there is no consensus about the optimal management of this condition, because we may have mild, moderate, or severe degree of dysplasia. There are many treatment options, such as LASER therapy, total esophagectomy, repeated endoscope with biopsies,...etc. The risk of adenocarcinoma is very high when compared to the general population. The incidence of esophageal adenocarcinoma is rising at a global level.

8) e.

Esophageal adenocarcinoma results from Barrette esophagus. The other stems are true, and patients may present with one of these complications rather than simply with heartburn and regurgitation.

9) c.

The history is highly suggestive of a pharyngeal pouch. Cystic hygromas, branchial cysts, and thyroid cysts are seen in young age groups.

10) c.

The tight lower esophageal sphincter protects against acid reflux and therefore no heartburn occurs. It is a pre-malignant condition, and the risk of developing malignancy persists even in those who have received successful treatment.

11) e.

The other etiologies behind the development of esophageal cancer are Barrette esophagus, post-cricoid web, post-caustic stricture, achalasia of the cardia, and chewing betel nuts.

12) e.

Pernicious anemia causes chronic atrophic gastritis; no acid, no ulcer! The other causes of acute gastritis are NSAIDS and aspirin, Herpes simplex (with CMV, are seen in HIV infected patients), alcohol ingestion, and severe stress (after skin burns or CNS trauma).

13) a.

In addition, there is no (or very weak) association with non-ulcer dyspepsia; therefore, in both (gastro-esophageal reflux disease and non-ulcer dyspepsia) there is no need for H. pylori eradication therapy.

14) c.

Rapid urease test on an antral biopsy specimen has a low sensitivity but a high specificity. Urea breath testing is based upon the hydrolysis of urea by H. pylori to produce CO_2 and ammonia. A labeled carbon isotope is given by mouth; H. pylori liberates tagged CO_2, which can be detected in breath samples. False positive results after doing urea breath testing are uncommon; however, false negative results may be observed in patients who are taking anti-secretory therapy, bismuth, or antibiotics. To prevent false negative results, the patient should be off antibiotics for at least four weeks and off proton pump inhibitors for at least two weeks.

15) d.

Age above 60 years is a risk factor for NSAID-induced peptic ulceration.

The most ulcerogenic NSAID is azapropazone and the least one is low-dose ibuprofen. For the treatment or prevention of NSAID-induced peptic ulceration, we may choose prostaglandin analogues (e.g., misoprostol) or proton pump inhibitors; both have the same efficacy but the side effect profile differs (e.g., diarrhea and abdominal pain are more common with the use of misoprostol).

16) a.

Diarrhea is very common and occurs in 30-50% of cases. The other stems are true and common.

17) c.

Cimetidine may result in confusion and diarrhea; it also potentiates warfarin. Omeprazole produces hypergastrinemia and diarrhea. Sucralfate binds digoxin and warfarin, and therefore, lessens their effect. Unfortunately, the main limitation for using misoprostol is the 20% incidence of diarrhea (which may be severe). In addition, it is abortifacient and is contraindicated in woman of childbearing age. In addition to blackening teeth, tongue, and stool, prolonged treatment with colloidal bismuth may produce bismuth toxicity.

18) e.

About 20-60% of cases are part of MEN type I. Gastrinomas are derived from multipotential stem cells of endodermal origin. Known as enteroendocrine cells, they arise mainly in the pancreas as well as the small intestine. They have a particular affinity for chromium and silver salts, which are useful markers for histopathologic diagnosis.

19) d.

In ZE syndrome, barium meal may reveal *thick* mucosal folds which occur due to the hypertrophic effect of gastrin on the gastric mucosa. The presence of ulcers at atypical sites should always prompt a search for ZE syndrome. Causes of thick gastric mucosal folds are (in decreasing order): chronic gastritis/lymphoid hyperplasia; benign tumors; gastric malignancy; Zollinger-Ellison syndrome; and Ménétrier's disease.

20) d.

The overall 5-year survival is 60-75%, which is a good figure. High-dose proton pump inhibitors produce profound symptomatic relief, heal ulcers, and alleviate diarrhea. The factors that cause diarrhea include:
1. The high rate of gastric acid secretion, resulting in a volume load that cannot be fully reabsorbed by the small intestine and colon.
2. The excess acid secretion, which exceeds the neutralizing capacity of pancreatic bicarbonate secretion. The exceptionally low pH of the intestinal contents inactivates pancreatic digestive enzymes, interferes with the emulsification of fat by bile acids, and damages intestinal epithelial cells and villi. Thus, both maldigestion and malabsorption may result in steatorrhea.
3. The extremely high serum gastrin concentrations may inhibit absorption of sodium and water by the small intestine, thereby adding a secretory component to the diarrhea.
Three tests are generally used to diagnose this syndrome; fasting serum gastrin concentration, secretin stimulation test, and gastric acid secretion studies.

21) e.

Tylosis (palmo-planter thickening) predisposes to esophageal squamous cell carcinoma. The other risk factors are smoking, alcohol, H. pylori, dietary association with smoked or pickled food, and Ménétrier's disease. Tylosis is keratoderma of the palms and soles and presents as a yellow, symmetrical, smooth bilateral thickening of the epidermis. The inherited type of tylosis (Howell-Evans syndrome) has been most strongly associated with squamous cell carcinoma of the esophagus. However, sporadic cases of tylosis have also been associated with Hodgkin's disease, leukemia, and breast cancer. This condition needs to be distinguished from the T-cell malignancy mycosis fungoides localized to the palms and soles.

22) a.

The incidence of gastric cancer is falling in the UK; the disease is extremely common in China, Japan and South East Asia. Pseudo-achalasia syndrome may result from involvement of Auerbach's plexus, due to local extension or to malignant obstruction near the gastroesophageal junction. Therefore, gastric cancer needs to be considered in the differential diagnosis for older patients presenting with achalasia

23) e.

Celiac disease is much more common in northern Europe with a prevalence of 1/300. The disease may be observed in up to 10% of first-degree relatives of patients. Glomerular IgA deposition is common in celiac patients, occurring in as many as $1/3^{rd}$ one-third of them; however, the great majority of affected patients have no clinical manifestations of renal disease, perhaps because there is no associated activation of complement.

24) c.

Rice, maize, and potatoes can be safely ingested and are satisfactory source of complex carbohydrates. The histological changes in the small bowel are reversible but the sensitivity is life-long.

25) b.

Esophageal squamous cell carcinoma and intestinal lymphoma can develop in celiac patients. Option d refers to dermatitis herpetiformis. Celiac crisis is rarely seen nowadays because of better patients' education. Some patients don't respond to gluten-free diet; patients with refractory sprue fall into two clinical categories:
1. Patients who have no initial response to a gluten-free diet.
2. Patients who experience initial clinical improvement on a gluten-free diet, but, after a period of remission, develop a disease that is refractory to gluten abstinence.
The cause of refractory sprue is unknown. It is possible that some patients with this condition develop sensitivity to a dietary constituent other than gluten; however, identification of the responsible antigens in most patients is difficult and unrewarding. As a result, treatment has focused on immune suppression, which has traditionally relied upon corticosteroids.

26) e.

In stem "e", serum IgA anti-endomysium would be falsely negative, but serum IgG anti-endomysium is positive. The other stems are correct. Serum IgA anti-endomysial and anti -tissue transglutaminase antibody testing have the highest diagnostic accuracy. The IgA and IgG antigliadin antibody tests have lower diagnostic accuracy with frequent false positive results, and are therefore no longer recommended for initial diagnostic evaluation or screening.

27) a.

Type I, not type II, diabetes is an association in 2-8% of celiac patients. The other associations are Down syndrome, sarcoidosis, selective IgA deficiency (3-4%), autoimmune hepatitis, microscopic colitis, and thyroid disease (5%).

28) e.

Tropical sprue is commonly seen in Malaysia and Indonesia, as well as southern India. The other stems are true. Considerable evidence supports the notion that tropical sprue is an infectious disease (however, no single infectious agent has been isolated and implicated). There are four major observations in support of this hypothesis:
1. In some patients, the disease follows an acute diarrheal illness typical of infectious enteritis.
2. Epidemics in households and communities have been documented, particularly in India where the disease is common. The majority of such patients recover, with only some developing tropical sprue.
3. Most patients have overgrowth of toxigenic coliform bacteria (Klebsiella, E. coli, and Enterobacter) in the proximal small intestine.
4. Treatment with broad spectrum antibiotics is usually curative.

29) c.

Bacterial overgrowth syndrome may also result from long-term use of proton pump inhibitors, partial gastrectomy, hypogammaglobulinema, scleroderma, diabetic autonomic neuropathy, intestinal strictures (e.g., in Crohn disease), gastric surgery (blind loop after Billroth II operation), jejunal diverticulosis, and enterocolic fistulae (e.g., in Crohn disease). The gold standard diagnostic investigation is duodenal and jejunal aspirate for bacterial studies and cultures; however, these are academic methods and are rarely done in clinical practice.
Serum folate is normal or high (and is a good clue) while that of vitamin B_{12} is low. The treatment of choice in most patients is tetracycline for 1 week, although many patients may require extension to 28 days or continuous rotational courses of antibiotics.

30) a.

Whipple disease results from infection with the Gram-positive bacilli Tropheryma whippelii.

The clinical manifestations are diverse and the presentation depends on the organ involved (with neurological, gastrointestinal, lung, musculoskeletal features,…etc.). After starting medical treatment, profound symptomatic improvement is seen within in a week, but the histological changes may take few weeks to revert to normal; however, 30% of patients will relapse and the relapsed site is usually the CNS.

31) d.

Anti-diarrheal agents are useful to control bile salt diarrhea of short bowel syndrome. The commonest causes of this syndrome in adults are Crohn's disease, mesenteric infarction, radiation enteritis, and volvulous.

32) e.

Cholestyramine is used when bile salt diarrhea follows terminal ileal disease and is very helpful in reducing the stool volume.

33) e.

Many gastrointestinal diseases can cause protein-losing enteropathy, such as celiac disease, Crohn disease, lymphoma, Whipple disease, tropical sprue, eosinophilic gastroenteritis, and radiation enteritis.

34) b.

In protein-losing enteropathy, lymphocytopenia is characteristic, and although hypogammaglobulinemia is present, the susceptibility to infections is not increased.

35) e.

The prognosis of eosinophilic gastroenteritis is generally good and almost all patients respond to corticosteroids and/or disodium cromoglycate. Dietary manipulations are rarely effective. The disease is characterized by inflammatory eosinophilic infiltration of the gut wall in the absence of parasitic infestation or eosinophilia of other tissues. Exclusion of other causes of eosinophilia must be part of the management. Eosinophilic ascites may occur when there is serosal involvement.

36) e.

Crohn disease is usually seen in active smokers; ulcerative colitis is more common in non-smokers and ex-smokers.

37) c.

The occurrence of daily bowel motions of more than 6 is considered a sign of severe ulcerative colitis relapse. The other features indicating severe attacks are: presence of frank blood in stool; pulse rate >90 beats/minutes; temperature >37.8 °C two days out of 4; frank blood seen in the gut lumen by sigmoidoscopy; and dilated bowel loops and/or mucosal islands on plain X-ray films.

38) c.

Crohn disease can target the whole GIT, from the mouth down to the anus.

39) e.

Sacroiliitis, ankylosing spondylitis-like picture, and primary sclerorsing cholangitis are independent of the inflammatory bowel disease activity.

40) e.

Ulcerative colitis can be cured by surgery (e.g., panproctocolectomy with permanent iliostomy), unlike Crohn's disease, in which surgery should be conservative as much as possible.

41) c.

Up to 50% of irritable bowel syndrome patients meet criteria of a psychiatric disease, ranging from anxiety and depression to frank panic attacks.

42) e.

Metaplastic (or hyperplasic), hamartomatous, and inflammatory polyps have no malignant potential. Up to 50% of the Western population, who are older than 60 years of age, have polyps; of those, 50% are multiple and are more common in the distal colon.

43) e.

Familial adenomatous polyposis (FAP) is an uncommon autosomal dominant disease, which is responsible for about 1% of all colorectal carcinomas. The duodenal polyps may undergo malignant changes; this is the usual cause of death in those who had successfully undergone total colectomy. Congenital hypertrophy of the retinal pigment epithelium (which is seen as dark round pigmented retinal lesions) when found in at risk patient is a 100% predictive for the presence of FAP. Gardener syndrome is characterized by the predominance of the extra-intestinal manifestations, especially osteomas. Turcot syndrome main clinical feature is malignant brain tumors. Desmoid tumors are benign tumors and are usually found in the mesentery or abdominal wall.

44) b.

Juvenile polyposis is an autosomal dominant disease in 20% of cases and carries a risk of colorectal carcinoma, which usually develops before the age of 40 years.

45) e.

Protective factors against the development of rectosigmoid cancer are diets rich in vegetables, folic acid, and calcium. Other risk factors for the development of rectosigmoid cancer are obesity and sedentary life style, colorectal adenomas, and long-standing extensive ulcerative colitis.

46) e.

Viral infections, such as mumps and coxsachie, are rare causes of acute pancreatitis. The other rare causes of acute pancreatitis are abdominal trauma, surgery, cardiopulmonary bypass surgery, prolonged hypothermia, and drugs (e.g., azathioprine and valproate sodium).

47) d.

Although serum amylase is used as a tool in the diagnosis of acute pancreatitis, its level has no prognostic value at all. The other predictive factors are: WBC >15000/mm^3; blood urea >16 mmol/L (after re-hydration); serum ALT >200 u/ L; and serum LDH >600 u/L.

48) e.

Because the kidneys very efficiently excrete the blood amylase, serum amylase is only useful in the first 24-48 hours of acute pancreatitis; otherwise, after that urinary amylase:creatine ratio is much more useful as a diagnostic tool. Necrotizing pancreatitis is better assessed by CT scan, which appears as failure of the inflamed pancreas to enhance after intravenous contrast CT scanning. Visible gases in the pancreatic tissue suggest the development of abscess formation; this calls for surgical drainage. Persistently elevated serum amylase suggests the development of pancreatic pseudocyst (due to disruption of the pancreatic ducts). Serial measurement of C-reactive protein is a very important indicator of the overall progression of the disease.

49) e.

Unfortunately, despite counseling and education, almost all chronic pancreatitis patients continue to drink alcohol (which actually produces pain relief) and up to 20% of patients become opiate-dependent. Wight loss is common and is due to anorexia, avoidance of food because of post-prandial pain, diabetes, and malabsorption.

50) e.

Pancreatic enzymes supplements are also useful as an analgesic by resting the pancreas. Percutaneous or endoscopic celiac nerve blocks (with either alcohol or steroids) have had only limited success in chronic pancreatitis and should be considered to be an unproven therapy.

51) d.

About 15% of the liver is composed of cells other than hepatocytes, e.g., Kupffer cells and Ito cells. Clearance of bacteria, viruses, and erythrocytes as well as removal of IgG complexes and cytokine production is a function of Kupffer cells. Ito cells take up and store vitamin A as well as being involved in the synthesis of extra-cellular matrix and synthesis and release of collagenases and metalloprotease inhibitors. Vitamin K and folic acids are stored in small amounts, and therefore, their deficiency states ensue relatively rapidly when there is no supply of them. During fasting, the liver also releases glucose.

52) e.

Stem "e" favors hepatocellular damage over cholestasis; the reverse is true in biliary obstruction. Carbamazepine induces the synthesis of gamma glutamyl transferase enzyme.

53) e.

Cimetidine and acute alcohol ingestion are liver enzyme inhibitors. Enzyme inducers are meprobamate, isoniazid, phenytoin, primidone, and phenobarbitone.

54) e.

The main limitation of hepatic ultrasonography is that small focal lesions <2 cm will be missed; it is less useful in diffuse parenchymal diseases, in addition.

55) d.

Liver biopsy is contraindicated in agitated and confused patients (the patient should be cooperative), platelets $<100000/mm^3$, severe anemia, hepatic hydatid cyst, hepatic hemangioma, and bile duct obstruction.

56) d.

Up to 4 mg of urobilinogen is excreted in urine every day. The other stems are true.

57) d.

Megaloblastic anemia of vitamin B_{12} deficiency results in indirect hyperbilirubinemia due to ineffective erythropoiesis and premature destruction of red cells in the bone marrow. The jaundice of Wilson disease may be hemolytic (unexplained Coombs negative hemolytic anemia should prompt you search for Wilson disease) or hepatocellular (from parenchymal damage). Gilbert syndrome and Crigler–Najjar syndrome (type I and II) result in plasma elevation of the indirect bilirubin, while Rotor syndrome and Dubin- Johnson syndrome cause direct hyperbilirubinemia. Major ABO incompatibility reaction (and other causes of intravascular hemolysis) is another cause of indirect hyperbilirubinemia.

58) e.

We are dealing with causes of *direct* hyperbilirubinemia. A palpable gall bladder may result from pancreatic cancer (which has resulted in obstructive jaundice). Bile duct stricture may be an aftermath of a previous biliary surgery. Stem "c" reflects hepatic metastases. A palpable mass in the upper mid-abdomen may suggest tumor or masses (including lymphoma and choledochal cyst) with secondary biliary obstruction. Skin scratch marks and "polished" nails simply indicate itching from cholestasis (but without providing any additional clue about the underlying culprit).

59) e.

Apart from stem "e", the other stems are local measures to arrest esophageal variceal bleeding in the acute setting; terlipressin (triglycyl lysine vasopressin) is given intravenously (i.e., a systemic measure, not a local one). The other drugs used in the acute bleeding are vasopressin and octreotide. Propranolol is used in the primary and secondary prophylaxis against bleeding, but not to stop an active hemorrhage.

60) d.

Esophageal trans-section is used to arrest acute bleeding episodes in certain situations; it has no place in the secondary prophylaxis regimen.

61) e.

Esophageal trans-section can stop bleeding efficiently, but it should be done only when there is failure of endoscopic hemostasis *and* TIPSS is not available or cannot be carried out. It carries a risk of future esophageal stricture. It is done by using a stapling gun.

62) e.

TIPSS should be done by an experienced operator. It is the placement of a stent between the hepatic vein and the portal vein in the liver under radiological control. The objective is to produce porto-systemic shunting with secondary reduction in the portal venous blood pressure. However, this may precipitate or worsen hepatic encephalopathy; this necessitates iatrogenic shunt narrowing to lessen the shunt.

All coagulation defects should be corrected beforehand. Upper gastrointestinal re-bleeding may occur and this may indicate in-stent narrowing; venography and angioplasty should be carried out.

63) c.

About 30% of spontaneous bacterial peritonitis (SBP) patients have no abdominal signs and symptoms and may therefore present with fever only or as an acute sudden hepatic encephalopathy in a previously well-managed and compensated patient. The infection is mono-microbial (poly-microbial infection should cast a doubt on the diagnosis and should be regarded as visceral perforation until proved otherwise). The commonest organisms are enteric Gram-negative, but the source of infection is usually unknown (hence the designation spontaneous). Successfully treated patients are at risk of future development of SBP; this can be prevented by giving daily 400 mg of norfloxacin.

64) d.

Oral neomycin is used as a bowel decontaminant in hepatic encephalopathy and is not a precipitating factor for a decompensation. The other precipitating factors:
1. Uremia: which may be spontaneous or diuretic-induced.
2. Gastrointestinal bleeding (upper or lower).
3. Surgery and trauma.
4. Surgical or spontaneous porto-systemic shunts.
5. Large volume abdominal paracentesis.
6. Hypokalemia.
7. Sedatives and hypnotics.

65) e.

Impaired sensorium in cirrhotic patients may result from a multitude of causes, which may be additive. This may include hepatic decompensation, hypoglycemia, subdural hematoma, renal failure, electrolyte disturbances, and side effects of medications. These should be addressed and managed accordingly.

66) c.

The commonest causes of acute liver failure are viral hepatitides (especially hepatitis E infection in pregnancy) and medications/drugs. Encephalopathy marks the disease appearance; its absence should suggest an alternative diagnosis. Although the liver is massively damaged, jaundice could be entirely absent; it is characteristically absent in Rye syndrome. Acute hepatic failure has a long-list of complications, all of which should be anticipated and treated. Meanwhile, the hepatic transplant unit should be contacted as early as possible.

67) b.

Although hepato-renal syndrome is an intrinsic renal disease, the fractional Na^+ excretion is <1 and the Na^+ concentration in urine is less than 10 mmol/L, i.e., a pre-renal failure-like picture. Renal-dose dopamine may produce slight improvement. The improvement in renal function depends entirely on improvement of the liver function.

68) e.

Macrovesicular steatosis is much more common than the microvesicular variety and is a benign condition. The microvesicular type usually implies sinister diseases, e.g., fatty liver of pregnancy. Amiodarone, iron, and minocycline may result in macrosteatosis, while didanosine, valproate, and ketoprofen can cause microsteatosis. Wolman disease is lysosomal acid lipase deficiency, which is responsible for 3% of childhood adrenal failure.

69) e.

Hemochromatosis causes chronic liver disease, cirrhosis, and hepatocellular carcinoma. Wilson disease can cause acute hepatitis (which may be recurrent), chronic active hepatitis, and cirrhosis.

70) e.

EBV causes acute hepatitis, which never progresses to cirrhosis or chronic hepatic carrier state. Liver disorders, such as neonatal hepatitis, cirrhosis (both in children and adults), and hepatocellular carcinoma are associated with some α1-antitrypsin (ATT) deficient phenotypes.

Approximately 10 to 15% of ATT newborns at risk have some form of hepatic disease, and approximately 10 to 15% of adults develop hepatic disease.

The liver disease of ATT is caused by pathologic polymerization of the variant AAT, resulting in intra-hepatocyte accumulation of AAT molecules, rather than a proteolytic mechanism. Necrotizing panniculitis, which is characterized by inflammatory lesions of the skin and subcutaneous tissue, is the major dermatologic manifestation of AAT deficiency. The other extra-pulmonary manifestations of ATT are glomerulonephritis (proliferative or IgA nephropathy), and vascular diseases (abdominal and intracranial aneurysms and arterial fibromuscular dysplasia; this is mainly observed in the PI*ZZ type).

71) d.

Skin hyperpigmentation is a feature of hemochromatosis and prolonged biliary obstruction; therefore, pallor can easily escape detection. Spider nevi are common during the 2nd and 3rd trimesters of pregnancy and are detected in up to 20% of the general population; otherwise, if you find them, investigate the liver. Bilateral parotid gland enlargement and prominent gynecomastia point out towards an alcoholic etiology. Ascites indicates an advanced cirrhosis, and ascites per se may be an indication for hepatic transplantation in cirrhosis. Low grade fever and finger clubbing are non-specific signs of cirrhosis.

72) e.

Hepatitis E viral infection in pregnancy carries a mortality rate of 20%. Up to 80% of hepatitis C infections become chronic; acute infections are usually asymptomatic. Approximately 20% of chronically infected patients will develop cirrhosis after 20 years of infection, a figure that rises up to 50% after 30 years. This is more likely to happen if patients misuse alcohol as well. Once cirrhosis is established, 2-5% of patients per year will develop hepatocellular carcinoma.

73) e.

The first 4 stems predict poor response to interferon alpha therapy. Criteria for initiating interferon alpha treatment in chronic hepatitis B infections are:
1. Raised serum aminotransaminases.
2. Chronic active hepatitis picture on liver biopsy.
3. The infection was not acquired transplacentally.
4. HIV-negative status.

74) e.

Hepatocellular carcinoma as a complication of autoimmune hepatitis is surprisingly uncommon despite the changes in liver histology. The disease occurs in exacerbations and remissions and eventually will end up with cirrhosis. Corticosteroids are effective in the treatment of acute flare-ups and in the prevention of future attacks, but they do not prevent the progression to frank cirrhosis. The disease has many associations (especially type I disease), e.g., ulcerative colitis, nephrotic syndrome, Coombs positive hemolytic anemia, autoimmune thyroid disease,…etc. Type I autoimmune hepatitis is ANA and anti-smooth muscle antibodies positive, while type II is anti-LKM antibodies positive.

75) e.

Early in the course of primary biliary cirrhosis, there is proliferation of endoplasmic reticula (and mitochondrial swelling), which can be seen only with the aid of electron microscope.

76) c.

Although xanthelasmas develop and hypercholesterolemia is common in primary biliary cirrhosis (PBC), the latter is predominantly of HDL-type and hence is cardio-protective! Liver transplantation is the only curative option. Xanthoma is a late manifestation that occurs in less than 5% of patients (xanthelasmas are more common and occur in approximately 10%). Anti-mitochondrial antibodies are the serologic hallmark of PBC. Antinuclear antibodies (ANA) are found in up to 70% of patients with PBC. Antinuclear antibodies have clinical significance in PBC for two reasons. First, their presence can cause confusion with autoimmune hepatitis or autoimmune hepatitis/PBC overlap. Second, ANA may be associated with more rapid progression of disease and a poorer prognosis.

77) e.

Corticosteroids and immune-suppressants have no value at all to arrest or reverse the pathological changes of primary sclerosing cholangitis. Liver transplantation is the only curative treatment. The majority of patients with primary sclerosing cholangitis are asymptomatic at the time of diagnosis, although some may have advanced disease.

The disease should be considered in patients with inflammatory bowel disease who have otherwise unexplained abnormal liver biochemical tests, particularly an elevation in serum alkaline phosphatase. Complications of primary sclerosing cholangitis include systemic and metabolic complications related to cholestasis, cholangitis, cholelithiasis, cholangiocarcinoma, and colonic cancer.

78) e.

The fibrolammellar variant of primary hepatocellular carcinoma has a good prognosis and is not associated with hepatitis B or C infections or cirrhosis, and serum alpha-fetoprotein is usually normal. When hepatocellular carcinoma is less than 5 cm in maximum diameter, confined to one liver lobe, and occurs on the background of cirrhosis, it can be treated by liver transplantation. The following are paraneoplastic manifestations of primary hepatocellular carcinoma: hypoglycemia (typically mild and produces no or mild symptoms), polycythemia, hypercalcemia, watery diarrhea (because of secreting vasoactive intestinal polypeptide, gastrin, and peptides with prostaglandin-like immunoreactivity), and a variety of skin diseases (dermatomyositis, pemphigus foliaceus, pityriasis rotunda, and porphyria cutanea tarda).

79) e.

Patient should be asymptomatic or has very mild symptoms. Gallstone patients with chronic cholecystitis are at risk of developing gall bladder cancer. The presence of the following factors (besides gallstones) further increases the development of gall bladder cancer and/or cholangiocarcinoma: choledochal cysts; Caroli disease; anomalous pancreatic ductal drainage (in which the pancreatic duct drains into the common bile duct duct); gallbladder adenomas; and porcelain gallbladder. Gall bladder stasis predisposes to gall stones formation; this can result from prolonged total parenteral therapy, post-vagotomy, octreotide therapy, somatostatinoma, diabetes mellitus, and spinal cord injury.

80) e.

Pregnancy and oral contraceptive pills are risk factors for cholesterol stone formation. Medications that increase the risk of gallstone formation are: estrogen and oral contraceptive pills; octreotide; clofibrate; and ceftriaxone.

Three non-surgical methods can be used to treat (dissolve) gallstones and these are: oral bile salt therapy (with ursodeoxycholic acid); contact dissolution (using methyl tert-butyl ether, MTBE; a potent cholesterol solvent); and extra-corporeal shock-wave lithotripsy.

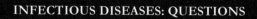

Chapter 8
Infectious Diseases and Genitourinary Medicine

Questions

1) A 35-year-old farmer complains of a 1-month history of fever, myalgia, lassitude, and back pain. Serum Rose-Bengal test is strongly positive. Regarding brucellosis, all of the following statements are correct, *except?*
  a) Splenomegaly usually indicates chronic brucellosis or severe acute infection
  b) The gold standard investigation is blood culture
  c) Neuro-brucellosis may presents just like multiple sclerosis
  d) Spinal tenderness is highly uncommon
  e) It is difficult to differentiate a new infection from a relapsing one in a patient who was treated in the past

2) A 29-year-old Asian man, who has arrived from India, presents with a 2-week history of stepladder fever, cough, headache, and myalgia. The spleen is just palpable. Widal test turns out to be positive. You consider typhoid fever. With respect to typhoid fever, which one of the following statements is the *wrong* one?
  a) Human being is the only reservoir of this infection
  b) Constipation is seen in the first week while diarrhea occurs in the second week
  c) Rose spots are found in a minority of patients
  d) Blood cultures have the highest yield in the first week of symptomatic infection
  e) Bone marrow cultures are useless in those who were partially treated with antibiotics

3) A 47-year-old businessman has recently visited tropical Africa. He develops attacks of fever, rigors, and sweating. The spleen is enlarged and he has anemia. Thin and think blood films reveal Plasmodia organisms. All of the following statements with respect to malaria are correct, *except?*
  a) Thick blood film is mainly used for rapid diagnosis
  b) Thin blood film is used to confirm the diagnosis and to speciate the organism
  c) Severe infections should be treated with intravenous quinine
  d) Mefloquine is contraindicated in epilepsy and lactation
  e) Steroids are very useful in treating cerebral malaria of falciparum infections

4) A 21-year-old man presents with fever, jaundice, and renal failure. Liver function tests are abnormal. You suspect leptospirosis. With respect to this infection, choose the *incorrect* statement?

a) The occupation or hobby of the patient is relevant
b) Early phase is due to leptospiremia and blood cultures have high positive results during this period
c) Hemorrhagic manifestations are common
d) The second phase is an immune-mediated one
e) There is peripheral blood lymphocytosis

5) A 15-year-old schoolgirl presents with fever, vomiting, diarrhea, diffuse skin redness, renal impairment, and hypotension. Toxic shock syndrome comes into your mind. Which one of the following statements with respect to toxic shock syndrome is the *incorrect* one?
a) About $2/3^{rd}$ of cases are due to infected tampons
b) Blood cultures are almost always negative for Staphylococcus aureus
c) Serum creatine phosphokinase is commonly raised
d) The mortality rate is 90%
e) The resulting shock is usually resistant to fluid therapy

6) A 27-year-old man was brought to the Emergency Room. He developed vague behavioral and personality changes, 5 days after a stray dog bite. You suspect rabies. Regarding rabies, all of the following statements are correct, *except?*
a) Caused by an RNA rhabdoviral infection
b) Transmitted through animal saliva
c) Once symptomatic, it is almost virtually fatal
d) May present as an isolated ascending flaccid paralysis state
e) The incubation period is very short in all cases

7) A 46-year-old farmer from West Africa presents with fever and diffuse body pain. You find bradycardia. Urine is positive for protein and you detect raised serum ALT and AST, leucopenia, and thrombocytopenia. You consider hemorrhagic fever, and Lassa fever is one of your differential diagnoses. Regarding Lassa fever, all of the following statements are correct, *except?*
a) Results from arena virus and is transmitted through rat urine
b) Usually presents with fever, myalgia, and pharyngitis
c) Renal and hepatic failure may occur
d) The mortality rate may reach 50%
e) No drug therapy against this infection is effective till now

8) A 38-year-old man is referred to you for further management. His face is flushed and there are conjunctival suffusion, bradycardia, and leucopenia.

His friend states that the patient has been exposed to many insects' bites during his recent visit to Africa. You suspect yellow fever. Which one of the following statements about yellow fever is *correct*?

a) Caused by an RNA togavirus infection and is transmitted from cows through fleas' feces
b) Has a long incubation period of 3-6 months
c) There may be intense neutrophil leukocytosis and even a leukemoid reaction
d) Usually presents with fever, join pain, muscle pain, and headache
e) There is a favorable response to lamivudine

9) Your intern asks about the management of dengue, as one of your ward patients is suspected of having this infectious illness. With respect to dengue, which one of the following statements is *incorrect*?

a) Mosquito-borne infection
b) There are fever, lymph node enlargement, and skin rash
c) All individuals in endemic areas should be vaccinated against it
d) There are intense headaches and backaches
e) Has an incubation period of 5-6 days

10) A 29-year-old farmer presents with fever, headache, cough, and jaundice. Liver biopsy reveals granulomatous hepatitis. There is a four-fold rise in serum antibodies against Coxiella burnetii on compliment fixation testing. All of the following statements about Q-fever are correct, *except*?

a) There may be a history of exposure to animals
b) May present as an atypical pneumonia
c) One of the causes of culture-negative infective endocarditis
d) Demonstrates poor response to tetracycline
e) Liver dysfunction occurs

11) A 38-year-old man has recently visited the western area of the United States. After returning to the UK, he develops fever and diffuse body pain. A maculopapular skin rash appears on day 4, and is prominently seen in the wrists and ankles. You find hepatosplenomegaly. The Weil-Felix reaction is positive, and there is a 4-fold rise in the titer. Which one of the following statements about typhus fevers is *incorrect*?

a) Are arthropod-borne infections
b) Rickettsia organism infects and damages the endothelium of small blood vessels
c) The mortality is variable and depends on the infecting agent

d) All infections result in gastroenteritis, which is prominent and common
e) The treatment of choice is tetracycline

12) A 41-year-old reporter has multiple cranial nerve palsies, arthritis of knees, and cardiac conduction defects. He gives a history of visiting the United States 3 months ago and remembers multiple tick bites. These bites were followed by a strange-looking red skin rash. With respect to Lyme disease, all of the following statements are correct, *except*?
a) History of tick bite may not always be obtained
b) The characteristic early skin rash may be not seen
c) Bilateral facial palsy may occur
d) Caused by spirochetes
e) Joint involvement usually occurs in the form peripheral symmetrical small joint arthropathy

13) A 36-year-old woman presents with fever, upper right abdominal pain, and generalized skin rash involving the palms and soils. There are multiple fleshy moist plaque-like skin lesions in the vulval area. You suspect secondary syphilis. Regarding this form of syphilitic infection, which one of the following statements is the *wrong* one?
a) The skin rash is non-itchy and characteristically spares the face
b) Lymphocytic meningitis occurs and is usually self-limiting or asymptomatic
c) Serum VDRL is always positive
d) Snail-tract oral mucosal ulcerations are characteristic
e) Even if not treated, not all patients develop a tertiary phase in the future

14) Your junior house officer asks you about the differences between typhoid and paratyphoid fevers. In comparison to typhoid fever, which one of the following statements about paratyphoid fever is *false*?
a) Has a shorter incubation period
b) Usually presents with acute onset of vomiting and diarrhea
c) The skin rash is much more prominent and is more commonly observed
d) Less frequent intestinal complications
e) Higher incidence of chronic carrier state

15) A 28-year-old male presents with severe painless watery diarrhea. There is no fever. He denies history of contact with patients who have diarrhea.

He has just returned from India. Regarding cholera, which one of the following statements is wrong?
a) May cause death even before the diarrhea appears
b) Usually occurs in epidemics
c) The organism is motile
d) Tetracycline is used to shorten the duration of shedding of the organism in stool
e) May result in severe abdominal pain that may be diagnosed as a medical acute abdomen

16) A 23-year-old college student presents with a small ulcer on his penis. He denies unprotected sex, oral ulcers, or abdominal symptoms. Which one of the following does *not* result in genital ulceration?
a) Certain Chlamydia strains infection
b) Secondary syphilis
c) Genital Herpes infection
d) Infection with some Hemophillus strains
e) Donovania (Calymmatobactrium) granulomatis infection

17) A 17-year-old schoolboy presents with burning urethral discharge, few days after unprotected sex with a stranger. All of the following statements with respect to gonorrhea are correct, *except?*
a) Has a short incubation period of 2-5 days
b) Purulent urethral discharge and dysuria are the main features
c) Calls for oral tetracycline therapy for at least 21 days
d) In comparison to males, females tend to be more asymptomatic
e) Recurrences and re-infections are common

18) A 23-year-old man is about to travel to an African country where falciparum malaria is endemic. He was advised about receiving mefloquine prophylaxis for malaria. You educate him well about the methods of malaria prophylaxis. Which one of the following statements about infectious chemoprophylaxis is *wrong?*
a) Rifampicin is useful in Hemophillus infection prophylaxis in close contacts
b) Rifampicin helps prevent meningococcal infection in close contacts
c) Erythromycin is useful in preventing Legionnaires' disease in close contacts
d) Erythromycin is useful in diphtheria prophylaxis in close contacts
e) Ampicillin is part of prophylactic programs against infective endocarditis in high-risk patients undergoing dental surgery

19) A 27-year-old man presents with fever, sore throat, generalized lymph node enlargement, oral ulcers, and aseptic meningitis. He has recently visited Africa and had unprotected sex there. You suspect primary HIV seroconversion illness. With respect to HIV infection, choose the *wrong* statement?
- a) The env viral gene encodes for the envelope proteins, which are important for the attachment and entry of the virus into cells
- b) Tuberculous chest infection is much more common than Pneumocystis carinii infections in Africans
- c) AIDS-dementia complex may be the presenting feature
- d) Cryptosporidium diarrhea can prove fatal in AIDS patients
- e) Kaposi sarcoma of the gastro-intestinal tract is prominently symptomatic

20) A 28-year-old homosexual man complains of abdominal obesity and progressive facial and limbs wasting. He has been taking highly active anti-retroviral therapy over the past 6 months. There is hypertriglyceridemia and impaired oral glucose tolerance test. You think of protease inhibitor-associated lipodystrophy. Which one of the following statements about anti-HIV medications is the *wrong* one?
- a) Zidovudine can result in ragged red fibers on muscle biopsy
- b) Zalcitabine may produce esophageal ulceration
- c) Protease inhibitors inhibit the P450 cytochrome enzymatic system
- d) Non-nucleoside reverse transcriptase inhibitors can result in severe skin rashes
- e) Nail pigmentation is characteristically observed with squanavir

21) A 32-year-old HIV-positive homosexual man presents with fever, dry cough, weight loss, and shortness of breath. He is not taking his anti-HIV medications and he is on no prophylactic regimen against infections. CD4 positive T-cell count is $110/mm^3$. Trans-bronchial biopsy reveals Pneumocystis carinii. All of the following are poor prognostic factors in Pneumocystis carinii pneumonia (PCP) in HIV-positive patients, *except?*
- a) If it develops in a patient on no PCP prophylaxis
- b) High serum LDH
- c) High hypoxemia ratio
- d) Extensive chest radiographic changes
- e) Coexistent chest infection

22) A 37-year-old HIV-positive bisexual male, who has a CD4-positive cell count of $270/mm^3$, is referred to you for further evaluation of chronic cough and bloody sputum. Chest X-ray reveals bi-apical fluffy opacities.

HIV-positive patients have a greater risk than the normal population of all of the following, except?
   a)   Reactivating latent tuberculous infection
   b)   Acquiring tuberculosis from an infectious tuberculosis contact
   c)   Developing progressive primary tuberculous disease
   d)   Developing extra-pulmonary tuberculosis
   e)   Developing mild brain/meningeal tuberculous infection

23) A 52-year-old female presents with dry irritative cough, fever, and exertional shortness of breath. She says that her serological tests for HIV infection proved positive, 2 years ago. Her up-to-date CD4 positive T-cell count is $160/mm^3$ and there is raised serum LDH level. You plan to do a chest X-ray. Which one of the following is *not* an "atypical" chest X-ray appearance of Pneumocystis carinii pneumonia in HIV-positive patients?
   a)   Bilateral peri-hilar ground glass infiltrates
   b)   Upper lobe infiltrates
   c)   Unilateral lung involvement
   d)   Lobar cavitations
   e)   Focal consolidation

24) A 39-year-old homosexual man presents with rapidly progressive right-sided vision. Fundoscopy reveals hemorrhagic exudates along the retinal vessels. He is turned out to be HIV positive and his CD4 positive count is 42 cells/mm^3. You suspect CMV retinitis. Regarding CMV retinitis in HIV-infected patients, which one of the following statements is *false*?
   a)   There is 90% fall in the incidence after the introduction of highly active anti-retroviral therapy
   b)   In about 30% of patients, both retinas are affected
   c)   Should be differentiated from acute retinal necrosis syndrome
   d)   Vision improves gradually in almost all cases
   e)   Involvement of the macular area is rare

25) A 41-year-old heterosexual pub owner developed severe large volume watery diarrhea and unintentional weight loss over the past 2 months. Stool microscopy using acid-fast stain revealed Cryptosporidium öocysts. With respect to Cryptosporidium diarrhea in HIV-infected patients, which one is the *wrong* statement?
   a)   Responsible for 20% of diarrhea in HIV-infected patients
   b)   Becomes chronic when the CD4+ count falls below 200 cells/mm^3
   c)   Can cause malabsorption
   d)   Immuno-fluorescence stains can detect the organism in stool

e) May respond to interferon alpha injections

26) A 42-year-old female, who is HIV-positive, presents with fever, night sweats, weight loss, chronic diarrhea, vomiting, and abdominal pain. She is on no medications. Her CD4 positive T-cell count is $27/mm^3$. You conduct further investigations and the final cause behind this presentation turns out to be Mycobacterium avium intracellulare. Regarding Mycobacterium avium intracellulare infections in HIV-positive patients, which one is the *incorrect* statement?
   a) It is an environmental organism found in food and water
   b) All organs can be affected by heavy infiltration of the organism
   c) Anemia and raised serum alkaline phosphatase are commonly found
   d) The presence of hepatosplenomegaly suggests an alternative diagnosis
   e) Can be diagnosed by blood culture

27) You attend a symposium about filariasis and have learned that infection with the filarial worms Wuchereria bancrofti and Brugia malayi is associated with a range of clinical outcomes ranging from subclinical infection to hydrocele formation and elephantiasis. With respect to filariasis, all of the following statements are correct, *except?*
   a) This infection results in very high peripheral blood eosinophil count
   b) Microfilariae survive for 2-3 days only
   c) Wuchereria bancrofti is transmitted through the bite of Culex mosquito
   d) Calabar swelling is observed in Loiasis
   e) Onchocerciasis results in severe visual impairment

28) A 32-year-old male from central Africa was found to have a high blood eosinophil count during routine pre-employment assessment. Stool examination reveals hookworms. Which one of the following tropical infections is associated with peripheral blood eosinophilia?
   a) Malaria
   b) Trypanosomiasis
   c) Leishmaniasis
   d) Giardiasis
   e) Ascariasis

29) A 43-year-old male has recently emigrated from Papua New Guinea. Today he visits you because of certain complaints. You examine him and think he has Löeffler syndrome. Which one of the following statements about this syndrome is *incorrect?*

a) This is the trans-pulmonary passage of certain helminthes' larvae
b) Dry cough and wheezes occur
c) There is burning substernal discomfort which is aggravated by coughing or deep breathing
d) Chest X-ray films show fixed pulmonary infiltrates with pleural effusion
e) Blood-tinged sputum containing eosinophil-derived Charcot-Leyden crystals may be present

30) You are reviewing flash cards about tropical infections and their causative organisms. The following infections correctly match their corresponding clinical disease, *except?*
a) Onchocerciasis-diffuse dermatitis
b) Fascioliasis-urticarial skin rashes
c) Strogyloidiasis-meningitis
d) Trichinosis-myositis
e) Cysticercosis-malabsorption

31) A 32-year-old refugee from Somalia has developed a bizarre skin lesion. You refer him to the tropical diseases department for further evaluation. Regarding tropical infective skin lesions, which one of the following is the *incorrect* one?
a) Myiasis can cause subcutaneous swellings
b) Buruli ulcer results from Mycobacterium ulcerans infection
c) Larva currens is produced by Strongyloidiasis
d) Oriental sore is a manifestation of Treponema infection
e) Calabar swelling is caused by Loa loa

32) A 32-year-old man underwent renal transplant surgery 1 month ago. He takes daily cyclosporine, azathioprine, and prednisolone. Today, he develops fever and rigor. There is no change in his urine output and his graft is not tender. Blood cultures are taken and you order chest X-ray and urine examinations. During the 1st month following solid-organ transplantation, the following infections are common, *except?*
a) Herpes simplex
b) Wound infections
c) Bacterial pneumonias
d) Cather-related bacteremia
e) Nocardia brain abscess

33) A 41-year-old diabetic nephropathy patient underwent renal transplant operation before 4 months. The early post-operative period was uneventful and he was doing well on cyclosporine, azathioprine, and prednisolone. Recently, he has been experiencing headache, vomiting, and blurred vision. Nocardia brain abscess is diagnosed. During the 2nd to 6th months following solid-organ transplantation, the following infections are common, *except*?

a) Listeria monocytogenes
b) Cytomegalovirus
c) Aspergillus
d) Cryptococcus
e) Staphylococcus aureus

34) A 32-year-old male presents with abdominal cramps, bloody diarrhea, and fever. Stool examination reveals pus and RCBs, but no organism is seen; cultures are pending. All of the following infections demonstrate fecal leukocytes on stool examination, *except*?

a) Vibrio cholerae
b) Shigella
c) Salmonella enteritidis
d) Campylobacter jejuni
e) Vibrio parahemolyticus

35) A 9-year-old girl from Brazil presents with a swelling on her right eye after a reduviid bug bite. You think that this is Romaña's sign. Which one of the following organisms is the cause of Chagoma?

a) Toxoplasma gondii
b) Leishmania donavani
c) Hemophilus species
d) Trypanosoma cruzi
e) JC virus

36) A 23-year-old pregnant woman visits the physician's office to ask about the risks of having a small cat at home. All of the following are cat-related infections, *except*?

a) Toxacariasis
b) Toxoplasmosis
c) Strogyloidiasis
d) Dermatomycosis
e) Skin candidiasis

37) A 43-year-old woman demonstrates positive serum VDRL during routine pre-medical insurance testing. She denies any sexual contact with strangers or with anyone with a known sexually transmitted disease. No history of genital ulceration is obtained. All of the following are causes of biologic "chronic" false positive rapid plasma reagin testing, *except?*
   a) Systemic lupus erythematosus
   b) Lepromatous leprosy
   c) Intravenous drug addiction
   d) Pregnancy
   e) Rheumatoid arthritis

38) A 19-year-old college student presents with fever, sore throat, and splenomegaly. He develops diffuse skin rash after receiving ampicillin. Blood film reveals atypical lymphocytes. There are raised serum AST and ALT levels. You suspect an infectious disease and you order serum monospot test. All of the following can cause positive monospot test, *except?*
   a) Hepatitis viruses
   b) Lymphoma
   c) Serum sickness
   d) Infectious mononucleosis
   e) Streptococcus skin infections

39) A 32-year-old soldier presents with lockjaw, opisthotonus, and repetitive generalized muscle spasms 2 weeks after sustaining a deep thigh injury. You diagnose tetanus. All of the following can result in lockjaw and trismus, *except?*
   a) Phenothiazines
   b) Jaw sarcoma
   c) Strychnine poisoning
   d) Guillain-Barre syndrome
   e) Ludwig angina

40) A 43-year-old bisexual male develops prominent groove at both inguinal areas. Past notes reveal many sexually transmitted diseases. In which sexually transmitted disease this "groove sign" appears?
   a) Lymphogranuloma venereum
   b) Granuloma inguinale
   c) Secondary syphilis
   d) Gonorrhea
   e) Herpes simplex type I infection

Infectious Diseases and Genitourinary Medicine

Answers

1) d.

In brucellosis, generalized lymphadenopathy may occur. Blood culture is the gold standard in diagnosing this infection; unfortunately, this is time-consuming and needs special culture media with a special environment as well as being hazardous to the lab workers. Neurological manifestation of brucellosis may be in the form of encephalopathy, meningoencephalitis, pseudotumor cerebri-like picture, and peripheral neuritis. Spinal tenderness is very common. Usually, it is clinically difficult to differentiate a new infection from a relapsing one in a patient who was treated in the past; this is especially seen in highly endemic areas and 2-mercapto ethanol test is used as a differentiating tool which helps distinguish between IgG and IgM responses.

2) e.

Unlike other Salmonella species, human being is the only reservoir of Salmonella typhi. During the second week of symptomatic infection, the spleen becomes palpable and rose spots appear. Although these rose spots are observed in a minority of patients, but they are highly suggestive of typhoid. About 90% positivity is the yield of blood culture in the first week; this gradually declines, week after week. The main indication for doing bone marrow culture is to isolate the organism from those who were exposed to systemic antibiotics. In the first week, the diagnosis may be challenging, because in this invasive stage with bacteremia, the symptoms are those of a generalized infection without localizing features.

3) e.

Thin and thick blood films should be ordered and examined whenever malaria is a clinical suspicion. The thin blood film has the advantage of speciating the Plasmodia parasite, as well as counting the resulting parasitemia. The intravenous preparation of quinine may cause hypoglycemia, cardiac toxicity, and cinchonism. Mefloquine is contraindicated in the 3rd trimester of pregnancy. Systemic steroids add no benefit to the management of cerebral malaria; actually they should be avoided as they may impart a deleterious effect. Chemoprophylaxis for malaria should ideally begin 1 week before entering the malarious area and is continued for 4 weeks after leaving it. Fansidar should not be used for chemoprophylaxis, as deaths have occurred from agranulocytosis or Stevens-Johnson syndrome. Mefloquine is useful in areas of multiple drug resistance, such as East and Central Africa and Papua New Guinea.

4) e.

Risk factors for leptospirosis are occupational exposure (farmers, ranchers, abattoir workers, trappers, veterinarians, loggers, sewer workers, rice field workers, military personnel, laboratory workers), recreational activities (fresh water swimming, canoeing, kayaking, trail biking), and household exposure (pet dogs, domesticated livestock, rainwater catchment systems, and infestation by infected rodents). Therefore, knowing the patient's occupation or hobbies is part of the disease assessment. The early phase is due to leptospiremia and blood cultures during this period have high positive results; urine culture becomes gradually positive in the second week. There is blood neutrophil leukocytosis, which is a very important clue in differentiating leptospirosis from severe viral hepatitis infections. Leptospira aseptic meningitis is classically associated with L. canicola infection; this illness is very difficult to distinguish from viral meningitis.

5) d.

Toxic shock syndrome is not an invasive or bacteremic illness; it is an exotoxin-mediated syndrome. The other sources of this exotoxin (TSST-1) are skin abscess and conjunctivitis. Raised serum level of creatine phosphokinase is one of the diagnostic criteria of toxic shock syndrome. The disease confers a mortality rate of 10-20%. The resulting shock is usually resistant to fluid therapy and necessitates the use of pressers.

6) e.

Rabies is caused by RNA rhabdoviral infection; usually through a bite of an infected animal. The virus is present in the saliva although it infects the CNS. The disease is fatal; very few exceptions were documented worldwide (yet such patients are left with severe neurological disability). Rabies may present as an ascending paralysis state (so-called dumb-type) in 20% of cases; the remaining 80% of patients present with the classical furious type of many characteristic phobias. The incubation period is highly variable. The incubation period varies in humans from a minimum of 9 days to many months but is usually between 4 and 8 weeks. Severe bites, especially if on the head or neck, are associated with shorter incubation periods.

7) e.

Lassa fever results from infection with an RNA arena virus and is transmitted through rat urine. The incubation period is 1- 3 weeks. Renal and hepatic failures may occur, as well as circulatory collapse. Intravenous ribavirin can be used in the treatment and this may decrease the mortality rate if given within the 1 st week. Ribavirin has been used as prophylaxis in close contacts of Lassa fever patients but there are no formal trials of its efficacy.

8) d.

Yellow fever is a mosquito-borne infection (Aedes africanus in Africa and the Hemagogus species in America are the vectors). The incubation period is 3-6 days. Leucopenia occurs. The usual presentation is with fever, join pain, muscle pain, and headache; a highly non- specific picture. No specific treatment is available; only supportive. A single vaccination with the 17D non-pathogenic strain of virus gives full protection for at least 10 years.

9) c.

Until now, there is no dengue vaccine. The principal vector is Aedes aegypti, which breeds in standing water; collections of water in containers and tyre dumps are a particular risk in large cities. Aedes albopictus is a vector in some Southeast Asian countries. There are four serotypes of dengue virus, all producing a similar clinical syndrome; homotypic immunity is life-long but heterotypic immunity between serotypes lasts only a few months.

10) d.

Q-fever is mainly seen in farmers, slaughterhouse workers, and vets. The treatment of Q- fever endocarditis is usually prolonged; may reach 1 year. Tetracyclines are the drug of choice. Liver granulomas may develop, but they are usually asymptomatic.

11) d.

Typhus fever may be mite- borne, flea-borne, tick-borne and louse-borne (depending on the species). Rickettsia organism infects and damages the endothelium of small blood vessels in the skin, brain, and lung. There is no gastroenteritis as a common feature of all Rickettsiae. Tetracycline or chloramphenicol are the treatment of choice.

Routine blood investigations are unhelpful. The diagnosis is mainly clinical; however, it should be differentiated from malaria, typhoid fever, meningococcal sepsis, and leptospirosis.

12) e.

Many patients may not be able to remember the tick bite (of Ixodes tick, which is the vector for the spirochete Borrelia burgdorferi); on the other hand, the skin rash may be mild and disappear before being noticed by the patient. Therefore, the diagnosis can easily be overlooked if the physician does not have a high index of suspicion. Multiple cranial nerve palsies may develop; bilateral facial weakness is the commonest one. Peripheral joint involvement usually takes the form of asymmetric large joint oligoarthritis.

13) c.

The rash of secondary syphilis characteristically involves the palms and soles. Lymphocytic meningitis occurs and is usually self- limiting or asymptomatic; rarely, it is severe. Serum testing for VDRL does not approach 100% positivity; FTA- ABS and TPHA are much more sensitive and specific. Serum VDRL and rapid plasma reagin (non-treponemal tests) are mainly used for screening and for following up patients on medical therapy; the treponemal tests (e.g., TPHA) are used to confirm the diagnosis. The oral mucosa could have snail-tract ulcers while the genital areas may demonstrate condylomata lata; genital ulceration is not part of secondary syphilis. Secondary syphilis usually starts 6-8 weeks after the appearance of the primary ulcer (chancre) due to dissemination of treponemes to produce a multi-systemic disease.

14) e.

In general, Salmonella paratyphi infection is milder and shorter illness when compared to Salmonella typhi. The incidence of chronic gallbladder carrier state is much lower in the paratyphi group.

15) e.

The so-called cholera sicca can result in large fluid loss and sequestration within the bowel and death may ensue rapidly without the clinical appearance of diarrhea; however, this is rare. The disease usually occurs in large epidemics and a common infecting source is usually present.

The cholera organism is motile and comma-shaped; this can be seen with the use of dark field or phase-contrast microscopy of stool. Vibrio cholerae is not an invasive pathogen; the resulting painless watery diarrhea is exotoxin-mediated.

16) b.

Stem "a" refers to lymphogranuloma venereum which results from infection with Chlamydia trachomatis serovars L1, L2, and L3. Genital ulceration does not occur in secondary syphilis; the primary chancre is the hallmark of the primary phase. Genital Herpes simplex infection (usually type II virus) results in recurrent grouping of painful genital ulcers. Chancroids are usually multiple and painful ulcers that result from infection with Hemophilus ducreyi. Calymmatobactrium granulomatis infection is the cause of granuloma inguinale.

17) c.

The hallmark of symptomatic gonorrhea is purulent urethral discharge and painful micturition; however, these may be mild. The disease favorably responds to single doses of antibiotics, such as 500 mg of ciprofloxacin. Because many patients are co- infected with Chlamydia, oral tetracycline (or doxycycline) is co-administered for 10 days. Sexual partners should always be contacted and investigated. Most female patients are asymptomatic and this population is the usual source of spreading the infection to males.

18) c.

Erythromycin is useful in the treatment of Legionnaires' disease but it has no place in the chemoprophylaxis against this form of chest infection; there is no human-to-human transmission of this infection.

19) e.

The env viral gene encodes for the envelope proteins (such as gp160 and gp45), which are important for the attachment and entry of the virus into cells. Tuberculous chest infection is much more common than Pneumocystis carinii infections in Africans, unlike the Western population, in which P. carinii is much more common.

AIDS-dementia complex may be a presenting feature (and it is an AIDS-defining illness); the progression can be slowed by zidovudine. Cryptosporidium diarrhea is almost always a self-limiting infection in healthy persons; it can be fatal in AIDS patients. Gastrointestinal Kaposi sarcoma is usually asymptomatic, in contrary to pulmonary Kaposi sarcoma, which results in prominent dyspnea, cough, pleural effusion,...etc.

20) e.

Similar to mitochondrial cytopathies, zidovudine results in myopathy with ragged red fibers. Many non-nucleoside reverse transcriptase inhibitors (NRTIs) can cause liver dysfunction, lactic acidosis, and pancreatitis; some may cause severe skin reactions which necessitate stoppage of the medication. Protease inhibitors (PIs) results in many types of lipodystrophy; fat redistribution occurs in 20- 30% of patients on protease inhibitors by 2 years. PIs result in a multitude of metabolic disturbances; insulin resistance develops in 60% of patients, abnormal glucose tolerance affects 35% of them, while overt diabetes mellitus occurs in 8%. In 30- 50% of patients, PIs raise serum levels of total cholesterol, LDL cholesterol, and triglycerides, while those of HDL cholesterol are lowered; coronary artery disease is much more common in this population. Nail pigmentation is characteristically seen with zidovudine.

21) c.

Low hypoxemia ratio portends a poor outcome in Pneumocystis carinii pneumonia. The other predictors of a poor outcome are delayed diagnosis, low CD4 positive count, and low serum albumin.

22) e.

In addition, HIV patients have a greater risk of developing military and disseminated tuberculosis and of developing a second episode of tuberculosis from exogenous infection as demonstrated by isolate typing. Tuberculous meningitis in HIV patients is an aggressive illness which confers a high morbidity and mortality.

23) a.

The appearance of bilateral peri-hilar ground glass infiltrates is the typical picture of Pneumocystis carinii pneumonia (which is seen in 80% of cases); later, an ARDS-like picture may develop.

Atypical appearances (e.g., upper lobe infiltrates, unilateral lung involvement, lobar cavitations, and focal consolidation) are encountered in 20% of cases; a nodular appearance of opacity may also be seen.

24) d.

Prior to the introduction of highly active anti-retroviral therapy (HAART), about 25% of HIV patients with a CD4 positive count of <100 cells/mm^3 developed this grave sight -threatening complication. Some patients may be entirely asymptomatic and discovered by routine fundoscopic examination. CMV retinitis should also be differentiated from Toxoplasma retinitis, progressive outer retinal necrosis syndrome, syphilis, and Pneumocystis infection. No recovery of vision occurs in the affected areas and there is always a risk of future retinal detachment because of retinal necrosis. Fortunately, macular involvement is rare (but it is the most sight-threatening type).

25) e.

Cryptosporidium is a highly contagious zoonotic protozoal enteric pathogen. By infecting the small bowel mucosa, it results in severe large volume watery diarrhea, which is usually painful. The organism rarely causes acalculous cholecystitis, sclerosing cholangitis, and pneuomonitis. Immuno -fluorescence stains or modified acid-fast satin on stool specimens can identify the organism in 90% of cases. Cryptosporidia are also readily identifiable on histopathological examination of duodenal biopsies; this may be necessary, as the organism's öocysts are intermittently (or never) seen in the stool of 10% of patients. It may respond to paromomycin or azithromycin.

26) d.

Acquisition of Mycobacterium avium intracellulare is probably through respiratory or gastrointestinal tracts, where colonization occurs and precedes the disseminated disease in 2/3rd of cases. In theory, any body organ can be infiltrated by the organism; however, the reticulo-endothelial system bears the major burden of the infection. CT scans usually reveal an enlarged intra-abdominal and mediastinal lymph nodes, as well as hepatosplenomegaly. The organism can be isolated by doing blood cultures (or positive cultures of liver biopsy or bone marrow aspirates) . Therapy must consist of multiple agents, and a combination of ethambutol, azithromycin, rifabutin and ciprofloxacin is effective.

27) b.
Microfilariae survive for 2-3 years while the adult worms live for 10-15 years, in general. Loa loa is transmitted by the day-biting fly Chrysops. The two classic clinical presentations of loiasis are localized subcutaneous swellings that appear transiently (Calabar swelling) or the pathognomonic migration of the adult worm across the eye. "River Blindness" or onchocerciasis is transmitted by a bite of Simulium fly.

28) e.
Ascariasis can cause blood eosinophilia. Tropical infections that are *not* associated with eosinophilia are leprosy, tuberculosis, tapeworms (but cysticercosis causes eosinophilia), typhoid fever, brucellosis, and amoebiasis. With the exception of the eosinophilia that may develop following scarlet fever, acute bacterial or viral infections characteristically produce eosinopenia. Even in patients with eosinophilia due to parasitic or allergic diseases, the development of an inter-current bacterial or viral infections leads to suppression of blood eosinophilia until the superimposed acute infection has resolved. The major causes of marked (>3000 cells/mm^3) peripheral blood eosinophilia are: Ascariasis and hookworm infestation (during their early trans-pulmonary larval migration; often, this is absent when the parasite matures); Angiostrongyliasis costaricensis; Strongloidiasis; visceral larva migrans (this is especially in children); Gnathostomiasis; Filariasis (tropical eosinophilia syndrome, onchocerciasis, and Loiasis); and flukes (during their early phase; this includes Katayama fever).

29) d.

Löeffler syndrome is the trans-pulmonary passage of certain helminth larvae (Ascaris and hookworms). Ascariasis is the most common cause of Löeffler syndrome worldwide. Fever occurs in many patients, but infrequently exceeds 38.3°C. About 15% have urticaria during the first four to five days of the illness. The pulmonary opacities are non-fixed and fleeting; these appear as round or oval infiltrates ranging in size from several millimeters to several centimeters in both lung fields. These lesions are more likely to be present when blood eosinophilia exceeds 10%. The infiltrates are migratory and may become confluent in perihilar areas; they usually clear completely after several weeks

30) e.

Cysticercosis does not produce a malabsorption state. Cysticercosis can involve the CNS or extra-neural tissues. Cysticerci can develop in almost any body site, but tend to have a predilection for muscle or subcutaneous tissues. Cysticerci at these sites are usually asymptomatic, but the patient may notice subcutaneous, pea-like or walnut-sized nodules.

31) d.

Oriental sore is caused by Leishmania tropica infection via the bite of a female sand-fly. In Buruli ulcer, the initial lesion is a small subcutaneous nodule on the arm or leg. This breaks down to form a shallow, necrotic ulcer with deeply undermined edges, which extends rapidly. Healing may occur after 6 months but the accompanying fibrosis causes contractures and deformity. Myiasis is due to skin infestation with larvae of the South American botfly, Dermatobia hominis, and the African Tumbu fly, Cordylobia anthropophaga. The larvae develop in a subcutaneous space with a central sinus. This orifice is the air source for the larvae, and periodically the larval respiratory spiracles protrude through the sinus.

32) e.

Infections during the 1st month of solid-organ transplantation generally are not due to immune suppression; they are infections that are commonly seen in post-operative patients. Nocardia infections reflect a prominent underlying immune suppression and develop within 2-6 months following solid-organ transplantation.

33) e.

Staphylococcus aureus infections are characteristically seen during the 1st 4 weeks post-transplantation in the form of skin wound, catheter -related, and chest infections. During the 2nd to 6th months following solid-organ transplantation, the infections are of those seen in immune -compromised patients: EBV, varicella zoster, papova viruses (JC and BK), adenoviruses, Toxoplasma, Nocardia, and Pneumocystis carinii.

34) a.

The presence of fecal leukocytes on stool examination indicates that an invasive pathogen is infecting the bowels; cholera is not an invasive infection and no fecal leukocytes are found on stool examination.

35) d.

Chagoma forms as a result of local replication of the organism Trypanosoma cruzi at the bite site of reduviid bug. After the entrance of the organism into the skin, a dark erythematous firm swelling appears and this is accompanied by enlargement of the regional lymph nodes. The conjunctival location of the lesion is less common but is more characteristic; the unilateral firm reddish swelling of the lids may close the eye and constitutes "Romaña's sign." Young children are most commonly affected.

36) e.

Having a cat at home does not confer an increased risk of skin candidiasis. The other cat-related infections are Pasteurellosis (usually a bite wound), cat-scratch disease, tularemia, hookworms, and rabies.

37) d.

Both treponemal and non -treponemal serological tests can have false positive results (acute or chronic; sometime called biological false positive testing). Causes of acute (within 6 months) false positive rapid plasma reagin (RPR) test are pregnancy, recent immunization, and febrile illnesses; the causes are usually self-limiting and transient. Chronic (>6 months) false positive RPR test are chronic infections (especially hepatitis C and HIV), autoimmune diseases (especially SLE), and intravenous drug addicts. In false positive testing, the titer is usually <1:8. On occasion, biologic false positive results may also be of high titer; thus the level of the titer does not help the clinician differentiate between a true or false positive result.

38) e.

Heterophile antibodies react to antigens from phylogenetically unrelated species. They agglutinate sheep red blood cells (the classic Paul-Bunnell test), horse red blood cells (used in the monospot test), and ox and goat erythrocytes.

The monospot is a latex agglutination assay using horse erythrocytes as the substrate. The first three stems are the cause of false positive monosopt (heterophile antibodies) test while stem "d' is the cause of truly positive test. Rare false-positive heterophile tests have been reported in patients with leukemia, lymphoma, pancreatic cancer, systemic lupus erythematosus, HIV infection, and rubella. In addition, the patient may have previously had infectious mononucleosis, since heterophile antibodies can persist at low levels for up to one year. Stem "e" has nothing to do with monospot testing.

39) d.

Guillain-Barre syndrome is a cause of acute inflammatory demyelinating polyradiculopathy; the weak muscles are flaccid (there is no lockjaw). Causes of lockjaw (apart from tetanus) are any inflammatory lesion in the mouth, pharynx, cheeks, or external auditory canal (such as Ludwig's angina, peritonsillar abscess, tooth abscess); malignancies, such as sarcoma of the jaw and squamous cell carcinoma of the oral cavity; conversion disorders (so-called hysterical tetanus); mechanical problems (jaw dislocation, jaw ankylosis); phenothiazines (as part of dystonic reactions); strychnine poisoning (lockjaw is a late sign); and encephalitides.

40) a.

The "groove sign" of the inguinal syndrome (of the secondary phase of lymphogranuloma venereum) is due to adenopathy above and below the inguinal ligament; this is said to pathognomonic of lymphogranuloma venereum in the appropriate clinical setting. An anorectal syndrome can also occur during the secondary phase of the disease, which causes inflammatory reaction and mass in the rectum and retroperitoneum. Those patients usually present with painful rectal discharge, constipation, fever, and tenesmus. Overt hemorrhagic proctocolitis and hyperplasia of intestinal and perirectal lymphatic tissue may occur; this can be mistaken for inflammatory bowel disease. Chronic colorectal fistulas and strictures may develop in the long-term.

Chapter 9
Neurology, Psychiatry, and Ophthalmology

Questions

1) A 55-year-old man is due to undergo a battery of investigations because of certain neurological complaints. Which one of the following statements regarding neurological investigations is the *correct* one?
   a) Brain CT scan is a week tool to diagnose acute subarachnoid hemorrhage
   b) The CSF imparts a black signal on the brain MRI T2-weighted images
   c) Diffusion-weighted MRI is useful to detect early cerebral ischemia
   d) PET scan is contraindicated in seizure disorders
   e) Conventional cerebral angiography does not increase the risk of contrast nephropathy

2) The MRI lab technician phones you about a 44-year-old woman, who was planned to do brain MRI. He says he cannot do it because MRI is contraindicated in this patient. Which one of the following is *not* a contraindication for doing MRI scans?
   a) Cochlear implants
   b) Eye metallic foreign body
   c) Cardiac pacemaker
   d) Older model intracranial aneurysmal clips
   e) Coronary stents

3) Your intern intends to do lumbar puncture for a 21-year-old man. You disagree with him and tell him that the patient has a contraindication. Which one of the following is *not* a contraindication for doing lumbar puncture?
   a) Platelets count of $10000/mm^3$
   b) Large brain hemispheric glioma
   c) Cellulitis at the site of entry of the needle
   d) Patient with suspected acute subarachnoid hemorrhage, papilledema, and unremarkable brain CT scan
   e) Suspected case of pyogenic meningitis with left-sided weakness

4) A 54-year-old man developed recurrent nocturnal seizures. You arranged for electroencephalography (EEG) and started anticonvulsant therapy. EEG is useful in supporting the diagnosis of all of the following, *except?*
   a) Herpes simplex encephalitis
   b) Subacute sclerosing panencephalitis
   c) Creutzfeldt-Jacob disease
   d) Petit mal epilepsy
   e) Multiple sclerosis

5) A 53-year man presented with proximal muscle weakness. The pelvic girdle muscles are tender. You suspected inflammatory myositis and arrange for electromyography (EMG). Which one of the following is *not* a recognized EMG finding of muscle diseases?
   a) Prolonged insertional activity
   b) Presence of fibrillation potentials at rest
   c) Complex repetitive discharges
   d) Pseudomyotonic discharges
   e) High amplitude, long-duration, and polyphasic motor unit potentials

6) You are giving a short lecture about localization in clinical neurology to undergraduate medical students. Which one of the following is *correct*?
   a) The skin overlying the medial malleolus is the 1st sacral dermatome
   b) Big toe dorsiflexion represents 5th lumbar myotome
   c) Symmetrical weakness of the pelvic/shoulder girdle is consistent with peripheral neuropathy
   d) Pyramidal weakness of the right upper limb is translated into weakness of shoulder adduction and elbow/wrist flexion
   e) Dissociated sensory loss means loss of join position with preserved vibration sense

7) A 65-year-old man was diagnosed with ischemic stroke. He developed sudden aphasia and right-sided weakness. All of the following statements about ischemic stroke is incorrect, *except*?
   a) Ischemic strokes are less common than hemorrhagic strokes
   b) Thrombotic strokes are rare while embolic ones are very common
   c) The volume of infracted tissue in lacunar stroke is <5 cm^3
   d) Aspirin is the only useful option in treating asymptomatic carotid artery stenosis
   e) r-tPA thrombolytic therapy is contraindicated in patient who presents within 1 hour of symptoms

8) A 69-year-old woman is brought to the Emergency Room. You examine the patient carefully and find Wernicke's aphasia. Which one of the following statements about this type of aphasia is *correct*?
   a) The speech output is non-fluent
   b) Comprehension is intact
   c) Naming is relatively normal
   d) Repetition is impaired
   e) Neologism suggests an alternative form of aphasia

9) A 16-year-old female was diagnosed with epilepsy after developing 3 non-provoked generalized seizures. Which one of the following is the *correct* statement regarding seizure disorders?
   a) Petit mal epilepsy usually develops before the age of 4 years
   b) Complex partial seizures usually last more than an hour
   c) Lamotrigine is safe during pregnancy
   d) The commonest cause of status epilepticus is CNS infections
   e) EEG is normal in 10% of inter-ictal recording of idiopathic generalized tonic-clonic epilepsy

10) A 67-year-old woman developed idiopathic trigeminal neuralgia. With respect to trigeminal neuralgia, which one is the *correct* statement?
   a) The idiopathic variety usually develops in young males
   b) The area of the ophthalmic division of the 5th cranial nerve is the commonest involved area
   c) Bilateral involvement is common
   d) Responds poorly to carbamazepine
   e) Brain MRI is normal in the idiopathic variety

11) A 17-year-old female was diagnosed with migraine before 1 week. She visited the doctor's office to ask about her headaches and their long-term outcome. With respect migraine, which one of the following is the *correct* statement?
   a) Migraine with aura is the commonest subtype
   b) Family history is usually absent
   c) The disease usually targets males
   d) Sumatriptan is useful as a prophylactic agent
   e) Chocolate food can precipitate migraine attacks

12) A 21-year-old man desperately asks for relief of his cluster headaches. With respect to this form of cephalgia, which one is the *correct* statement?
   a) The headache usually attacks at noon
   b) Bilateral headache is common
   c) Dryness of the eyes is a characteristic feature during headache
   d) Aggravated by intranasal capsaicin
   e) Alcohol ingestion precipitates these clusters

13) A 22-year-old woman has had pancephalic headache for the past 6 weeks, which is turned out to be due to idiopathic pseudotumor cerebri. All of the following are incorrect with respect to this disease, *except?*

a. There is intermittent obstruction of the CSF flow at the level of arachnoid granulations
b. Double vision is not consistent with the diagnosis
c. Eventually leads to primary optic atrophy if not treated
d. CSF mononuclear pleocytosis does not exceed 100 cell/mm^3
e. Responds to oral prednisolone

14) A 44-year-old man undergoes brain imaging because of certain complaints. The imaging is suggestive of acoustic neuroma. All of the following statements are incorrect regarding this tumor, *except?*
   a) Bilateral tumors are the hallmark of neurofibromatosis type I
   b) Loss of hearing is a late feature
   c) The tumor lies in the inter-peduncular cistern
   d) Results in a very high CSF protein level
   e) The best clinical screening test is the exaggeration of jaw jerk

15) A 71-year-old man scores 21 on mini-mental status examination. His final diagnosis is normal pressure hydrocephalus. Which one of the following statements regarding normal pressure hydrocephalus is *correct?*
   a) Some patients display past history of subarachnoid hemorrhage
   b) The optic discs are swollen
   c) The CSF opening pressure is elevated but does not exceed 60 cm water
   d) The resulting dementia is cortical in type
   e) Large volume CSF aspiration should be discouraged

16) A 63-year-old woman presents with rapidly progressive dementia and startle myoclonus. You suspect Creutzfeldt-Jacob disease. Which one is the *correct* statement about this disease?
   a) The infecting prion protein has a human-human mode of transmission
   b) The full-blown dementia syndrome appears after 2-3 years of the occurrence of myoclonus
   c) Cerebellar ataxia suggests an alternative diagnosis
   d) The characteristic EEG findings may be transient and occur in 65% of patients
   e) Brain biopsy is contraindicated

17) A young man brings his 72-year-old father to the physician's office. His father has memory problems and you consider Alzheimer's disease after interviewing him. With respect to Alzheimer's disease, which one is the *correct* statement?

a) Most cases are autosomal dominant
b) Early loss of remote memory is the core finding
c) Atrophy of the frontal and temporal lobes is the usual brain imaging feature
d) Most patients die from inanition after an average of 8 years of symptoms
e) Donepezil is useful to treat moderate-severe dementia

18) A 66-year-old man is referred by his physician as a difficult-to-manage case of Lewy-body dementia. With respect to this type of dementia, choose the *correct* statement?
a) The presence of formed visual hallucinations is against the diagnosis
b) Rigidity and hypokinesia indicate an alternative diagnosis
c) Loss of recent memory is not that conspicuous early in the course of the disease
d) The cognitive decline is fixed and does not fluctuate
e) Chlorpromazine is useful in controlling psychotic features

19) A 21-year-old woman was diagnosed with relapsing-remitting multiple sclerosis. Her neurologist plans to prescribe some form of therapy. Which one of the following about multiple sclerosis is *correct*?
a) The primary progressive form should be differentiated from high spinal cord compression
b) Optic neuritis is usually of papillitis type
c) Patients with clinically isolated syndrome of demyelination should not receive interferon beta therapy
d) Brain MRI shows peri-ventricular plaques in 30% of relapsing remitting multiple sclerosis
e) The CSF mononuclear pleocytosis usually exceeds 100 cells/mm^3 during relapses

20) A 43-year-old man is referred from a rural hospital for further management of Wernicke's encephalopathy. All of the following are correct with respect to Wernicke's encephalopathy, *except*?
a) May result from hyperemesis gravidarum
b) The presence of seizures is incompatible with the diagnosis
c) The resulting ataxia is primarily of gait
d) Confusion is more common than coma
e) Facial weakness is usually bilateral

21) A 52-year-old man develops sudden severe thunderclap headache. His final diagnosis turns out to be acute subarachnoid hemorrhage. Regarding this form of stroke, which one is the *correct* statement?
   a) Most cases are due to rupture of cerebral hemispheric arteriovenous malformation
   b) Focal signs at presentation are very common
   c) Berry aneurysms are more common in the anterior circulation than in the posterior circulation
   d) Re-bleeding usually occurs in the 1st 24 hours
   e) The complicating focal cerebral vasospasm responds well to oral hydralazine

22) A 50-year-old man is brought to the Emergency Room by a policeman who states that the patient has lost his mind. The patient keeps saying "Where am I?" After you examine the patient thoroughly and do brain CT scan, you diagnose transient global amnesia. All of the following are incorrect with respect to this type of acute amnesia, *except?*
   a) Usually lasts for few months
   b) Recurrent attacks is the rule
   c) Treated by electroconvulsive therapy
   d) The amnesia is mainly for remote memories
   e) Hysterical amnesia is a differential diagnosis

23) Because of medication-refractory headache, a 43-year-old woman undergoes brain imaging. The result is consistent with supra-tentorial meningioma. Which one of the following regarding this type of tumors is *correct?*
   a) Arises from cortical neurons
   b) Calcification never occurs within these tumors
   c) Usually multiple
   d) The tumor mass weakly and heterogeneously takes the contrast material on brain imaging
   e) Prior cranial irradiation is a risk factor for meningioma development

24) A physician is in a dilemma over a case he has referred to you. He is not able to control seizures in a 21-year-old woman. After careful evaluation, you diagnose non-epileptic attacks (pseudo-seizures). Which one of the following is *incorrect* with regard to pseudo-seizures?
   a) Should be suspected if patients have minimal to no movement during a seizure with unresponsiveness that lasts more than a few minutes

b) Some patients keep their eyes tightly closed during the seizure and, particularly, if passive opening is actively resisted
c) Abrupt return of normal alertness after a generalized motor seizure or prolonged loss of consciousness suggest pseudo-seizures
d) During generalized motor pseudo-seizures, the mouth is partially or tightly closed or clenched
e) Urinary incontinence always favors genuine seizures over pseudo-seizures

25) A 33-year-old man has developed fever and confusion. He was admitted to the neurology ward and a battery of investigations was conducted. The final diagnosis was Herpes simplex encephalitis. With respect to this brain infection, which one of the following is the *correct* statement?
a) Herpes simplex type II is the usual culprit
b) The CSF analysis is normal in 30% of cases
c) EEG is not helpful in the work-up
d) The virus is isolated from brain biopsy specimens, not from the CSF
e) Brain CT scan is superior to MRI

26) A 32-year-old immigrant from India develops progressive somnolence and fever. His neck is stiff. Brain CT scan is unremarkable. After conducting proper investigations, you diagnose tuberculous meningitis. You start anti-tuberculous therapy. All of the following statements about this form of CNS infection are incorrect, *except?*
a) Cranial nerve palsies suggest another diagnosis
b) About 5% of patients have military tuberculosis
c) Chest X-ray films show evidence of tuberculosis in 10% of patients
d) Negative skin tuberculin testing virtually excludes the disease
e) May present as acute rapidly progressive meningitic syndrome, suggesting pyogenic meningitis

27) A 65-year-old man presents with rapid onset of dense left-sided weakness and drowsiness. Brain CT scan reveals intra-cerebral hemorrhage (ICH). All of the following statements about ICH are wrong, *except?*
a) Hypertension is a rare cause of spontaneous ICH
b) Hypocholesterolemia protects against the development of ICH
c) Enlargement of the hematoma is rare in the 1st 24 hours
d) Thalamic hemorrhage may decompress itself into the 3rd ventricle
e) Cerebellar hemorrhage usually starts in the midline cortical area

28) A 61-year-old man, who is hypertensive and diabetic, presents to the Emergency Department dizzy. He has nausea and vomiting. You examine the patient and arrange for brain imaging. His diagnosed is lateral medullary syndrome. Which one of the following statements regarding this form of ischemic stroke is *true*?
   a) Results from ischemic damage to the upper part of the brainstem
   b) Nystagmus is unusual
   c) Seizures occur
   d) Both planter responses are flexors
   e) Horner syndrome is observed on the contralateral side

29) A 32-year-old man is referred from a primary care center because of weakness of all limbs. You suspect Guillain-Barre syndrome (GBS). Regarding GBS, which one is the *correct* statement?
   a) The resulting weakness usually progresses over 3 months
   b) Marked asymmetrical weakness is the usual presenting feature
   c) Loss of pain and temperature sensations in the hands and feet usually develops in the 2nd week
   d) Axonal degeneration may be primary, or secondary to severe demyelination
   e) Preceding Campylobacter infection confers a good prognosis

30) A 69-year-old man presents a 2-year history of unsteady stance and gait. He is pale and you notice a tinge of jaundice. Serum vitamin B12 is very low. Which one of the following is the *correct* statement about subacute combined degeneration of the cord?
   a) Anemia must be present to consolidate the diagnosis
   b) The resulting lower limb weakness is flaccid
   c) Romberg's sign is detected
   d) Fecal and urinary incontinence are early features
   e) Peripheral neuropathy and cognitive dysfunction should not accompany this syndrome

31) A 61-year-old man is brought by his wife to see you. He has memory disturbance and talks little with his wife. She is afraid that he has developed Alzheimer's disease and she is desperate for your help. You interview the patient and examine him thoroughly. You tell his wife that he has depressive pseudo-dementia. Which one of the following is *inconsistent* with depressive pseudo-dementia?
   a) The onset of cognitive decline is more or less abrupt
   b) The cognitive dysfunction is progressive

c) The patient is aware of his cognitive deficits and may exaggerate them
d) There are prominent vegetative symptoms
e) Somatic complaints and hypochondriasis are common

32) A 76-year-old man presents with headache and scalp tenderness. Your intern suspects giant cell arteritis and he plans to prove it by doing some tests. Which one of the following statements regarding giant-cell (temporal) arteritis is *true?*
a) The location of headache is never occipital
b) High fever occurs in 50% of patients
c) Some patients have trismus-like symptoms rather than jaw claudication while eating
d) Most patients eventually develop bilateral blindness
e) The presence of aortic aneurysm suggest an alternative diagnosis

33) A general practitioner refers a 32-year-old female to your office as a suspected case of myasthenia gravis. She has had certain symptoms for 3 months. All of the following statements with respect to myasthenia gravis are incorrect, *except?*
a) The resulting weakness is relatively fixed and rather painful
b) Isolated distal muscle weakness never develops
c) Thymoma is found in 70% of patients
d) Motor neuron disease patients may demonstrate false positive Tensilon® test
e) Seronegative myasthenia gravis does not respond to plasma exchange

34) A 65-year-old smoker man develops weakness. His chest CT scan reveals a mass that is most likely lung cancer. You ascribe his weakness to Lambert-Eaton myasthenic syndrome (LEMS). Which one of the following about LEMS is *correct?*
a) About 30% of small-cell lung cancer patients develop this syndrome
b) Respiratory muscles weakness is the usual cause of death
c) Dry mouth is the commonest symptom of autonomic dysfunction
d) Anti-P/Q type voltage-gated potassium channel anti-bodies play a central role in the pathophysiology of LEMS
e) Low frequency EMG repetitive stimulation results in decremental response

35) You intend to present a poster about muscular dystrophies during a national conference about muscle diseases.

You surf the internet and collect many helpful information to create your work. Which one of the following you have *included* in your poster?

a) Duchenne's muscular dystrophy is a cause of hypertrophic cardiomyopathy

b) Becker's muscular dystrophy patients become wheelchair-bound by the age of 15 years

c) In facioscapulohumeral muscular dystrophy, the deltoids are relatively preserved

d) Extra-ocular muscle weakness occurs early in limb-girdle muscular dystrophy

e) The presence of early contractures should cast a doubt on Emery-Dreiffus muscular dystrophy

36) A 16-year-old male is referred by your ophthalmology colleague to your office. The boy has bilateral ptosis and ocular weakness. You examine the patient and think his diagnosis is Kearns-Sayre syndrome. With regard to this syndrome, choose the *correct* statement?

a) In spite of the prominent ophthalmoplegia, diplopia is rare

b) Fundoscopy may reveal old choroiditis

c) Heart block is uncommon and is rarely the cause of death

d) CSF protein should be normal

e) Hypoglycemia is one of the core features of this syndrome

37) A 13-year-old boy is brought to the Emergency Room in the early morning. The boy demonstrates flaccid weakness of all limbs. Breathing is good and he does not seem to be in distress. He can swallow quite well. After conducting some tests, you diagnose familial periodic paralysis (FPP). All of the following statements regarding FPP are incorrect, *except?*

a) Mutations in the dihydropyridine L-type calcium channels result in hyperkalemic periodic paralysis

b) The hypokalemic variety may be associated with eye closure myotonia

c) The hyperkalemic type may be associated with hyperthyroidism

d) Progressive muscular hypertrophy occurs in between attacks of the hypokalemic type

e) Acetazolamide helps prevent attacks of hypokalemic FPP

38) A 22-year-old man is brought by his sister to the Acute and Emergency Department. He is confused and rigid, and has high fever. There are tachycardia and fluctuating blood pressure. The patient's sister says that his GP prescribed him a medication 1 week ago. You consider neuroleptic malignant syndrome (NMS). All of the following can result in NMS, *except?*

a) Haloperidol
b) Risperidone
c) Amitriptyline
d) Withdrawal of long-term L-dopa therapy
e) Dantrolene

39) Because of severe hyperthermia, a 32-year-old man is brought to the Emergency Department. He takes daily fluoxetine for major depression. Yesterday, he ingested 20 capsules of fluoxetine. Which one of the following is the *correct* statement about serotonin syndrome?
a) Can occur with therapeutic doses of certain medications
b) The planter reflexes should be flexors in severe cases
c) Severe constipation can result in subacute intestinal obstruction
d) Hypokinesia causes recurrent falls
e) Dantrolene is the drug of choice to treat this syndrome

40) A 60-year-old man presents with fluctuating consciousness and left-sided weakness. Brain CT scan reveals subacute subdural hematoma. Which one of the following statements with respect to subacute subdural hematoma is *correct*?
a) The source of bleeding is arterial
b) History of head trauma must be present
c) Bilateral hematomas are usually fatal
d) Epileptic patients are at risk of developing this type of hematoma
e) All hematomas should be evacuated

41) The psychiatry department consults you about this 65-year-old retired writer who looks confused. However, you suggest Alzheimer's disease. All of the following suggest acute confusional state rather than dementia, *except*?
a) Fluctuated level of consciousness
b) Rapid development
c) Visual hallucination
d) The presence of fever
e) Gradual loss of short-term memory over time

42) A 16-year-old schoolgirl is brought by her patents to see you because of severe global wasting. The patient denies any weight loss and she thinks that she is chubby. The peripheries are cold and lanugo hair is found on her back. Regarding anorexia nervosa, all of the following statements are correct, *except*?
a) The blood levels of prolactin, cholesterol, and glucagon are raised
b) Bradycardia and hypotension are found
c) Prominent sexual and physical retardation occur

d) Amenorrhea is usually present for more than 3 months
e) Usually starts around adolescence

43) A 26-year-old female has unexplained hypokalemic alkalosis. You find severe dental caries and erosions. Her roommate states that the patient induces vomiting several times a day, after having a good meal. Which one of the following is *correct* regarding bulimia nervosa?
a) Weight loss is prominent
b) Usually starts around puberty
c) Hospital admission is required to treat it
d) Loss of self-control and binge eating behavior is present
e) Grossly disorganized perception of body image and weight

44) A 32-year-old man presents with auditory hallucinations. He says that he hears unknown voices talking about him and commenting on his daily actions. He has poor work records and marital strife. The patient denies substance abuse. His father has a psychiatric disease affecting his personality and social life, as he said. You are thinking of schizophrenia. Which one of the following is *not* considered a good prognostic factor in schizophrenia?
a) Acute onset of schizophrenia
b) Good job records
c) Presence of clear-cut precipitating event
d) Normal pre-morbid personality
e) Absence of affective symptoms

45) A 37-year-old man presents with low mood. He thinks that he is hopeless and helpless, and he says that there is no reason to live for. His wife died 1 year ago because of breast cancer. He has some weight gain, but no interest in sex. Every day, he awakes at 4 AM. He denies alcohol or drug abuse. Regarding major depression, all of the following statements are correct, *except?*
a) The presence of low self-esteem is prominent
b) Some patients lose weight
c) There is loss of sense of pleasure
d) The presence of irritability suggests an alternative diagnosis
e) Some patients have early insomnia

46) A 38-year-old schizophrenic woman takes long-term haloperidol. She complains of nipple discharge and weight gain. You tell her that these are side effects of her medication. Regarding antipsychotic medications, all of the following statements are correct, *except?*

a) The newer generation, such as olanzapine, is more effective at the negative symptoms of schizophrenia
b) In general, no drug is superior in efficacy over the others
c) Extra-pyramidal side effects are uncommon in antipsychotics which have prominent anti-cholinergic effects
d) May be used in the control of acute mania
e) Some may be used in the control of nocturnal enuresis

47) A 43-year-old male receives daily imipramine for severe endogenous depression. He complains of dry mouth and some difficulty urination. He also has some blurring of vision which interferes with reading. Regarding tricyclic antidepressants, all of the following statements are correct, *except?*
a) Mainly inhibit the reuptake of noradrenalin and serotonin at receptor sites in the brain
b) Should be used with caution in elderly people
c) May result in postural hypotension
d) Their clinical effect in terms of depression improvement is usually seen after 2-4 weeks of starting therapy
e) First-line agents in severe depression with attempted suicide

48) A 43-year-old vagrant was found unconscious in the street. He was then found to have Wernicke's encephalopathy, and he was treated accordingly. After recovery, he admits to drinking alcohol heavily, every day. Alcohol abuse is suspected in all but one of the following?
a) Unexplained atrial fibrillation
b) Unexplained hypertension
c) Unexplained diarrhea
d) Unexplained gout in a pre-menopausal female
e) Unexplained maculopapular skin rash

49) A 33-year-old female is brought to A&E with a failed attempt to suicide after stressful conversation with her boyfriend. She is 4 months pregnant and the boyfriend wants to end their relationship. The risk of success after an attempted suicide is increased in all of the following, except:
a) Being a young female rather than a male
b) Living alone
c) Presence of an underlying malignant disease
d) Using a gun rather than self-poisoning
e) Being divorced rather than happily married

50) A 43-year-old male, who has severe endogenous depression, will receive electroconvulsive therapy (ECT) as a rapid mode of effective treatment. Which one of the following is *not* an indication for using ECT in depression?
   a) Severe depression with failed suicidal attempt using a gun
   b) Severe depression interfering with eating and resulting in nutritional problems
   c) Mildly depressed patient intolerant to oral tricyclics
   d) Severe depression with prominent psychotic symptoms
   e) Agitated depression

51) A 32-year-old man was referred to the psychiatry outpatients' clinic. He has hypochondriasis. Hypochondriasis occurs in all of the following, *except*?
   a) Separation anxiety
   b) Panic disorder
   c) Obsessive-compulsive disorder
   d) Major depression
   e) Acute amnesic disorders

52) A 25-year-old woman visits the physician's office. She is anxious and tells the physician that her general practitioner told to her that she has somatization disorder. She insists that her symptoms are real and she is not mad. You explain to her what the term somatization means. All of the following about somatization disorders are incorrect, *except*?
   a) In most patients, the duration of complaints covers 1-2 months
   b) The patient usually has many undiagnosed diseases
   c) Most symptoms start after the age of 30 years
   d) The majority of patients have a multitude of negative investigations at the time of diagnosis
   e) This disorder is a subtype of conversion disorders

53) A 67-year-old man presents with visual impairment. Which one of the following is not a common cause of visual impairment in elderly people?
   a) Glaucoma
   b) Age-related macular degeneration
   c) Presbyopia
   d) Cataract
   e) Retinal melanoma

54) A 66-year-old man was found to have age-related macular degeneration after developing visual impairment. Which one of the following is the *incorrect* statement regarding age-related macular degeneration?

a) There is degeneration of the central portion of the retina
b) Smoking increases the risk of this disease
c) Patients with the wet type usually complains of acute distortion of vision, usually in one eye
d) The dry type results in gradual loss of vision in one or both eyes
e) Ranibizumab is contraindicated in the wet variety

55) A 31-year-old man was referred to the neurology outpatients' clinic for further assessment of optic nerve head swelling. However, you find optic nerve head drüsen and you tell him he has pseudo-papilledema. Regarding optic nerve head drüsen, which one is the *correct* statement?
a) Drüsen is calcified hemorrhagic spots
b) Drüsen tends to be large in children and gradually become buried in adults
c) About 90% are unilateral
d) Usually result in prominent visual impairment
e) Disc drüsen appears as a lumpy mass with refractile bodies

56) A 61-year-old woman develops progressive loss of vision in both eyes. His final diagnosis turns out to be primary open-angle glaucoma (POAG). Which one of the following is the *correct* statement regarding this type of glaucoma?
a) The incidence of POAG decreases with age
b) Optic cup that is >25% of the vertical disc diameter points to POAG
c) The angle of the anterior eye chamber is distorted
d) Confrontational field testing is not useful in the screening of glaucoma
e) Measurement of intra-ocular pressure is the gold standard investigation

57) You've just diagnosed central retinal artery occlusion (CRAO) in this 71-year-old man, who has developed sudden visual loss. Which one of the following with respect to CRAO is *correct*?
a) Carotid artery atherosclerosis is a rare cause
b) Most patients present with acute but mild loss of vision in both eyes
c) Visual acuity can be normal in 15% of patients
d) Spontaneous clinical improvement is very common
e) Unlike central retinal vein occlusion, CRAO is not associated with late retinal neovascularization response

58) A 60-year-old man is due to undergo cataract surgery. He asks about the possible post-operative complications this procedure may incur.

All of the following are well-recognized post-operative complications of this form of eye surgery, *except?*
   a)   Bullous keratopathy
   b)   Retinal detachment
   c)   Cystoid macular edema
   d)   Posterior capsular opacification
   e)   Acute pre-septal cellulitis

59) A 37-year-old woman presents with left-sided peri-ciliary flush and painful visual blurring. Your final diagnosis is acute uveitis. Which one of the following uveitis syndromes is *not* primarily confined to the eyes?
   a)   Birdshot chorioretinopathy
   b)   Sympathetic ophthalmia
   c)   Fuchs heterochromic iridocyclitis
   d)   Immune recovery (reconstitution) uveitis
   e)   Vogt-Koyanagi-Harada syndrome

60) A 41-year-old woman presents with acute red eye. You intern suggests episcleritis, after he has examined her. Which one of the following statements is *correct* with respect to episcleritis?
   a)   Mostly afflicts males
   b)   Could be cystic or atrophic in type
   c)   Associated with inflammatory reaction in the underlying uveal tract
   d)   Most cases are not associated with systemic diseases
   e)   The vast majority of cases resolve within 3 months of proper treatment

This page was intentionally left blank

Neurology, Psychiatry, and Ophthalmology

Answers

1) c.

The sensitivity of non-contrast brain CT scan is about 90% in the first 24 hours of acute subarachnoid hemorrhage, a figure that drops down to 40% after few days. Therefore, it is very helpful in the diagnosis if patients presents acutely; it is convenient to be done in confused and critically ill patients, unlike MRI scans. The CSF appears white (hyper-intense) on the MRI T_2-weighted images, while the T_1-weighted and FLAIR images produce a black (hypo-intense) CSF signal. Diffusion-weighted MRI measures the relative flow of blood though the brain tissues; this is done either by injection a contrast medium or using an endogenous technique (where the patient's own blood is used as a contrast material). This would detect cerebral blood flow abnormalities and demonstrate early reperfusion of tissue after doing revascularization techniques. It can pick up areas of very early ischemia and can differentiate reversible from non - reversible ischemic damage. One of the tools that are used to assess epileptic patients for brain surgery is PET; it can for example detect hypo-metabolic area(s) in the temporal lobe in medically-refractory temporal lobe epilepsy which may be suitable for resective surgery. PET is useful to differentiate between various dementia and degenerative brain diseases (such as Alzheimer's disease, progressive supranuclear palsy,
Huntington's disease,…etc.). Differentiating brain radiation-induced necrosis from tumor recurrence can be achieved through the use of PET. The contrast material of conventional angiography (renal, coronary, or cerebral) carries a risk of contrast nephropathy.

2) e.

The risks of MRI scanning in patients with cardiac pacemakers (or other implanted electronic devices) are related to possible movement of the device, magnetically-induced programming changes, electromagnetic interference, and induced currents in lead wires leading to heating and/or cardiac stimulation. It is currently considered inadvisable for patients with pacemakers or other intracardiac wires to undergo MR imaging. Nerve stimulators, cochlear implants, and other implanted electronic devices also may be affected by MRI and are considered unsafe. Any ferromagnetic object within the body represents a potential hazard when exposed to the large magnetic field of an MR imaging system. The hazard primarily reflects the possibility of deflecting the foreign body sufficiently to injure vital structures; ocular metallic foreign bodies and the older intracranial aneurysm clips have this risk.

It appears to be safe to perform an MRI at any time after placement of coronary artery stents of any type, and it is also safe to scan most prosthetic cardiac valves because, at most, they experience only a mild torque (the only exception is the old generation of Starr-Edwards valves).

3) c.

Any form of uncorrected coagulopathy represents a contraindication for doing lumbar puncture (LP). For example, clotting factors deficiency and platelets count $<20000/mm^3$ should be corrected before doing lumbar puncture. Local infections at the site of needle insertion may result in extension of the infection to the meninges; in such cases, cervical or cisternal puncture should be done.

Any intracranial mass lesion (brain tumors, abscess, large granuloma,...etc.) precludes this form of the investigation; fatal trans -tentorial herniation can ensue rapidly. The presence of focal or lateralizing signs in acute pyogenic meningitis calls for doing brain CT scan first to exclude mass lesions (e.g., brain abscess that has burst out and resulted in acute meningitis). Acute subarachnoid hemorrhage rapidly elevates the intracranial pressure; this elevation is diffuse and is equal in all compartments of the brain (unlike the case in brain masses, where the pressure unequally distributed between compartments); LP would not result in sudden decompression of one compartment or consequent herniation. Actually, stem "c" calls for LP to see the grossly hemorrhagic CSF (and to look for xanthochromia, 8 -12 hours after the ictus). LP should be done in patients with suspected spinal masses only if myelography is contemplated.

4) e.

Some neurological disorders result in characteristic but non-specific EEG abnormalities. Multiple sclerosis (MS) is not an epileptic disorder; it is a disease of white matter and therefore EEG is not helpful in suggesting the diagnosis of MS. The presence of repetitive slow-wave complexes in one or both temporal lobes may indicate Herpes simplex encephalitis. A patient with rapidly progressive dementia and periodic EEG complexes may have subacute scelrosing panencephalitis or Creutzfeldt-Jacob disease. The finding of 3-Hz per second spike-and-wave EEG discharges in a child with frequent staring is highly suggestive of petit mal epilepsy.

5) e.

EMG is a helpful tool in the diagnosis of inflammatory myositis, in the appropriate clinical setting. Many changes can be seen; none is diagnostic per se, as they are encountered in many muscle diseases (metabolic, toxic,…etc.). On the other hand, EMG testing is completely normal in 11% of polymyositis/dermatomyositis cases. The first 4 stems reflect increased muscle fiber membrane instability (because of the inflammatory process). Low amplitude, short duration, and polyphasic motor unit potentials are the myopatic EMG changes; stem "e" refers to neurogenic muscle disease with chronic denervation/reinnervation.

6) b.

The skin area of the medial malleolus is the L4 dermatome, while the lateral malleolus area represents the 1st sacral dermatome. Big toe dorsiflexion is the 5th lumbar myotome; foot inversion is the 4th lumbar myotome while eversion of the foot is the 1st sacral myotome. The 5th lumbar segment of the cord has no deep tendon reflex representation in the lower limb. Dissociated sensory loss means loss of pin and temperature sensation while join position and vibration senses are intact. Pyramidal weakness of the upper limbs is translated into weakness of anti-gravity muscles, and that is weakness of should abduction and elbow/wrist extension. Lower limbs' pyramidal weakness indicates weakness of hip/knee flexion and foot dorsiflexion/eversion. The weakness pattern in peripheral neuropathy is distal, i.e., hand and foot weaknesses. Proximal weakness (of the shoulder/pelvic girdle) is consistent with muscle disease, i.e., myopathy.

7) d.

Ischemic stroke comprises 80-85% of all strokes; thrombotic stroke (40%) is more common than embolic (20%) strokes. Lacunar strokes constitutes about 20% of all strokes; the lacunar area volume is <1 cm^3. One of the inclusion criteria of receiving r-tPA infusion is that the patient should present within a 3 to 6-hour window period. The best treatment of asymptomatic carotid artery stenosis is anti- platelets (e.g., aspirin). The leading causes of death after stroke are pneumonia, pulmonary embolism, and coronary artery disease.

8) d.

Wernicke's aphasia is a fluent type of aphasia that results from damage to the upper posterior part of the dominant temporal lobe. Naming and repetition are impaired. Comprehension is severely impaired; this is the core feature of this type of aphasia. Those patients have neologism and paraphasic errors; they invent words and sounds as they go along and string word together in non-grammatical meaningless fashion.

9) c.

Lamotrigine can be given for women taking oral contraceptive pills (there is no risk of contraceptive failure) and during pregnancy (as it poses no fetal risks). Complex partial seizures last for 1- 2 minutes. Petit mal seizures usually starts between the ages of 4-9 years. The EEG recording is normal in 50% of cases of generalized epilepsy, interictally. Sudden withdrawal of anti-epileptics still ranks first on the list of causes of status epilepticus.

10) e.

The idiopathic variety usually starts in middle-aged women. The maxillary area is the commonest one to be involved. Examination is otherwise unremarkable apart from triggering zones of pain; brain imaging is also normal. Occurrence in young age, bilateral/alternating involvement, presence of focal/lateralizing signs should cast a strong doubt on the "idiopathic" variety and should prompt you search for an underlying disease in the posterior fossa. The idiopathic one responds favorably to carbamazepine; if this fails, phenytoin is a useful alternative.

11) e.

The majority of patients with migraine headache present before the age of 20 years. Most patients are females and a family history of the same disease is usually positive. Migraine without aura is responsible for about 70% of migraines. The migraine headache can be precipitated by emotional stress, sleep deprivation/excessive sleep, fasting, undue physical exertion, menstruation, changes in barometric pressure and environmental changes, and diet rich in chocolate. Triptans are abortive therapies; they should never been given as prophylactic treatments; propranolol, sodium valproate, amitriptyline, and verapamil are migraine prophylactic therapies.

12) e.

Cluster headache is a relatively uncommon form of headache; the prevalence is less than 1%. Men are affected more commonly than women (in contrast to migraine headache), with a peak age of onset of 25 to 50 years. The pain of cluster headache is strictly unilateral and begins quickly without warning and reaches maximal intensity within a few minutes. The headache is usually deep, excruciating, continuous, and explosive in quality; occasionally, it may be pulsatile and throbbing (i.e., migraine-like). Most attacks occur at night. Eye redness and excessive lacrimation together with nasal stuffiness and Horner syndrome are the usual accompaniments. Although alcohol precipitates these headaches, this does not mean that the patient is alcoholic. Attacks of headache can be aborted by high flow, high concentration nasal oxygen, intranasal capsaicin or lidocaine, ergotamine, and triptans.

13) e.

Apart from unilateral or bilateral abducens palsy and bilateral papilledema (or secondary optic atrophy), the neurological examination should be unremarkable in idiopathic pseudo-tumor cerebri. The presence of other cranial nerve palsies or focal signs should prompt a search for an alternative diagnosis. Brain imaging is normal apart from slit-like ventricles (because of diffuse brain edema). The CSF parameters should be normal; the only abnormal finding is the elevated CSF opening pressure. It can be treated with prednisolone, acetazolamide, and frusemide. Stem "a" is the proposed mechanism behind the developing normal pressure hydrocephalus.

14) d.

Acoustic neuroma is an 8th nerve schwannoma, which arises from the cochlear branch of vestibulocochlear nerve; the term vestibular schwannoma is a misnomer, therefore. Progressive hearing loss is an early feature. Facial weakness occurs as an advanced feature or as an early sign in patients who had undergone surgical resection of the tumor. The tumor expands the internal auditory canal and gradually fills in the cerebellopontine angle. The best clinical screening test is the loss of corneal reflex while brainstem auditory evoked potential is the best para-clinical screening tool. Bilateral tumors are the hallmark of neurofibromatosis type II. The 3 causes of very high CSF protein are acoustic neuroma, tuberculous meningitis, and Guillain-Barre syndrome.

15) a.

The intermittent obstruction of the CSF absorption at the level of arachnoid granulation is thought to be the cause of this type of hydrocephalus. The CSF pressure is normal and is below 18 cm water. Most cases are idiopathic; however, it may occur as a secondary phenomenon to a prior head trauma, meningitis, or subarachnoid hemorrhage. The optic discs are normal. Clinical improvement after aspirating 30-50 ml of the CSF is thought to predict a favorable response to planned shunting surgery. Normal pressure hydrocephalus results in a subcortical type of dementia.

16) d.

Human-human transmission of the prion protein of Creutzfeldt-Jacob disease is very rare and mainly results from corneal transplantation. The disease is sporadic; however, about 10% of patients have a familial predisposition. The spongiform changes in the brain may be patch; therefore, multiple brain biopsies are required for histopathological diagnosis and detection of mutant form of PrP^{Sc}. Cerebellar involvement is common. The disease is usually fatal within 1 year of symptoms. In the appropriate clinical setting, the presence of periodic complexes on EEG is highly suggestive of the diagnosis; they are not diagnostic, however. These changes are usually transient and appear in 65% of patients.

17) d.

Autosomal dominant mutations in pre-senilin 1 and 2 (on chromosomes 14 and 1, respectively) are responsible for about 10% of cases of Alzheimer's disease (early-onset). Alzheimer's disease is the commonest cause of dementia; this is followed by vascular dementia and Lewy-body dementia. It is a cortical type of dementia; cortical signs are common. Early loss of recent memory with relative preservation of remote memory and immediate recall is the cardinal feature; as the disease progressive all 3 components of memory become impaired. Donepezil (and other central acetylcholinesterase inhibitors) is prescribed for mild dementia. Most patients die after an average of 8 years from inanition; advanced cases are deaf-mute and bed-ridden.

18) c.

Lewy-body dementia patients have a fluctuating cognitive dysfunction, Parkinsonism, and formed visual hallucinations.

The loss of recent memory is not that prominent early in the course of the disease, unlike Alzheimer's disease. Those patients are very sensitive to anti-Parkinsonian side effects of conventional neuroleptics; chlorpromazine should be avoided at all possibility.

19) a.

The primary progressive form of multiple sclerosis usually presents as spastic paraparesis in middle-aged men; compressive myelopathy is a differential diagnosis. Clinically isolated syndromes of demyelination (e.g., an isolated attack of optic neuritis but abnormal brain MRI) should be given long-term interferon beta therapy (similar to the relapsing-remitting form). The optic neuritis of multiple sclerosis is usually of the retro-bulbar type; papillitis is uncommon and usually occurs in children. The peri- ventricular plaques of multiple sclerosis are observed in 90- 95% of patients; therefore, brain MRI scanning is negative in 5-10% of cases. During disease relapse, the CSF protein is raised (usually between 60- 80 mg/dl; it does not exceed 100 mg/dl) and CSF mononuclear pleocytosis is mildly elevated (usually <50 cells/mm^3); the presence of a single neutrophilic cell in the CSF strongly questions the diagnosis of multiple sclerosis.

20) e.

Thiamine deficiency causes Wernicke's encephalopathy; this vitamin deficiency results from malnutrition (alcoholism, hyperemesis gravidarum, cancer and its treatment, and famine). Clouded consciousness usually takes the form of agitated confusion with impaired registration and immediate recall; progression to coma is rare. The ataxia is primarily of gait; limb ataxia is uncommon. Nystagmus and double vision are additional features. The disease is not epileptogenic. The presence of unilateral or bilateral facial weakness is not part of Wernicke's encephalopathy.

21) c.

Rupture of cerebral Berry aneurysm is responsible for most cases of acute non-traumatic subarachnoid hemorrhage; rupture of cerebral arteriovenous malformation ranks next. Berry aneurysm is found in 2-5% of the adult population and these aneurysms are multiple in 20% of cases; most of them actually don't rupture. They are more common in the anterior circulation; about 20% are in the posterior circulation.

Focal signs after rupture are uncommon, and may suggest intracranial extension of the blood jet or that the rupture has caused by intra-cerebral arteriovenous malformation. Mycotic aneurysms are mainly found in the distal segments of middle cerebral artery and are usually multiple. Re-bleeding and focal cerebral vasospasm usually occur 3-12 days after the ictus. Vasospasm may be treated by the triple -H therapy (hemodilution, induced hypertension, and hypervolemia); resistant cases may respond to direct intra-arterial infusion of papaverine or nicardipine, or intrathecal nitroprusside. Angioplasty can also be successful.

22) e.

Transient global amnesia is a syndrome characterized by the acute onset of severe anterograde amnesia, which is accompanied by variable degree of retrograde amnesia, but without other cognitive or focal neurologic impairment. The attack of memory loss usually resolves spontaneously within 24 hours; there is no specific treatment. Most patients are middle-aged or older adults with a prior history of ischemic stroke (usually in the posterior circulation). Episodes are usually not recurrent; rare patients have infrequent attacks that recur over several years. Hysterical amnesia, occult head trauma, and complex partial seizure disorder are the usual differential diagnoses.

23) e.

Patients at risk of developing intracranial meningioma are those who have prior cranial irradiation and history of breast cancer. Meningiomas are multiple in neurofibromatosis type II. These tumors arise from the arachnoid meningothelial cells. Meningioma takes the contrast homogenously and intensely. Most meningiomas show a characteristic marginal dural thickening that tapers peripherally; the dural tail sign. Secondary reactions in the adjacent bone (reactive sclerosis, invasion, or erosion) are uncommon with cerebral convexity meningiomas; however, these occur in 50% of skull base meningiomas. About 15% of meningiomas, show atypical features; they display cystic changes, focal areas of necrosis, and hemorrhage. Calcification sometimes occurs.

24) e.

Certain behaviors were thought to be rare in pseudo-seizures, but have been recently found to be relatively common, and these are: incontinence (usually urinary one); self -injury (surprisingly common in non-epileptic events; burn injuries, however, may be confined to patients with true seizures); and tongue biting (patients with non-epileptic seizures are more likely to bite the tip of the tongue, rather than its side, or the inside of the cheek or lip as occurs with true seizure).

25) d.

Herpes encephalitis is the most common cause of fatal sporadic (not epidemic) encephalitis in the Western World. It is typically caused by type I virus; Herpes simplex type II produces more global form of encephalitis with profound neurologic impairment. Brain MRI is superior to CT scanning in showing the temporal lobe abnormalities; these lesions are predominantly unilateral and may have associated mass effect. The CSF is totally unremarkable in 2% of cases; otherwise, it typically shows raised opening pressure, mononuclear pleocytosis, and elevated protein. The CSF could be hemorrhagic (increased number of RBCs is found in 84% of cases). Low CSF glucose is actually uncommon, and if profound, an alternative diagnosis should be searched for.
Although the viral genome can be detected in the CSF by PCR, the virus itself cannot be isolated form the CSF; it is usually isolated from brain biopsy specimens. Prominent intermittent high-amplitude slow waves (and sometimes continuous periodic lateralized epileptiform discharges) in the affected region can be detected by doing EEG in at least 80% of victims.

26) e.

Most patients with tuberculous meningitis presents with subacute onset of fever, headache, nausea and vomiting, and clouded consciousness; however, some patients have acute meningitic syndrome, suggesting pyogenic meningitis, or presents very slowly over months (or even years) as slowly progressive dementia illness. Focal signs and cranial nerve palsies are common. Around half of patients have military tuberculosis; fundoscopy may reveal choroidal tubercles. Tuberculin skin testing is usually positive; a negative test, however, does not refute the diagnosis. Chest X-rays usually show evidences of tuberculosis, ranging from focal lesions to subtle miliary patterns.

27) d.

Hypertension is the commonest cause of spontaneous non -traumatic ICH. Congophilic angiopathy is an independent risk factor (even in the presence of other risk factors) for the development of lobar hemorrhage. Hypertensive hemorrhages occur in the territory of penetrator arteries that branch off major intra-cerebral arteries, often at 90° angles with the parent vessel.

Hypocholesterolemia is associated with increased risk of ICH, while hypercholesterolemia decreases this risk; hypocholesterolemia resulting from statin therapy, however, does not increase this risk. Thalamic hemorrhages may extend horizontally to compress the posterior limb of the internal capsule, downward to put pressure on the tectum of the midbrain, or may rupture into the third ventricle and decompress themselves. Cerebellar hemorrhage usually starts in the deep dentate nuclei and may enlarge to fill the cerebellar hemisphere, decompress into the 4th ventricle, and possibly involving the pontine tegmentum. About 40% of hematomas enlarge in the 1st 24 hours; this results in further neurological deterioration and increased mortality.

28) d.

This syndrome results from ischemic damage to a wedge- shaped area in the posterio-lateral part of medulla oblongata. The descending sympathetic fibers course there; their damage results in ipsilateral Horner syndrome. The ischemic damage of the central vestibular apparatus causes nystagmus, nausea, vomiting, diplopia, and oscillopsia. Nuclei of the 9th and 10th cranial nerves are also involved; dysphagia ensues. The ascending spinothalamic tracts and the trigeminal sensory tract are destroyed; this would result in contralateral loss of pain and temperature sensation in the upper and lower limbs and ipsilateral loss of the same sensory modalities in the face, respectively. The inferior cerebellar peduncle is involved; this would cause limb ataxia. The pyramidal tracts are anteriorly located; their damage is not part of this syndrome and, therefore, the planter reflexes are flexors.

29) d.

Guillain-Barre syndrome (GBS) usually develops 1-3 weeks following upper respiratory tract infection or a diarrheal illness; a preceding Campylobacter infection portends poor prognosis.

The other poor prognostic indicators are a rapid downhill course, need for assisted ventilation, and axonal degeneration on neurophysiological studies (which could be primary or secondary to severe wide-spread demyelination). Some degree of asymmetry in the weakness may be observed in 9% of cases at the time of diagnosis; marked asymmetry or persistence of this asymmetry strongly questions the diagnosis of GBS. The disease usually progresses over 1-2 weeks and then a plateau occurs with gradual improvement; a stepwise progression over 3 months is totally inconsistent with GBS. Many patients have dull aching pain in the thighs but their sensory examination is normal; prominent sensory deficits suggest an alternative diagnosis.

30) c.

Subacute combined degeneration of the cord is caused by vitamin B_{12} deficiency (not from folate deficiency). The resulting defect in myelin synthesis affects the posterior and lateral columns. This vitamin deficiency also produces axonal peripheral neuropathy (that affects the legs more than the arms) and cognitive dysfunction (ranging from vague personality changes to frank dementia). Some patients have normal MCV and hematocrit levels; therefore, their absence does not exclude the diagnosis. Degeneration of the pyramidal tracts results in spastic paraparesis while degeneration of the dorsal column tracts causes sensory ataxia and positive Rombergism. Sphincter disturbances indicate an advanced stage.

31) b.

Depressive pseudo-dementia should be differentiated from dementia; both, dementia and depression may coexist, however, making the distinction challenging. There is a plateau of cognitive dysfunction in pseudo-dementia and the neurological examination and lab studies should be normal. The cognitive dysfunction in dementia is insidious in onset and progressive. Dementia patients typically are unaware of their deficits and they deny memory problems (pseudo-dementia patients frequently complain of memory loss). Somatic complaints are uncommon in dementia and patients their mood and affect are variable rather than being frankly depressive; they show minor or no vegetative symptoms and their cognitive deficit worsens at night. Their neurological examination and lab studies may be abnormal.

32) c.

About 70% of patients have a new onset headache, which is usually located in the temple; a frontal or occipital location, however, may be noted. The headache may be mild or severe and is usually variable. Some patients have progressive headache while others may have resolution of headache before medical treatment is begun. Half of all patients suffer jaw claudication on eating; however, trismus-like symptoms in the jaw may occur in some. Although most patients have low-grade fever, high fever may develop and may wrongly suggest an infectious process. The visual loss could be partial or complete and starts in one eye, typically. The other eye can be involved within 1-2 weeks if no treatment is given. Complete visual loss in both eyes is rare, though. In about 3-15% of patients, the major branches of the aortic arch can be involved with gradual narrowing; if the subclavian and axillary artery is involved, arm claudication results. The development of aortic aneurysm is a late manifestation.

33) d.

The myasthenic weakness is typically fluctuating and the fatigue is painless. Symptoms in the eyes (diplopia and ptosis) and proximal weakness are the usual presenting features; less common modes of presentation are isolated distal weakness, isolated neck weakness, and isolated respiratory muscles weakness. About 5% of patients presents with proximal weakness alone. Thymoma is found in 10-15% of patients. False positive Tensilon® (edrophonium) test can be noted in motor neuron disease, compressive cranial neuropathies, and brainstem tumors. Seronegative myasthenia gravis refers to negative serological testing for anti-acetylcholine receptor (AChR) antibodies, and comprises about 15 -20% of all cases; anti-MuSK antibodies are detected in 50% of those patients and these antibodies are directed against the muscle specific receptor tyrosine kinase. MuSK is a receptor tyrosine kinase that mediates agrin-dependent AChR clustering and neuromuscular junction formation during development. Seronegative myasthenia gravis responds to plasma exchange and immune suppressive therapy.

34) c.

LEMS occurs in about 3% of small -cell lung cancer patients. The anti-P/Q type voltage-gated calcium channel anti -bodies play a central role in the pathophysiology of LEMS by acting on the pre-synaptic membrane of the motor end-plate.

The result is reduction in the pre-synaptic release of acetylcholine. About 95% of the functioning calcium channel receptors are of the P/Q type; this is the main immunologic target of LEMS. Unlike myasthenia gravis, respiratory failure is rare and is usually an advanced feature. Dysautonomia is very common; reduced salivation results in dryness of the mouth, which is the commonest autonomic failure symptom and may be the presenting feature in some. The characteristic incremental EMG response is noted after doing low frequency (2-3 Hz) repetitive nerve stimulation; the decremental response is seen in myasthenia gravis.

35) c.

Cardiomyopathy of Duchenne muscular dystrophy is a dilated type with conduction abnormalities and arrhythmia. Cardiac involvement occurs also in Becker muscular dystrophy, facioscapulohumeral muscular dystrophy, limb-girdle muscular dystrophy, and Emery-Dreifuss muscular dystrophy. Symptoms of Becker muscular dystrophy usually begin by the age of 15 years and gradually extend into adulthood (i.e., at the time when Duchenne patients are already wheelchair-bound); in the absence of positive family history, the distinction between Becker and limb-girdle muscular dystrophy can be very difficult. In facioscapulohumeral muscular dystrophy, the muscles around the shoulder are atrophied with relative preservation of the deltoids; the distal musculatures are spared, but foot drop occurs in the scapuloperoneal variety. Facial weakness with bilateral ptosis is noted. A highly characteristic feature of limb-girdle muscular dystrophy is the preservation of facial and extra-ocular muscles. Contractures occur early in Emery -Dreifuss muscular dystrophy and may be the presenting feature; it can be X-linked recessive or autosomal dominant.

36) a.

Kearns-Sayre syndrome is usually a sporadic disease and rarely inherited maternally; there is large (deletion of 9 to 50 base pairs) mutation in the mtDNA. This results in bilateral progressive external ophthalmoplegia with ptosis (because of the symmetrical onset and very slow progression, double vision is rare). Patients must present before the age of 20 years. There are retinitis pigmentosa (not old chroiditis), sensori- neural deafness, raised CSF protein, and heart block (which is responsible for 20% of deaths). Diabetes is the cardinal feature of all mitochondrial cytopathic diseases. Serum LDH is raised and serum lactic acid may be elevated. Short stature may occur. The treatment is symptomatic.

37) e.

Mutations in the dihydropyridine L-type *calcium* channels results in the hypokalemic variety of FPP, while the hyperkalemic variety is caused by mutations in the SCN4A *sodium* channels. The same mutation in the SCN4A gene can result in normokalemic periodic paralysis, paramyotonia congenita, and potassium aggravated myotonia. In between the attacks of the hypokalemic variety, progressive weakness of muscles occurs; the disease may also be associated with hyperthyroidism (especially in Asians).

The hyperkalemic type may be associated with myotonia of eyelid closure, chewing and swallowing, and hand gripping. Acetazolamide helps prevent attacks of flaccid weakness of the hypokalemic and hyperkalemic FPPs. Mutations in *potassium* channels may produce episodic ataxia-myokemia, long QT syndrome, and Anderson syndrome (a very rare type of periodic paralysis with ventricular dysrhythmia and dysmorphic facial features).

38) e.

Dantrolene and bromocriptine are used in the treatment of NMS; they reduce the duration of hyperthermia. NMS results from blockage of dopaminergic receptors in the nigrostriatal pathways in the brain; this would cause skeletal muscles excitation-contraction uncoupling and excessive heat production. There is no clear-cut genetic explanation/mutation of this syndrome. It occurs in <1% of patients receiving conventional neuroleptics; haloperidol is the commonest culprit. The syndrome typically develops in the 1st 30 days of starting therapy and it is subacute. NMS may be precipitated by phenothiazines, thioxanthenes, tricyclic antidepressants, MAO inhibitors, and atypical neuroleptics. Sudden withdrawal of amantadine and L-dopa may also precipitate NMS. The mortality rate is 10-20% and those who recover carry a 30% risk of recurrence.

39) a.

Serotonin syndrome results from increased serotonergic activity in the CNS; it may occur with therapeutic use of serotonergic agents, intentional poisoning, or inadvertent reactions between these medications. The presentation ranges from benign tendency to startle easily to death.

There are hyperthermia (in moderate-severe syndrome), excessive sweating, tachycardia, hypertension, hyperkinesia, ocular clonus, lower limb clonus and extensor planters, tremor, and hypertonia. The diagnosis of serotonin syndrome is clinical; there is no diagnostic lab test. The syndrome must be differentiated from neuroleptic malignant syndrome, anti-cholinergic intoxication, sympathomimetic poisoning, and CNS infections. The treatment is supportive and diazepam is useful to sedate patients. Antipyretics, dantrolene, and bromocriptine are not useful in the treatment.

40) d.

Collection of blood in the subdural space forms subdural hematoma, which could be acute (day 0 to 3), subacute (day 3 -21), or chronic (of more than 3 weeks duration) . The source of bleeding is venous. Patients who have atrophied brain (elderly, alcoholics) are at risk of developing subdural hematoma when they sustain head trauma. The other risk factors are epilepsy, warfarin therapy (and coagulopathy), and tendency for recurrent falls (such as Parkinson's patients). Bilateral subacute hematomas are found in $1/6^{th}$ of cases; because of the bilateral compression of the brain and the isodensity of subacute hematomas, the brain CT imaging appears "hyper-normal" in elderly patients. Small hematomas with very minor symptoms may resolve spontaneously, although may fluctuate in size. However, large hematoma or very symptomatic ones should be removed surgically. About 20% of elderly patients with these hematomas do not recall head trauma.

41) e.

Dementia patients are at risk of developing confusion from many causes, such as fever, infections, electrolyte disturbance, medications' side effect,…etc.; on the other hand, chronic subdural hematoma may result in fluctuated level of consciousness and may gradually end up with dementia.

42) c.

In anorexia nervosa, serum levels of TSH, FSH, and LH are reduced. The extremities are cold and lanugos hair usually covers the back. The presence of hallucination (or psychotic features) should cast a strong doubt on the diagnosis. The loss of weight must be more than 25% of the original weight, or the current weight is blow 25% of the normal one, for age and sex. Anorexia nervosa usually starts around puberty; features of bulimia usually start later on. Despite the presence of global wasting, physical retardation is not present.

43) d.

Unlike anorexia nervosa patients (who lose weight), bulimic patients usually maintain their weight. The disease typically starts in the post-pubertal period (but later than anorexia nervosa 's peak). In contrast to anorexia nervosa patients, inpatient management is rarely required to treat bulimic patients. A characteristic feature is the loss of self-control and binge eating behavior; this is followed by self-induced vomiting, diuretic or laxative abuse, or the patient may enter a phase of prolonged dieting after these recurrent binge eating episodes. Grossly disorganized perception of body image and weight suggests anorexia nervosa.

44) e.

The chronic insidious onset of schizophrenic features portends a bad outlook, as is the presence of poor job records. The absence of positive family history and the presence of catatonic variety have a good prognosis. Patients who have pre-morbid schizoid personality have a gloomy outcome. The absence of affective symptoms confers a bad prognosis.

45) d.

Depressed patients feel worthless, helpless, and hopeless, as they demonstrate low self-esteem. Weight gain or loss may occur as part of the disease vegetative symptoms. A characteristic feature is the loss of pleasure sensation (anhedonia). Although most patients have psychomotor slowing, some are agitated (so-called agitated depression). In severe depression, psychotic symptoms may be prominent and must be differentiated from schizophrenia with affective symptoms. Insomnia could be early (difficulty falling asleep) or late.

46) e.

The older (conventional) generation of antipsychotics (e.g., chlorpromazine) is more effective at the positive symptoms (hallucinations and delusions) of schizophrenia. Antipsychotics which produce weak anti-cholinergic side effects (e.g., trifluoperazine) confer more risk of extra-pyramidal features. Haloperidol can be used in the acute control of mania. Some tricyclics (e.g., imipramine), can be used to control enuresis in children.

47) e.

The main limitation of tricyclics is the cardio-toxicity in elderly people. They are contraindicated in prostatism and glaucoma; both are common age -related diseases. They may also cause postural hypotension; this increases the risk of falls. Stem "e" calls for immediate electro-convulsive therapy.

48) e.

Alcohol abuse should always be included in the list of causes of unexplained forms of cardiomyopathy, atrial fibrillation, diarrhea (and malabsorption, through chronic pancreatitis), gout in premenopausal women (interferes with the tubular uric acid excretion), and unexplained rib fractures. Alcohol does not cause a specific skin rash; repeated unexplained skin bruises may be a clue to repeated falls from alcohol intoxication.

49) a.

Although the risk of suicide is much more common in females, the success rate is higher in males, especially elderly men. Those who write a will or prepare a plan for their suicide are more likely to succeed with their attempt. The presence of malignancy and chronic illnesses increase the patient's chances of ending his/her life. Many females use self-poisoning as parasuicide to have others' attention. Divorced and separated people (from their spouses) are more prone to have their life ended vigorously.

50) e.

ECT is usually applied under general anesthesia, twice or thrice weekly, for a total of 3 weeks. Some patients may need maintenance ECT in the long- term. ECT has many contraindications: aortic aneurysm, recent fractures, recent myocardial infarction, unstable cervical spine, and brain tumors with raised intracranial pressure. The agitated form of depression does not call for this form of therapy.

51) e.

Hypochondriasis is a preoccupation with the fear of having a serious disease (not a symptom) based on a misattribution of bodily symptoms or normal functions. It may also be seen as part of generalized anxiety and separation disorders.

Hypochondriasis can be viewed as a disturbance of cognition, perception, or interpersonal relationship. Hypochondriacal patients may have anxious and insecure interpersonal attachments. Unlike somatization disorder, which is more common in females, hypochondriasis has an equal gender distribution. Hypochondriasis is not part of acute amnesic disorders.

52) d.

Somatization disorder has a chronic and fluctuating course over many years. Patients' symptoms start in early adult life (usually before the age of 30 years) and are more common in women. Patients complain of a variety of symptoms. All of the following are present at any time during the course of illness: 4 pain symptoms; 2 gastrointestinal tract symptoms; 1 sexual symptom; and 1 pseudo-neurologic symptom. The commonest complaints are pain, nausea and vomiting, headache, dizziness, menstrual irregularities, and sexual dysfunction. By the time the patient is referred to a psychiatrist, there is usually a multitude of negative investigations and unhelpful operations (usually in the form of hysterectomy and cholecystectomy).

53) e.

Visual impairment in elderly people has a major adverse impact on the patient's quality of life. The commonest causes are presbyopia, cataract, age-related macular degeneration, primary open-angle glaucoma, and diabetic retinopathy. Ocular tumors are rare causes of impaired vision in elderly individuals.

54) e.

Age-related macular degeneration (ARMD) results from a multi -factorial degenerative process involving the macular portion of the retina. As a result, central vision is lost; central vision is required to do activities such as driving, reading, watching television, and performing activities of daily living. ARMD is of 2 types; the dry (atrophic) and wet (exudative/neovascular) ones. Smoking increases the relative risk of the dry and wet types (2-4 folds when compared with non-smokers), and seems to increase the risk of progression of the disease from early to advanced stages. The dry type patients usually complains of progressive visual impairment in both eyes, while patients with the wet variety reports acute distortion of visual images (mainly of central lines) or acute central visual loss in one eye, although the disease is present in both eyes. Vascular endothelial growth factor (VEGF) is a potent mitogen and vascular permeability factor that plays a pivotal role in neovascularization.

Ranibizumab is recombinant humanized monoclonal antibody with specificity for VEGF. This medication is given as monthly intra-vitreous injections of 0.5 mg to treat the wet ARMD.

55) e.

Drüsens are remnants of calcific axonal degeneration and found in 2% of the general population. An autosomal dominant inheritance is suggested. They are usually not that obvious in children (as they are buried within the head of the optic nerve) and gradually become exposed in adults. They are bilateral in 85% of patients. Although they are asymptomatic, careful examination usually reveals abnormalities in the form of enlargement of the physiological blind spot, some sort of field constriction, and inferior nasal defects. An afferent pupillary defect may be present if the condition is unilateral or asymmetric. Orbital CT scanning and ultrasonographic examination can secure the diagnosis of drüsen by showing calcium deposits in the optic nerve head that are not visible on visual examination. Disc drüsen appears (on fundoscopy) as a lumpy mass with refractile bodies within.

56) d.

The incidence of POAG increases with age and the prevalence of the disease is 4 times more common in African Americans than other races. Systemic hypertension, diabetes mellitus, hypothyroidism, and myopia are possible risk factors for POAG. Topical ocular and systemic glucocorticoids also increase the risk of glaucoma development; nasal steroids raise the intraocular pressure in those who already have glaucoma. The angle of the anterior chamber is open and is normal-looking. Measuring the intra-ocular pressure (IOP) alone is not a satisfactory tool to diagnose POAG; about 30-50% of patients with established POAG and severe field defects have an IOP <21 mmHg and on the other hand, many normal adults with IOP >21 mmHg have no visual felid defects. Using the examiner's fingers in the bed-side confrontation visual field testing is not useful in the detection of glaucoma; this form of testing lack sensitivity and specificity and is "operator-dependent". Ideally, POAG should be diagnosed before the occurrence of any significant visual loss.

57) c.

Carotid artery atherosclerosis disease is the commonest cause of central retinal artery occlusion (CRAO). The usual presentation is acute severe loss of vision which is painless.

However, 15% of patients demonstrate a normal visual acuity because they have a cilio-retinal artery feeding the retina, thereby protecting this critical area from sudden ischemia. In most patients, restoration of the patency of the central artery usually occurs spontaneously within hours and sometimes days; however, improvement in vision is extremely rare.

Experimentally, the retina can survive and resists ischemia for only 90 minutes after surgically ligating its feeding artery. A neovascular response may occur 2-3 months following this event if the retina survives under hypoxic conditions; it may involve the retina, iris, and anterior chamber. Central retinal artery occlusion is a rare event and its subtype, branch retinal artery occlusion, is even much less common. Victims of CRAO have increased risks of cardiovascular and cerebrovascular diseases as well as shortened life expectancy.

58) e.

Acute bacterial endophthalmitis occurs in 0.13% of patients who do cataract surgery, and results in marked reduction in vision. Acute pre-septal cellulitis is not a complication of cataract surgery. Opacification of the posterior aspect of the capsule develops in about 20% of patients and can be treated by YAG LASER capsulotomy. Around 0.7% of cases develop retinal detachment which calls for immediate intervention; highly myopic eyes are especially susceptible for this complication, which results in marked loss of vision. Clinically apparent cystoids macular edema is found in 1.5% of patients. Bullous form of keratopathy is seen in 0.3% of surgeries. Malposition or dislocation of the implanted intra-ocular lens is a complication in 1% of patients.

59) e.

Vogt-Koyanagi-Harada syndrome is the second leading cause of uveitis after Behçet disease in Japan. In this form of bilateral posterior uveitis, fluid gradually accumulates beneath the retina leading to elevation and eventual detachment of the retina. The extra-ocular manifestations are vitiligo, recurrent aseptic meningitis, and alopecia. Immune recovery (reconstitution) uveitis occurs after the use of highly active anti-retroviral therapy in AIDS patients which results in recovery of CD4 positive count; this would enhance the immune responses and lead to immune reconstitution inflammatory syndrome. Sympathetic ophthalmitis (ophthalmia) is an autoimmune reaction that leads to ocular inflammation following penetrating trauma to the contralateral eye.

60) d.

Episcleritis must be differentiated from scleritis, conjunctivitis, and keratitis. It has 2 forms; diffuse and nodular ones. Episcleritis usually presents as an acute onset of eye redness, irritation, and excessive lacrimation. Pain is highly unusual; however, chronic or nodular cases may be painful. Vision is never affected. Most recover completely with 3 weeks, with or without treatment. The vast majority of cases occur as an isolated disease without being associated with a systemic illness. Bilateral occurrence is observed in 50% of patients. About 70% of patients are females. Scleritis is painful and impairs vision and is more likely to be associated with a systemic disease. Some cases are atypical and cannot be designated as episcleritis or scleritis; in such cases, the application of topical phenylepherine results in rapid but transient disappearance of the episcleral redness permitting a better examination of the underlying sclera.

Chapter 10
Nephrology

Questions

1) The kidney is one of the body organs that secretes many important hormones and participates in acid-base balance maintenance, water homeostasis, red cell production, and blood pressure control. Regarding the normal kidney, all of the following statements are correct, *except*?
   a) Erythropoietin is secreted by peri-tubular cells in response to hypoxia
   b) Hydroxylates 1-hydroxycholecalciferol to its active form
   c) Renin is secreted from the juxta-glomerular apparatus
   d) Locally produced prostaglandins have a very important role in maintaining renal perfusion
   e) About 90% of erythropoietin comes from the kidneys and the rest is from the liver

2) You are reviewing the anatomy of the kidney and its overall structure in order to give a lecture during you annual nephrology association meeting. With respect to normal adult kidney, all of the following statements are correct, *except*?
   a) Kidney length is 11-14 cm
   b) Both kidneys rise and descend several centimeters during respiration
   c) Each kidney contains approximately 100 million nephrons
   d) Both kidneys receive 20-25% of the cardiac output
   e) The right kidney is usually few centimeters lower than the left kidney

3) A 54-year-old man complains of passing large amount of clear urine every day, for the past few weeks. He denies chronic illnesses or abusing diuretics. Polyuria may be result from all of the following, *except*:?
   a) Excessive fluid intake
   b) Hyperglycemia
   c) Early stage chronic renal failure
   d) Tubulo-interstitial diseases
   e) Heavy smoking

4) A 16-year-old female presents with severe right-sided flank pain that radiates to her right labia majora. Abdominal ultrasound discloses the presence of multiple bilateral renal stones. Regarding renal ultrasound examination, all of the following statements are correct, *except*?
   a) The disadvantage is that it is highly operator-dependent
   b) Quick, cheap, and non-invasive, and often the only required method of renal imaging
   c) It can show the renal size, position, dilatation of the collecting system and other abdominal pathologies
   d) In chronic renal failure, the density of the renal cortex is decreased

e) By utilizing the Doppler techniques, more information can be gained, such as the resistivity index

5) A 52-year-old man visits the physician's office. He has been experiencing bilateral flank heaviness and low back pain over the past 2 months. He denies abdominal trauma, bloody urine, or using medications. Intravenous urography reveals bilateral hydronephrosis, non-dilated ureters, and normal urinary bladder. The ureters seem to be very close to each other near the midline. All of the following statements are correct with respect to intravenous urography (IVU), *except*?
a) Risky in diabetes mellitus
b) Can have adverse effects in multiple myeloma
c) Patients with pre-existent renal disease may develop complications
d) The risk of contrast nephropathy can be reduced by avoiding dehydration and giving diuretics beforehand
e) The risk of contrast nephropathy can be reduced by using less hyperosmolar contrast media

6) A 44-year-old woman is referred to your office for further evaluation of her stag-horn calculi. She has renal impairment, in addition. She declines doing IVU because she thinks that this investigation will result in many fearful complications. Which one of the following is *not* a disadvantage of intravenous urography?
a) Time-consuming investigation
b) Needs an injection
c) Dependence upon adequate renal function for good images to be obtained
d) There is a risk of exposure to contrast media, such allergic reactions and nephrotoxicity
e) Poor definition of the collecting system on the anterio-posterior films

7) A 67-year-old man, who was reasonably well and healthy, visits the emergency Room. After taking a thorough history and conducting a proper examination, you consider acute obstructive uropathy. Abdominal ultrasound reveals bilateral hydronephrosis but the level of obstruction cannot be located. You are about to perform anterograde pyelography. Regarding this mode of imaging, which one of the following statements is *correct*?
a) It is the injection of a contrast media into the kidney through the bladder and ureters
b) Usually done blindly
c) Much more difficult and hazardous in a non-obstructed kidney

d) Usually used in cases of chronic glomerulonephritis
e) Poorly outlines the collecting system

8) Your junior house officer asks you about the clinical applications of micturition cystourethrogram and its complications. With respect to micturition cystourethrogram, which one of the following is the correct statement?
   a) Not used in the diagnosis and assessment of vesicoureteric reflux severity
   b) Usually used in conjunction with urodynamic studies
   c) Part of the last stage of intravenous urography
   d) Not indicated in patients with recurrent urinary tract infections
   e) Not indicated in patients with renal scars

9) A 7-year-old boy, who has recurrent urinary tract infections, is referred to you for further management. His renal ultrasound reveals bilaterally small and scarred kidneys. You consider another imaging study for the kidneys. What would you choose?
   a) Abdominal CT scan
   b) Kidney MRI
   c) Micturition cystourethrography
   d) Anterograde pyelography
   e) Plain abdominal films

10) A 44-year-old woman develops severe hematuria. Earlier that morning, she underwent renal biopsy procedure in order to diagnose the cause of unexplained proteinuria. She is due to undergo renal artery angiography as part of her management plan. Regarding renal angiography and venography, all of the following statements are correct, *except?*
   a) The main indication of renal angiography is the diagnosis of renal artery stenosis and renal hemorrhage
   b) Therapeutic intervention may be undertaken at the same time of doing renal angiography
   c) Unlike intravenous urography (IVU), there is a risk of cholesterol atheroembolization
   d) When compared to IVU, the risk of contrast nephropathy is much lower
   e) Renal venography is mainly used in the diagnosis of renal vein thrombosis and renal cell carcinoma extension

11) A 13-year-old child developed periorbital puffiness, leg edema, and frothy urine over the past 1 month. 24-hour urinary protein is 6.1 gram and is nonselective. Blood urea is 70 mg/dl and serum creatinine is 2.1 mg/dl. The blood pressure is 150/95 mmHg. You consider doing renal biopsy before starting medical treatment with immune suppressants for his nephrotic syndrome. Which one of the following is *not* an indications for renal biopsy?

a) Unexplained acute renal failure
b) Chronic renal failure with normal sized kidneys
c) Atypical childhood nephrotic syndromes
d) Isolated hematuria with normal looking urinary RBCs
e) Nephrotic syndrome in adults

12) A 16-year-old girl has steroid-resistant nephrotic syndrome and she is about to undergo renal biopsy. After a proper assessment, you think that there is a contraindication for doing this step. Which one of the following is *not* a contraindications for renal biopsy?

a) Severe hemophilia
b) Platelets count of $10000/mm^3$
c) Uncontrolled hypertension
d) Renal size less than 90% predicted
e) Single kidney

13) A 65-year-old man, who was diagnosed with benign prostatic hyperplasia 6 months ago, passes deep dark urine today. The urine is full of red blood cells. He denies burning sensation upon micturition. Urinary discoloration may be encountered in all of the following, *except?*

a) All cases of porphyria
b) Intervertebral discs calcification with dark ears
c) Parkinson's disease patient
d) Tuberculosis patient
e) Patient who sustained massive crushing trauma

14) A 22-year-old male is referred to you for further investigations after his GP has found excess protein in his urine. Regarding proteinuria, all of following statements are correct, *except?*

a) Standard urinary dip-sticks usually miss Bence-Jones protein
b) In myeloma, it is due to protein overflow rather than renal amyloidosis
c) The majority of the daily excreted protein is Tamm-Horsfall mucoprotein
d) Defined as albumin/creatinine ratio on a random urine sample of <3.5 in females and <2.5 in males

e)   Positive dipstick testing for urinary protein may occur in fever

15) A 23-year-old police officer sustained a bullet injury in the right thigh with subsequent femoral fracture and severe shock. Blood urea is 125 mg/dl, serum creatinine is 4 mg/dl, and serum potassium is 6.1 mEq/L. Regarding acute renal failure, which one is the *correct* statement?
a)   Pre-renal causes are uncommon
b)   About 85% of intrinsic renal causes of acute renal failure are due to acute tubular necrosis
c)   Under-perfusion causes of acute renal failure are usually irreversible
d)   Stones as a cause of acute obstructive uropathy are very common clinically
e)   Around 50% of post-renal acute renal failure cases are due to acute glomerulonephritis

16) A 43-year-old man is in the way of recovery from ischemic acute tubular necrosis. He has polyuria, hyponatremia, hypokalemia. With respect to the prognosis of acute renal failure, all of the following statements are correct, *except?*
a)   In uncomplicated renal failure, the mortality is low
b)   Serious infections complicating acute renal failure portend bad prognosis
c)   Multiple end-organ failure portends a gloomy outcome
d)   Complicated acute renal failure may have a mortality approaching 5-10%
e)   The outcome and prognosis is determined by the severity of the underlying disease and by complications rather than by the renal failure per se

17) A 54-year-old heavy alcoholic man developed acute renal failure after prolonged unconsciousness. Severe myoglobinuia ensued. Today, he demonstrates rapid respiratory rate. Rapid respiratory rate in acute renal failure may be due to all of the following, *except?*
a)   Acidosis
b)   Intravenous fluid overload and pulmonary edema
c)   Adult respiratory distress syndrome
d)   Chest infection
e)   Hyperkalemia

18) A 16-year-old girl developed post-streptococcal glomerulonephritis 3 weeks after tonsillitis.

Hemoglobin is 8.3 g/dl. Anemia in the setting of acute renal failure is very common and is usually multi-factorial. All of the following are causes of anemia in acute renal failure, *except?*

a) Hemolysis
b) Excessive bleeding
c) Profound suppression of erythropoiesis
d) Drug-induced
e) Hyperphosphatemia

19) General urine examination is one of the commonest investigations which are done in every-day clinical practice. With respect to this investigation, choose the wrong statement?

a) Elevated urinary concentration of ascorbic acid gives false negative result for bilirubin dipsticks
b) Elevated urinary concentration of ascorbic acid yields false negative result for glucose dipsticks
c) Gross hematuria can result in false positive results for protein
d) Significant glycosuria can falsely lower the specific gravity
e) MESNA produces false positive results for ketone sticks

20) Urine dipstick tests are commonly used in the medical ward side labs by nurses and interns. Dipsticks for nitrites can be used to detect urinary tract infections as a rapid diagnostic tool. All of the following are correct regarding this test, *except?*

a) False negative results for nitrite may be due to short bladder transit time
b) False negative results for nitrite may result from infecting organisms lacking nitrates and nitrate reductase
c) High urinary level of tetracycline produces false negative results for leukocyte esterase
d) High urinary level ascorbic acid gives false negative results for nitrite
e) Medications that discolor urine will cause false negative results for nitrite

21) Urinary specific gravity is measured in some clinical conditions, such as diabetes insipidus, but many physicians don't know how it is measured. It is measured by using all of the following methods, *except?*

a) Freezing point depression
b) Vapor pressure technique
c) Using a refractometer
d) Using a hydrometer

e) Calorimetric reagent strips

22) A 46-year-old man, who has long-standing history of type II diabetes and hypertension, has been recently found to have microalbuminuria. He is afraid that he will develop chronic renal failure and he is desperate for help. Regarding microalbuminuria, all of the following statements are true, *except?*
a) Defined as proteinuria of 30-300 mg/day
b) Defined as proteinuria of 20-200 µg/minute
c) Always protein dipstick-negative
d) Important in the follow-up of type II, but not type I, diabetes mellitus
e) Persistent proteinuria has been associated with the development of atherosclerosis

23) A 22-year-old man was displayed dipstick-positive proteinuria on routine pre-employment investigations. He denies any illness, fever, recent heavy exercise, or ingesting drugs. Regarding the daily excretion of urinary protein, all of the following statement are correct, *except?*
a) Up to 150 mg/day is normal
b) About 300-500 mg/day is expected to be dipstick test positive
c) More than 3.5 g/day is termed nephrotic-range proteinuria
d) If it is >2.5 g/day, a glomerular source is more likely than a tubular source
e) Between 0.5-2 g/day indicates a glomerular source of urinary loss of protein

24) A 54-year-old female developed severe hemorrhage after a gun shot. She was shocked and renal developed thereafter. All of the following are consistent with pre-renal failure, *except?*
a) History of excessive upper GIT bleeding
b) Bland urinary sediment
c) Progressive rise in blood urea and creatinine in congestive heart failure
d) Urine osmolality >500 mOsm/Kg
e) Fractional sodium excretion >3

25) A 64-year-old man presents with anorexia, nausea, and vomiting. He reports lassitude, decreased libido, and bone pains. You find hyperkalemia, hypocalcemia, and raised blood urea and serum creatinine. Abdominal ultrasound reveals bilaterally small kidneys. Regarding chronic renal failure, all of the following statements are correct, *except?*
a) The commonest causes worldwide are hypertension and diabetes mellitus

b) The presence of urea frost is a useful early clue
c) Itching is multi-factorial rather than due to hyperphosphatemia only
d) Hypotension and dehydration may occur
e) In clinical practice, about 4-18% of cases are of unknown or uncertain etiology

26) A 54-year-old man developed diabetic nephropathy and end-stage renal disease. Today, he visits the emergency Room. After a trivial fall, fractured his left femur. DEXA scan shows severe osteoporotic changes. With respect to endocrine abnormalities in chronic renal failure, all of the following statements are correct, *except?*
a) Hyperprolactinemia may occur but it does not respond to bromocriptine
b) The half-life of insulin is much shortened
c) Amenorrhea is common in females
d) Loss of libido in both sexes is very common
e) Grossly, patients' faces resemble hypothyroid faces

27) A 32-year-old man is referred to you for further management of his end-stage renal disease. You educate him about the various treatment options, including their pros and cons. Regarding the management of chronic renal failure, all of the following are true, *except?*
a) Hypertriglyceridemia is common and hypercholesterolemia is almost universal in those who have significant proteinuria
b) ACE inhibitors for hypertension have significantly been shown to retard the disease progression, especially in diabetics
c) Profound protein restriction is unwise, as this may produce malnutrition
d) Replacing sodium and chloride with high fluid intake should be avoided in all patients
e) Hypocalcemia is very common and should be corrected by vitamin D metabolites

28) A 37-year-old man was recently diagnosed with chronic renal failure of unknown cause. He asked you about the long term outlook of his disease. Regarding the prognosis of chronic renal failure, all of the following statements are true, *except?*
a) The commonest cause of death, in general, is vascular events
b) The 5-year survival of home hemodialysis patients is about 80%
c) The 5-year survival following renal transplantation approaches 80%

d) The 5-year survival of patients on regular hospital hemodialysis may reach 60%

e) The 5-year survival of continuous ambulatory peritoneal dialysis patients does not exceed 15%

29) A 35-year-old woman has developed acute renal failure and is due to receive renal replacement therapy in the form of hemodialysis using a temporary vascular shunt. All of the following are the treatment targets in this patient, *except?*

a) Maintain pre-dialysis blood urea concentration of <15 mmol/L
b) Adequate control of potassium
c) Adequate control of phosphate
d) Achieving normal extra-cellular fluid volume status
e) Each session of hemodialysis should be done every day in all cases

30) A 36-year-old female presents with resistant hypertension. She denies medication non-compliance. Apart from her anti-hypertensive medications, she takes daily tonics. You hear a bruit, about 2 cm above and lateral to the umbilicus, on the right side. Renal Doppler studies are suggestive of renal artery stenosis. Regarding renal artery stenosis, all of the following are true, *except?*

a) The commonest cause, in general, is atheromatous narrowing
b) It should be suspected when the blood pressure is severe, of rapid onset, or difficult to control
c) Fibromuscular dysplasia, as a cause, is more common in the young age group
d) Fibromuscular dysplasia does not usually cause complete occlusion and usually stabilizes once the patient stops growing
e) Surgical treatment is superior to medical treatment and angioplasty

31) A 16-year-old boy was referred to your office as a suspected case of Alport syndrome. His older brother has chronic renal failure. Regarding Alport syndrome, all of the following are true, *except?*

a) The second commonest inherited cause of renal disease
b) Usually autosomal recessive
c) Bilateral anterior lenticonus is the usual eye manifestation
d) Sensori-neural deafness, usually to high tones, is part of the syndrome
e) The pathological hallmark is progressive degeneration of the glomerular basement membrane

32) A 32-year-old man presents with heaviness in both flanks and infrequent dark urine. He reports passing small stones every now and then.

He has a sallow face with deep breathing. There are raised blood urea and creatinine serum levels. Abdominal ultrasound reveals enlarged kidneys full of cysts of various sizes. Regarding adult polycystic kidney disease, all of the following are correct, *except?*

a) About 85% of cases are due to mutations in PKD1 gene on chromosome 16
b) Mitral and/or aortic regurgitations are frequent but rarely severe
c) Around 30% of patients have associated hepatic cysts, but disturbances in hepatic function is rare
d) About 90% of patients develop subarachnoid hemorrhage
e) Colonic diverticulae and abdominal wall hernias are well-recognized associations

33) A female patient is being evaluated for vascular shunting surgery as part of her management plan of adult polycystic kidney disease. With respect to adult polycystic kidney disease, choose the *incorrect* statement?

a) The mean age of those who are heterozygous for PKD1 mutation to start dialysis is 57 years
b) About 50% of patients never need chronic dialysis
c) To screen patients' relatives, renal ultrasound is less reliable in the 10-18 years age group
d) Urinary tract infections should be treated aggressively
e) All patients develop hypertension at some point of their illness

34) A 57-year-old retired clerk received regular hemodialysis for end-stage renal disease, which was resulted from medullary sponge kidney. Regarding medullary sponge kidney, which one of the following is the *wrong* statement?

a) A sporadic disease, not an inherited one
b) Has a characteristic picture on intravenous urography
c) The cysts are confined to the proximal tubules
d) The prognosis is generally good
e) Nephrocalcinosis may be seen on the abdominal X-ray film

35) After developing polyuria and dehydration, you diagnose Fanconi syndrome in this young man. His tests reveal proximal renal tubular acidosis (RTA). Regarding Fanconi syndrome (renal tubular acidosis type II, proximal), which one of the following is *not* true?

a) Glycosuria is detected with normal blood sugar level
b) Aminoaciduria does not result in malnutrition
c) May be caused by hereditary fructose intolerance
d) Hypercalciuria is one of the core features of the syndrome

e) Large amounts of oral bicarbonate are need in the treatment

36) A 51-year-old female, who has primary Sjögren syndrome, is referred to you for further management of distal renal tubular acidosis. With respect to this form of renal tubular defect, choose the *wrong* statement?
a) May cause osteomalacia in adults, while children may develop rickets
b) Nephrocalcinosis is common
c) Hypokalemia is present besides normal anion-gap metabolic acidosis
d) Incomplete forms never occur
e) Inability to form very acidic urine in the context of systemic acidosis is the hallmark of the disease

37) A 45-year-old man presents with rapidly evolving renal impairment. There are raised blood urea and serum creatinine, hyperkalemia, and low serum C_3 and C_4 levels. Which one of the following does *not* result in hypocomplementemia in inflammatory nephritides?
a) Infective endocarditis
b) Systemic lupus erythematosus
c) Shunt nephritis
d) Post-infectious glomerulonephritis
e) Microscopic polyangiitis

38) Although chronic renal failure and end-stage renal disease are largely irreversible processes, but there are many "reversible factors" that may accelerate the course of this disease. All of the following are potentially reversible causes of worsening uremia, *except?*
a) Nephrotoxic medications
b) Renal artery stenosis
c) Hypotension due to drug therapy
d) Any infection
e) Normal blood pressure

39) A 32-year-old female developed reduced urinary output, leg edema, peri-orbital puffiness, pallor, and lassitude over the past week. Blood urea is 180 mg/dl and serum creatinine is 4.3 mg/dl. Rapidly progressive glomerulonephritis (RPGN) may result from all of the following, *except?*
a) Systemic lupus erythematosus
b) IgA nephropathy
c) Goodpasture syndrome
d) Post-infectious glomerulonephritis
e) Membranous nephropathy

40) A 45-year-old male developed shortness of breath, hemoptysis, and reduced dark urine output during the past 10 days. Serum ANA, ANCAs, and rheumatoid factor were negative. You found azotemia, active urinary sediment, and slightly enlarged kidneys. You considered Goodpasture syndrome. With respect to Goodpasture syndrome, which one is the *incorrect* statement?

a) Autoimmune disease against beta 1 chain of type III collagen
b) Linear IgG deposition in the glomerular basement membrane is observed on immune-fluorescence staining of renal biopsy specimens
c) Plasmapheresis may be used in the treatment
d) Lung hemorrhage is more common in smokers
e) Usually produces rapidly progressive crescentic form of glomerulonephritis

41) A 26-year-old SLE patient presented with rapidly progressive glomerulonephritis over the past 6 days. You started intravenous methyl prednisolone pulse therapy and you considered renal biopsy. Regarding renal biopsy with immunofluorescence staining, all of the following diseases are correctly matched with their immune-fluorescence findings, *except?*

a) Minimal change disease-no immune deposits
b) Focal segmental glomerulosclerosis-nonspecific immune trapping in focal scars
c) Membranous nephropathy-granular subendothelial IgG deposits
d) IgA nephropathy-mesangial IgA deposition
e) Type II membranoproliferative glomerulonephritis-intramembranous dense deposits

42) A 31-year-old man presents with dark urine, 1 day after developing pharyngitis. The urinary sediment is active and there is hypertension. Renal biopsy shows mesangial IgA deposits. Regarding IgA nephropathy, all of the following factors indicate bad prognosis, *except?*

a) Male gender
b) Presence of hypertension
c) Presence of hematuria
d) Presence of renal impairment
e) Persistent proteinuria

43) A 44-year-old man presents with fatigue and repeated vomiting for 1 month. He has bilateral small kidneys and mild elevation of blood urea and serum creatinine levels. He denies any toxic or infectious exposure, and does not admit to using regular medications.

There is no family history of note. Renal biopsy is consistent with non-specific chronic tubulo-interstitial disease. Chronic interstitial nephritis may be caused by all of the following, *except*?

a) Chronic exposure to ochratoxin
b) Chronic exposure to aristolochic acid
c) Wilson disease
d) Hanta virus infection
e) Chronic ingestion of phenacetin

44) A 36-year-old Korean man presents with acute renal impairment due to Hanta viral infection. The virus has resulted in acute interstitial nephritis. Regarding acute interstitial nephritis, all of the following statements are correct, *except*?

a) The commonest cause is drugs and medications
b) Blood eosinophilia occurs in 30% of cases
c) Should be suspected whenever patients develop non-oliguric acute renal failure
d) Predominant infiltration of the tubule-interstitial areas with eosinophils on renal biopsy is suggestive of a viral etiology
e) The majority of drug-induced acute interstitial nephritides recover following withdrawal of the offending drug

45) A 32-year-old sexually active female presents with urinary frequency and burning micturition over the past 2 days. Her urine is full of pus. Urine culture reveals *Enterococci*. This is the 3rd urinary tract infection (UTI) within the past 2 months. She asks if there is any way to prevent these infections. The following are prophylactic measures against recurrent UTI, *except*?

a) Fluid intake of at least 2 liters per day
b) Regular emptying of the urinary bladder
c) Local application of an antiseptic to the periurethral area before intercourse
d) Urinary bladder emptying before and after intercourse
e) Double micturition should be avoided in patients with reflux nephropathy

46) A 31-year-old male, who has developed recurrent urinary tract infections and kidney calculi, presents today with severe right-sided flank pain, radiating to the right scrotum. He has nausea and vomiting. After a proper assessment, you consider surgical interference to remove the stone. Indications for surgical intervention in renal calculi are all of the following, *except*?

a) If the patient becomes anuric
b) The presence of infection upstream
c) Large stone that is unlikely to pass outside spontaneously
d) Total obstruction of the pelvi-ureteric junction
e) The presence of a radiolucent stone

47) A 32-year-old male, who is a recurrent renal stone former, presents with severe attack of pyelonephritis. His kidney stones were removed surgically twice over the past 2 years. Blood and urine biochemical investigations have failed to detect any cause for these recurrent stone formation. Which one of the following is *not* a risk factor for renal stone formation?
a) Hypercalciuria
b) Hyperoxaluria
c) Hypercitraturia
d) Hyperuricosuria
e) Cystinuria

48) A 65-year-old heavy smoker male has been experiencing right flank pain and heaviness over the past 2 months. PCV is 61%. There is a palpable non-tender mass in the right flank. Abdominal ultrasound reveals 10 x 13 cm irregular complex mass in the right upper kidney pole. With respect to renal cell carcinoma, which one is the *incorrect* statement?
a) Hematuria is the commonest symptom
b) About 30% of patients present because of systemic metabolic effects of the tumor
c) Raised ESR is seen in 50% of cases
d) During surgical removal, the adrenal gland and local lymph nodes should be removed as well
e) Radiotherapy is very effective as a treatment modality

49) A 54-year-old female receives IL-II therapy for metastatic renal cell cancer. Her sister asks if the disease is inherited and whether the rest of the family may develop this form of cancer or not. All of the following statement are correct with respect to renal cell carcinoma, *except*?
a) More common in males
b) Adenocarcinoma is the commonest type
c) The tumor is vascular and may spread to the lungs and bones
d) The tumor may be multicentric and/or bilateral in some patients
e) The tumor may enlarge after administering progestins

50) A 45-year-old female complains of passing dark urine in addition to flank pain, 3 days after starting penicillin for sore throat. She denies reduction in the overall urine volume. You find eosinophilia and eosinophiluria. You suspect penicillin-induced allergic acute interstitial nephritis. Regarding drug and toxin-induced renal disease, all of the following associations are correct, *except?*

a) NSAIDS-minimal change nephropathy
b) Cyclosporine-chronic interstitial nephritis
c) Lithium-nephrogenic diabetes insipidus
d) Cisplatin-renal loss of sodium
e) Acyclovir-crystal formation inside tubules

51) Glomerulonephritides have many clinical, immunological, and pathological associations, besides displaying different treatment options and prognosis. With respect to glomerulopathies, all of the following associations are correct, *except?*

a) Minimal change disease is associated with HLA DR7, atopy, and certain medications
b) Membranous nephropathy is associated with HLA DR3, certain medications, and heavy metal poisoning
c) Association with liver disease has been documented in IgA nephropathy
d) Membrano-proliferative glomerulonephritis type I is associated with C_3-nephritic factor and partial lipodystrophy
e) Focal segmental glomerulosclerosis is associated with obesity, HIV infection, and heroin abuse

52) You are planning to give a lecture about body water physiology, distribution, and homeostasis in health and various diseases. Regarding body water, all of the following statements are true, *except?*

a) In a healthy 65 Kg male, it is about 40 liters in amount
b) About 70% of total body water is intracellular
c) Around 70% of extra-cellular water is in the interstitium
d) Water moves between different body compartments by an active process
e) Whole body extra-cellular water is about 12 liters

53) A medical student asks you about the pathophysiology of dehydration and its consequences, as well as its treatment. All of the following statements are correct, *except?*

a) The tonicity of plasma and interstitial fluids is determined by the concentration of sodium and chloride

b) The tonicity of intracellular fluid is determined by the concentration of potassium, magnesium, phosphate, and sulfate

c) The amount of hydrogen ion in the extra-cellular fluid is tiny, about 40 nmol/liter

d) Much of the extra-cellular hydrogen ions can be buffered by anionic proteins

e) The difference in the ionic composition of cells and interstitial fluid is important for normal cell function

54) A 42-year-old female was diagnosed with secondary Sjögren syndrome one year ago. Today she visits you for a scheduled follow-up. You find profound hypokalemia. 24-hour urinary potassium excretion is elevated. There is no gastrointestinal source of loss of this electrolyte. You are thinking of distal renal tubular acidosis as a cause of this potassium loss. Factors that increase potassium excretion in urine are all of the following, *except*?

a) Avid tubular sodium re-absorption

b) High urinary flow rates

c) Excess poorly absorbed anions, such as ketones and phosphates

d) A rise in intra-tubular potassium, as in alkalosis

e) A fall in intracellular potassium, such as in patients with acidosis

55) You are reviewing the physiology of nephrons and their role in electrolyte balance in health and disease states. Regarding proximal convoluted tubules in the healthy state, which one of the following is the *wrong* statement?

a) 90% of the filtered sodium is reabsorbed

b) 80-90% of the filtered potassium is reabsorbed

c) 90% of the filtered bicarbonate is reabsorbed

d) 99% of the filtered glucose is reabsorbed

e) 99% of the filtered amino acids are reabsorbed

56) A 54-year-old man asks you about the effect of ethanol on the urinary volume, because he states that he urinates a lot whenever he drinks. He thinks that his kidneys are diseased. You educate him in a simple way how the kidney handles water. Regarding the regulation of water excretion by the kidneys, all of the following statements are true, *except*?

a) In the presence of ADH, the collecting duct becomes more permeable to water

b) In the absence of ADH, the distal nephron is almost impermeable to water

c) About 95% of the filtered water is reabsorbed with an equivalent amount of sodium in the proximal tubule

d) ADH binds to V2-receptors in the distal nephron to enhance passive movement of water

e) In the thick ascending limb of loop of Henle, sodium and chloride are preferably absorbed without water

57) A 54-year-old man complains of leg swelling after starting some form of therapy. You tell him that certain medication may result in in this side effect because of water and sodium retention. Which one of the following does *not* result in sodium retention?
a) Corticosteroids
b) Licorice
c) Carbenoxolone
d) Estrogens
e) Ethacrynic acid

58) A 43-year-old male visits the physician's office for a scheduled follow-up. He has nephrotic syndrome. His leg edema is still prominent and Doppler studies are negative for deep venous thrombosis. He takes a daily diuretic and he denies non-compliance. You thinking of diuretic resistance. Which one of the following does *not* result in diuretic resistance?
a) Profound hypoproteinemia
b) Volume contraction
c) Reduced renal function
d) Secondary aldosteronism
e) Diuretic prescribed in high doses

59) A 32-year-old man develops severe travelers' diarrhea while visiting the Middle East. He has profound dehydration and hyponatremia. Causes of hyponatremia that is associated with low extra-cellular fluid volume are all of the following, *except?*
a. Salt-losing renal disease
b. Adrenal failure
c. Liver cirrhosis
d. Extensive burns
e. Cardiac failure

60) A 65-year-old man has small-cell lung cancer. Blood urea is 10 mg/dl and serum potassium is 3.0 mEq/L. You consider the syndrome of inappropriate secretion of ADH (SIADH) as the cause of these biochemical findings. SIADH may be result from all of the following, *except?*

a) Morphine
b) Cigarette smoking
c) Alcohol
d) Amitriptyline
e) Clofibrate

61) A 43-year-old epileptic man comes to see you as part of his regular follow-ups. Blood and urine studies are consistent with carbamazepine-associated syndrome of inappropriate secretion of ADH (SIADH). The following lab findings are consistent with SIADH, *except?*
a. Plasma osmolality of 260 mOsm/Kg
b. Serum sodium of 115 mmol/L
c. Urine osmolality of 100 mOsm/Kg
d. Blood urea of 2.5 mmol/L
e. Plasma potassium of 3.5 mmol/L

62) A 53-year-old man visits the physician's office for a check-up. He has high blood pressure, which is well-controlled using 2 anti-hypertensive medications. Serum potassium is 5.7 mEq/L but there is normal renal function. You suspect spironolactone as cause of this hyperkalemia. Hyperkalemia may result from all of the following, *except?*
    a. Digoxin toxicity
    b. Cyclosporine
    c. Heparin
    d. Beta agonists
    e. ACE inhibitors

63) A 55-year-old man, who has end-stage renal disease, has been found to have serum potassium of 6.4 mEq/L on a follow-up visit. He is on irregular hemodialysis program. The following statements are correct regarding the treatment of hyperkalemia, *except?*
    a) Bicarbonate infusion reduces serum potassium by 2 to 3.5 mEq/L
    b) Glucose and insulin infusion regimen reduces serum potassium by 0.6 to 1.2 mEq/L
    c) Calcium gluconate infusion does not reduce serum potassium
    d) Calcium resonium is not used in the treatment of hyperkalemia in the acute setting
    e) Beta agonist infusion may be additive or alternative to glucose and insulin regimen

64) A 43-year-old female undergoes hemodialysis because of uremia. Her serum phosphate has become low. Severe hypophosphatemia does *not* result in?
a) Elevated serum creatine phosphokinase
b) Respiratory muscle weakness
c) Intravascular hemolysis
d) Hypocalciuria
e) Cardiac dysrhythmia

65) You are reviewing the causes of hypophosphatemia with your interns while you are discussing the overall condition of one of your medical ward patient, who has this electrolyte disturbance. Hypophosphatemia does *not* result from?
a) Chronic alcoholism
b) Alcohol withdrawal
c) Peritoneal dialysis
d) Hemodialysis
e) Extra-cellular fluid contraction

66) Hypomagnesemia is usually overlooked in clinical practice. It has many causes and consequences. This low serum electrolyte is *not* caused by?
a) Gitelman syndrome
b) Post-obstructive diuresis
c) Acute pancreatitis
d) Protracted vomiting
e) Treatment with heparin

67) A 43-year-old man develops certain electrolytes abnormalities, which are consistent with normal anion-gap metabolic acidosis. Normal anion-gap metabolic acidosis can result from all of the following, *except?*
a) Medical treatment of glaucoma
b) After radical surgery of urinary bladder cancer
c) Ingestion of arginine hydrochloride
d) Renal tubular acidosis type IV
e) Advanced diabetic ketoacidosis

68) A 33-year-old man was brought to the Emergency Room. There is rapid and sighing breathing, which turns out to be due to high anion-gap metabolic acidosis. The following causes of high anion gap metabolic acidosis are correctly matched with their accumulating compounds, *except?*
a) Methanol poisoning-formic acid
b) Lactic acidosis-lactic acid

c) Ketoacidosis-acetoacetic acid and beta hydroxybutyrate
d) Ethylene glycol poisoning-formic acid
e) Chronic renal failure-phosphoric acid and sulfuric acid

69) A middle-aged man has developed lactic acidosis type A. Which one of the following is *not* responsible for this man's illness?
a) Septic shock
b) Severe anemia
c) Metformin
d) Cyanide poisoning
e) Respiratory failure

70) Because of respiratory alkalosis, a young child was referred to you. This type of alkalosis is *not* caused by?
a) Assisted ventilation
b) Salicylate poisoning
c) Hysterical over-breathing
d) Lobar pneumonia
e) Protracted vomiting

This page was intentionally left blank.

Nephrology

Answers

1) b.

Vitamin D is activated by a series of physiologic reactions. In the skin, the sun's ultraviolet rays convert 7-dehydroergosterol into vitamin D3. Vitamin D3 is hydroxylated in the liver to 25-hydroxycholecalciferol, and undergoes a second hydroxylation in the kidney to form 1,25-dihydroxycholecalciferol (calcitriol), which is the most active form of vitamin D. Calcitriol mediates the intestinal absorption of calcium, phosphorus, magnesium, and zinc. It also modulates bone mineralization and demineralization. In addition, vitamin D is important to maintain normal thyroid function. Vitamin D absorption may decrease with age. This kidney conversion process is impaired early in chronic renal failure; therefore, renal bone disease is almost always seen in established uremia.

2) c.

You should know the renal size when reading the ultrasound report. The size of the kidney could be normal (in health and in certain diseases, such as acute glomerulonephritis), increased (amyloidosis, diabetic nephropathy, and HIV nephropathy), or decreased (chronic reflux nephropathy and chronic glomerulonephritis). Both kidneys rise and descend several centimeters during respiration, but clinically, may not be that apparent. Each kidney has about 1 million nephrons. The right kidney is usually a few centimeters lower than the left, because of the site and size of the liver; this is important in reading the nephrogram phase of IVU.

3) e.

Smoking enhances the release of ADH (and this makes smokers urinate less). Polyuria is not a synonym to frequency. Frequency is the passage of frequent, yet small amounts of urine, but not more than 2 liters/day. Polyuria is the passage of >3-4 liters a day of a dilute urine. The other causes of polyuria are nephrogenic and cranial diabetes insipidus. One of the earliest abnormalities of chronic renal failure is polyuria and nocturia due to increased osmotic load per nephron and hyperactivity of the remaining normal nephrons.

4) d.

Renal ultrasound examination is highly operator- dependent and the other disadvantage is that the printed images convey only a fraction of the information gained by performing the investigation in real time.

However, it is rapid, cheap, and simple. Besides imaging the kidneys, it can show the rest of abdominal organs, such as liver, spleen, and pancreas. It differentiates renal cysts from solid masses. Chronic renal diseases usually impart an increased signal density with loss of corticomedullary differentiation. The resistivity index is the ratio of peak systolic and diastolic ratios, and is influenced by the resistance to flow through small intra-renal arteries and elevated in many intrinsic renal diseases, such as acute glomerulonephritis and renal transplant rejections. In addition, Doppler study may be used in the detection of renal vein thrombosis and renal artery stenosis.

5) d.

Diuretics should be stopped before doing contrast studies in at high risk patients. Contrast nephropathy is unfortunately a common iatrogenic mistake. Prior assessment of the risks with careful patient selection and appropriate application of certain precautions (such as good hydration, avoidance of diuretics, and stopping metformin in diabetics) is important in reducing the risk of contrast nephropathy.

6) e.

Intravenous urography provides an excellent definition and delineation of the collecting system and ureters. All patients should be assessed for the risk of contrast nephropathy.

7) c.

The contrast material is injected through the skin directly into the kidney(s) under ultrasound guidance. The procedure is dangerous in non-dilated kidneys. Anterograde urography is used to define the site of urinary tract obstruction. Excellent definition of the collecting system and ureters can be obtained.

8) b.

The main indication for using cystourethrography is vescicoureteric reflux, especially in children with recurrent urinary tract infections. It can also be useful in the assessment of urinary bladder emptying and urethral abnormalities. It is done by directly and retrogradely injecting the dye through a urethral catheter. The investigation is rarely used in patients with isolated renal scars.

9) c.

The overall clinical picture is suggestive of vesicoureteric reflux. Going next to micturition cystourethrogram would be the best next step.

10) d.

Renal angiography carries risks of contrast nephropathy in certain patients (long- standing diabetes, multiple myeloma,...etc.), of intra-arterial accessing (local hematoma or bleeding at the femoral entry site), and of intra-arterial manipulation (cholesterol atheroembolic disease). Therapeutic intervention may be undertaken at the same time of renal angiography, such as dilatation and stenting of renal artery stenosis and occluding an arteriovenous fistula.

11) d.

Isolated hematuria with *deformed* red blood cells (i.e., renal cause of hematuria) is an indication for renal biopsy. The other stems are true. Acute renal failure with an unexplained cause (e.g., absence of hemorrhagic shock or medication and toxin exposure as well as absence of urinary obstruction) calls for doing renal biopsy (which may show vasculitis for which immune suppressive therapy is given). In chronic renal failure, the kidney size is reduced; however, normal-sized kidneys necessitate renal histopathological examination. Atypical features of childhood nephrotic syndrome are the presence of hypertension, renal impairment, and active urinary sediment, as well as poor response to oral steroids; in such cases, kidney biopsy should be done. Nephrotic syndromes in adults are not commonly caused by minimal change nephropathy; more sinister causes are usually implied. All nephrotic syndromes in adults need renal histological examination.

12) d.

When the predicted renal size is <60% of the normal size, it is contraindicated to do renal biopsy. Transplant rejection is not a contraindication to perform renal biopsy (it may direct you what to do next). Biopsy from a single kidney is a *relative* contraindication and can be done safely by an experienced operator.

13) a.

Some porphyria types don't discolor urine. Alkaptonuria patients have inter-vertebral disc calcification and dark ears and urine.

L-dopa containing preparations (e.g., for Parkinson's disease) can discolor urine. Rifampicin imparts a dark red urine color. Massive crush trauma patients may develop myoglobinuria which results in red urine. The other causes of dark or discolored urine are beetroot, hemoglobinuria, hematuria, and medications (such as senna).

14) d.

The albumin/creatinine ratio (A/C ratio) of less than 2.5 in males and 3.5 in females is normal. Minor leaks of albumin in urine may occur after heavy exercise, fever, heart failure, exposure to extreme cold or heat weather, after general (especially abdominal) surgery, and extensive burns. Bence-Jones protein is positively charged while albumin is negatively charged; routine urinary dipsticks for protein detect negatively charged proteins. When the daily urinary loss of protein is high (say 4 g/day) but the albumin dipstick is negative, think of Bence-Jones proteinuria.

15) b.

Pre-renal etiologies rank first on the list of causes of acute renal impairment. Acute tubular necrosis (ischemic or toxic) is responsible for 85% of *intrinsic* causes of acute renal failure. Ischemic renal failure is usually reversible with proper management. Renal calculous disease and obstructive uropathy are uncommon causes of acute (post-) renal failure. About 5% of intrinsic acute renal failure cases result from acute glomerulonephritis and another 10% is caused by acute interstitial nephritis.

16) d.

Complicated acute renal failure may have a mortality approaching 50-70%, even at best centers. The other statements are correct.

17) e.

The respiratory rate in acute renal failure is rapid and deep and may result from acidosis, chest infection, pulmonary edema, adult respiratory distress syndrome, and lung aspiration (from impaired consciousness or following a seizure). Hyperkalemia can cause flaccid muscle weakness and paralysis (with resultant hypoventilation).

18) e.

Hypophosphatemia (severe and/or prolonged) may result in hemolysis and anemia, and is usually seen in patients with aggressive dialysis (whether hemodialysis or peritoneal dialysis). Hyperphosphatemia is more common than hypophosphatemia in renal failure, in general. Anemia is multi-factorial and should be addressed properly and managed accordingly.

19) d.

Dipstick analysis of urine is one of the commonest routine tests for inpatients and outpatients. These dipsticks look for protein, sugar, hemoglobin, bilirubin, infections,…etc. However, many factors interfere with the interpretation of these simple tests. Significant glycosuria and contrast media in urine can cause abnormally *high* specific gravity. The other stems are correct.

20) e.

When urinary tract infection is suspected clinically, a rapid and relatively cheap way of confirming this infection is the use of dipsticks for leukocyte esterase (which comes from WBCs lysis) and nitrite (a product of nitrate reductase on nitrate); yet many factors unfortunately affect the interpretation of these tests. Medications that discolor urine will give false positive results for nitrite.

21) e.

Calorimetric reagent strips are used for the detection of urinary proteins.

22) d.

Testing for urinary microalbuminuria is very important in the follow-up of both types of diabetes. Microalbuminuria is defined as persistent proteinuria of 30-300 mg/day or 20-200 μg/minute on 2 or more occasions (at least 6 months apart). ACE inhibitors have a very important role in the management, even in patients who have normal blood pressure. Microalbuminuria is a predictor for the future development of overt diabetic nephropathy and atherosclerosis. Neither the mechanism of the microalbuminuria nor an explanation for these associations has been found. Urinary protein of more than 300-500 mg/day is a frank proteinuria and is dipstick-positive.

23) e.

The finding of urinary protein loss of 0.5-2 g per day is considered as "source-equivocal," which may be glomerular or tubular; further investigations are required to solve this issue.

24) e.

Lab features suggestive of pre- renal failure are high specific gravity (usually >1.018), urine osmolality >500 mOsm/Kg (usually more than 600), urinary sodium <20 mEq/L (usually less than 10), fractional sodium excretion <1, and bland urine sediment (or clear hyaline casts). In established acute tubular necrosis, the urine specific gravity is <1.010, urinary osmolality <320 mOsm/Kg (usually around 280), urinary sodium >20 mEg/L, fractional sodium excretion >1, and the presence of muddy brown granular casts with tubular epithelial cells.

25) b.

Diabetes and hypertension are responsible for 65-70% of uremic cases. Although urea frost is a useful clue (pointing out towards a chronic process of renal impairment), but it is a relatively *late* sign of chronic renal impairment. Uremic patients commonly itch and the mechanism behind this wide-spread body pruritus is multi-factorial; raised PTH, high calcium-phosphate product, and unknown blood toxins are implicated. Hypokalemia and hyponatremia may occur due to diarrhea, vomiting, and salt-losing nephropathy. In up to 18% of chronic failure cases, no underlying cause can be identified.

26) b.

The half-life of insulin is prolonged in renal failure; the total daily doses of insulin should be reduced (otherwise, hypoglycemia can ensue easily).

27) d.

Patients who have tubulo-interstitial diseases, renal cystic disease, obstructive uropathy, and reflux nephropathy have what is called salt-losing nephropathy; water and salt loss may be profound and may further deteriorate the picture (fluids, sodium, and potassium should be replaced carefully).

28) e.

With the recent advances in renal failure management, the survival figures for dialysis and renal transplant patients are gradually becoming more and more optimistic. The 5 -year survival of continuous ambulatory peritoneal dialysis patients is around 50%.

29) e.

Hemodialysis may be done every day (usually in those with severe hypercatabolic state) or every other day, depending on the overall patient's condition and the hemodialysis center plans. There is no strict schedule to follow, generally speaking.

30) e.

There are three options in the treatment of renal artery stenosis:
1. Medical (with anti-hypertensive medications, low dose aspirin, and lipid lowering drugs).
2. Angioplasty (with/without stenting).
3. Surgical resection of the stenotic segment.
At present, there is no conclusive data to indicate the overall superiority of one approach over another. Each patient should be managed individually. MRA is now the *screening* method of choice for renal artery stenosis.

31) b.

Alport syndrome is usually X-linked recessive and most cases are due to abnormalities in the tissue specific type-IV collagen alpha 5 chain (which results from mutations in the gene *COL4A5* at Xq22).

32) d.

Although 15-40% of adult polycystic kidney disease patients harbor intracranial Berry aneurysms, only 10% of them develop subarachnoid hemorrhage, a figure that rises to 20% in patients who have a positive family history of ruptured intracranial aneurysm. MRA of the brain vascular tree is a good screening tool.

33) e.

Although hypertension is common (and should be treated aggressively), hypotension due to salt-losing nephropathy may occur and may call for fluid and salt replacement (some patients are dehydrated).

34) c.

The diagnosis of medullary sponge kidney can typically be done by IVU. The cystic dilatations lead to the appearance of a "brush" radiating outward from some or all of the calyces; enlargement of the pyramids and intra- ductal concretions are also often seen. Calcium stones, if present, are typically small and, almost pathognomonically, occur in clusters limited to the affected calyces. The cysts are usually confined to the papillary collecting ducts. The commonest causes of nephrocalcinosis are primary hyperparathyroidism, medullary sponge kidney, and renal tubular acidosis type I (not type II). Old healed tuberculosis and sarcoidosis are uncommon in clinical practice as a cause of nephrocalcinosis.

35) d.

Hyperphosphaturia occurs in Fanconi syndrome (and hypophosphatemia may follow). Hypercalciuria is a feature of type I (distal), not type II, renal tubular acidosis. The other causes of this syndrome are cystinosis and Wilson disease.

36) d.

Incomplete forms of renal tubular acidosis are well -documented, in which serum bicarbonate is normal but urine fails to acidify below pH of 5.3 upon ammonium chloride administration.

37) e.

The other causes of low serum complements are membranoproliferative glomerulonephritis (especially type II) and cryoglobulinemia. Systemic necrotizing vasculitides are pauci-immune and do not produce hypocomplementemia.

38) e.

The following factors should be looked for and removed (if possible), as these may accelerate the pace of chronic renal damage: urinary tract infection, urinary tract obstruction, nephrotoxic medications and drugs, any infection (a hypercatabolic state), hypertension, and reduced renal perfusion (due to drug-induced hypotension, renal artery stenosis, sodium and water depletion, and cardiac failure).

39) e.

There is progressive rise of blood urea and creatinine within few days-weeks. The kidneys are non- obstructed. The work-up of rapidly progressive glomerulonephritis must include serum antibodies (ANA, ANCAs, anti-GBM) and serum complements; renal biopsy should always be considered. Membranous nephropathy is not nephritic.

40) a.

Goodpasture syndrome is an autoimmune disease against alpha 3 chain of type IV collagen. It may present as rapidly progressive glomerulonephritis alone, with lung hemorrhages (dyspnea, cough,…etc.), or with both (i.e., a hepato-pulmonary syndrome). Serum IgG anti-GBM antibodies can be detected in the majority of patients.

41) c.

Membranous nephropathy shows granular sub-*epithelial* IgG deposition upon doing immune-fluoresce studies of renal specimens. Post-infectious and type I membranoproliferative glomerulonephritides display sub-endothelial deposits. Lupus nephritis kidney biopsies are always positive, showing various types of histopathological changes.

42) c.

Surprisingly, the absence of hematuria in IgA nephropathy portends a bad prognosis! The other stems are correct.

43) d.

Balkan nephropathy results from long-term exposure to ochratoxin, while aristolochic acid exposure causes Chinese herbs nephropathy. Wilson disease can produce chronic tubulo-interstitial disease, as well as proximal RTA and Fanconi syndrome. Hanta virus infection (as well as CMV and leptospirosis) is a cause of acute interstitial nephritis. Long-term ingestion of phenacetin has a risk of analgesic nephropathy.

44) d.

NSAIDS and antibiotics are the commonest causes of drug-induced (allergic) acute interstitial nephritis. Eosinophiluria can be detected in 70% of cases, while the peripheral blood eosinophil count is elevated in 30% of patients. Predominant infiltration of the tubule-interstitial areas with eosinophils on renal biopsy is suggestive of drug-induced etiology. The majorities are non-oliguric and simply respond to withdrawal of the offending agent; however, sometimes a tapering course of steroids is given to enhance recovery, although its effect is questionable.

45) e.

Double micturition is used in the management of reflux nephropathy (to lessen the amount of the refluxed urine and the incidence of urinary tract infection) and that is bladder emptying to be followed by another micturition 10-15 minutes later. It is usually applied at night.

46) e.

A radiolucent stone per se is not an indication for intervention. The presence of recurrent or intolerable pain is another indication to remove these stones.

47) c.

Hypercitraturia is protective against the development of renal stones, while hypocitraturia (e.g., in distal renal tubular acidosis) is a risk factor for renal calculi formation.

48) e.

Radiotherapy and chemotherapy are very weakly effective in the treatment of metastatic renal cell cancer. IL-II is used as a palliative measure in metastatic disease. Anemia (36%) is more common than polycythemia (4%), and hypercalcemia due to secretion of PTH-related peptide occurs in 5% of cases.

49) e.

Progesterone-like drugs may be used to slow the advancement of metastatic disease, although the impact on the overall prognosis is poor.

50) d.

Cisplatin induces loss of magnesium in urine through tubular dysfunction mechanism.

51) d.

Membrano-proliferative glomerulonephritis type II is associated with C_3-nephritic factor and partial lipodystrophy, while the type I disease is associated with hepatitis B infection, cryoglobulinemia (with/without hepatitis C infection), and bacterial infections. Goodpasture syndrome is associated with HLA DR15 (previously known as HLA DR2).

52) d.

Water moves between different compartments by a passive process. Plasma water is about 3 liters while the interstitial water (fluid) approximates 9 liters. The intracellular water is 27 liters. These amounts are approximate figures in a healthy 65 Kg male.

53) d.

Much of the extra-cellular hydrogen ions can be buffered by *cationic* proteins, such as albumin and hemoglobin.

54) e.

The first 4 stems increase the amount of potassium that is excreted in urine. This physiology should be kept in mind, which can explain the various causes of hypokalemia of renal tubular origin.

55) a.

About 65% of the filtered sodium is re-absorbed in the proximal tubule. Generalized dysfunction at the proximal tubular level explains the manifestations of Fanconi syndrome.

56) c.

About 65% of the filtered water is re-absorbed in the proximal tubule. The other stems are true.

57) e.

Ethacrynic acid is a loop diuretic. The other causes of salt retention are NSAIDS, calcium channel blockers, and vasodilators.

58) e.

Diuretic resistance is common in clinical practice. Profound hypoproteinemia would affect the diuretic serum protein-binding, while hypovolemia would affect the diuretic distribution. Some diuretics act after being excreted by the renal tubules into the intra-tubular fluid; therefore, renal impairment would impede this step. Aggressive diuresis results in secondary aldosteronism, and this would create a vicious circle. Thiazides are ineffective when the serum creatinine exceeds 1.5 to 2 mg/dl. Simply, giving a diuretic in a high dose does not mean it will not work, aggressive diuresis may ensue!

59) e.

The list of causes of hyponatremia with expanded (high) extra-cellular fluid (ECF) volume includes heart failure, renal failure, and SIADH (syndrome of inappropriate secretion of ADH). The following may result in hyponatremia with normal ECF volume: nephrotic syndrome, hypothyroidism, diuretics, and NSAIDS.

60) c.

Alcohol ingestion inhibits ADH secretion; therefore, urinary output rises considerably.

61) c.

Such urinary osmolality in stem "c" may indicate compulsive water drinking (causing water intoxication and dilute urine) or diabetes insipidus; urinary osmolality in SIADH is usually around 460 mOsm/Kg (i.e., is inappropriately high).

62) d.

Beta-blockers can result in hyperkalemia (beta-agonists are used in the management of hyperkalemia). The other causes of raised serum potassium include spironolactone, amiloride, and triamterene. Hypokalemia potentiates digoxin toxicity, while digoxin toxicity may result in a dangerous level of hyperkalemia.

63) a.

Bicarbonate infusion reduces serum potassium by 0.2-0.4 mEq/L; therefore, it is a weak agent alone against high serum potassium. Anyhow, it is used if severe acidosis co -exists (but we have to watch for circulatory overload). Calcium resonium is useful in the chronic setting to prevent hyperkalemia. Calcium gluconate infusion is a cardio-protective agent; it does not lower the serum potassium.

64) d.

Prolonged/severe hypophosphatemia can result in muscle fiber necrosis and even rhabdomyolysis; therefore, serum creatine phosphokinase rises. It can also produce skeletal muscle weakness; this may be observed as difficult weaning from ventilators in the ICU setting. Impaired red cell membrane ATPase causes red cell hemolysis (which is intravascular). It can affect other electrolytes handling and may produce hypercalciuria and hypermagnesuria. Cardiac dysfunction is a well-known complication and may appear as dysrhythmias (which are usually resistant to anti-arrhythmics).

65) e.

Volume expansion may be a cause of hypophosphatemia. The other causes of hypophosphatemia are nutritional recovery syndrome, prolonged parenteral nutrition, insulin infusion, glucose infusion, alkalosis (respiratory and metabolic), oral phosphate binders, and diabetic ketoacidosis (during treatment and recovery). Hypophosphatemia is usually multi-factorial and commonly coexists with other electrolyte imbalances.

66) e.

Drug-induced hypomagnesemia list includes cisplatin, gentamycin, and loop diuretics. Magnesium is very poorly absorbed orally; therefore, oral preparations are of no help in deficiency states. The other causes of low serum magnesium are chronic diarrhea, excessive lactation, hyperparathyroidism, primary and secondary aldosteronism, gastrointestinal fistulae, and protein-energy malnutrition.

67) e.

Acetazolamide (carbonic anhydrase inhibitor), for example when used in the treatment of glaucoma, results in normal-anion gap metabolic acidosis. This form of systemic acidosis can result from ureterosigmoidostomy (a form of ureteric diversion that was used in the past after radical removal of the urinary bladder), ingestion of arginine hydrochloride (because of its HCL content), and renal tubular acidosis (type I, II, and IV). Diabetic ketoacidosis (in the way of recovery) and early (not late) uremia may produce normal-anion gap metabolic acidosis. Stem "e" results in high-anion gap metabolic acidosis.

68) d.

In ethylene glycol poisoning, there is accumulation of oxalic acid and glycolic acid.

69) c.

Type-B lactic acidosis may be caused by metformin, sorbitol, isoniazid, and salicylates. Type-A lactic acidosis may occur in any severe shock, carbon monoxide poisoning, and in general, is caused by profound hypotension and/or severe anemia.

Type-B can be encountered in hepatic failure, severe infections, and ethanol and methanol poisoning, and in general, is caused by impaired mitochondrial respiration and increased lactate production.

70) e.

Respiratory alkalosis may result from excessive assisted ventilation, aspirin poisoning (due to direct stimulation of the respiratory center and metabolic acidosis), hysterical reactions (tetany with normal/increased PaO_2, high pH, and very low $PaCO_2$), and pulmonary embolism (resultant rapid shallow breathing). Loss of acid in vomitus (when protracted) causes metabolic form of alkalosis.

Chapter 11
Respiratory Medicine

Questions

1) You are discussing the anatomy of the lungs with your interns. A patient of yours is due to undergo bronchoscopic examination to uncover the nature of a suspicious-looking lung mass. With respect to the lungs, all of the following statements are correct, *except?*
   a) The right lung has 10 segments
   b) The left lung has 9 segments
   c) The visceral pleura is extremely sensitive to pain
   d) The parietal pleura is extremely sensitive to pain
   e) The left upper lobe has 4 segments

2) You examine a 53-year-old man. You find features that of chronic bronchitis and as well as cyanosis. You consider type II respiratory failure. Hypercapnia may result from all of the following, *except?*
   a) Myasthenia gravis
   b) Central sleep apnea
   c) Ankylosing spondylitis
   d) Motor neuron disease
   e) Acute uncomplicated pneumonia

3) A 44-year-old HIV-positive man was admitted to the hospital for suspected *Pneumocystis carinii* pneumonia. His pulse oximetry reveals oxygen saturation of 85%, which improved gradually when high flow O_2 is given. Hypoxemia that is corrected with O_2 therapy may result from all of the following, *except?*
   a) Large pulmonary arteriovenous malformation
   b) Acute pneumonia
   c) Pulmonary hemorrhage
   d) Pulmonary thromboembolism
   e) Acute asthma

4) You arrange CT scan of the chest for a 59-year-old patient, who has long-standing COPD. He presents with hemoptysis. A right hilar mass is found in the X-ray film. You intend to do further investigations. Which of the following statements is the *correct* one?
   a) MRI is usually used for parenchymal lesions
   b) High resolution CT scan is less helpful in interstitial lung diseases
   c) V/Q lung scan is contraindicated in atrial septal defects
   d) Bronchoscopy is useful in peripheral lung tumors
   e) Ultrasound examination of the chest is insensitive for the detection of small pleural effusions

5) Your junior house officer asks you about the causes of low DLCO and their implication in clinical pulmonology, and why some patients display normal KCO in spite of their dyspnea. Which one of the following does *not* result in low DLCO *with* normal or increased KCO?
   a) Right-sided pneumonectomy
   b) Severe ankylosing spondylitis
   c) Myasthenia gravis
   d) Poliomyelitis
   e) Fibrosing alveolitis

6) You assess a 66-year-old man, who was diagnosed with severe COPD before 3 years, whether he will get benefit from long-term domiciliary oxygen therapy or not. Which one of the following statements regarding lng-term domiciliary oxygen therapy in COPD is the correct one?
   a) Has been shown to decrease the mortality figure in selected patients with severe COPD
   b) Used for at least 15 hours/day, at a rate of 2-4 liters/minute
   c) The objective is to produce PaO_2 of >8 kPa without an unacceptable rise in $PaCO_2$
   d) The patient should stop smoking beforehand
   e) Indicated when the PaO_2 is below 9 kPa with any level of $PaCO_2$

7) Your colleague consults you about a 63-year-old man. He has severe COPD and has been deteriorating rapidly since 3 days. The presence of all of the following should prompt you consider acute exacerbation of COPD, *except?*
   a) Increase in the amount sputum production
   b) Increase in the purulence of sputum
   c) Some patients may present with features of fluid retention
   d) Increase in the chest tightness and wheeze
   e) Fever

8) A 66-year-old man, who was diagnosed with emphysema-predominant COPD, asked you about the possibility of using oral prednisolone in the management of his problem. You answered him that prednisolone has specific indications in COPD patients. Which one of the following is not an indication for using corticosteroids in chronic obstructive pulmonary disease?
   a) History of previous response to steroids
   b) During acute exacerbation s
   c) Early in the course to lessen progression
   d) Presence of concomitant asthma
   e) If the patients is already on steroids

9) A 57-year-old heavy smoker man presents with exertional shortness of breath and productive cough. You do some tests and diagnose COPD. He asks you about the disease's prognosis? Regarding the prognosis of COPD, all of the following statements are correct, *except?*

a) The best indicator of progression is the rate of decline in FEV_1 over time
b) The prognosis is inversely related to age
c) The prognosis is directly related to post-bronchodilator FEV_1
d) Atopy patients have bad prognosis
e) e- Pulmonary hypertension in COPD implies a gloomy outcome

10) A 52-year-old man, who is a known case of mild-moderate COPD, presents with increased amount of sputum production, which is associated with fever and increased shortness of breath. You admit him to the Emergency Room. You consider acute exacerbation of his chest disease. With respect to treatment of acute exacerbations of COPD, all of the following statements are correct, *except?*

a) Prophylactic subcutaneous heparin should be considered
b) If the blood pH is <7.26 and $PaCO_2$ is rising, consider ventilatory support
c) If ventilatory support is needed but not applicable because of poor quality of life or significant co-morbidity, intravenous doxapram should be considered
d) Diuretics should be avoided
e) O_2 should be given at a rate of 2 liters/minutes through nasal prongs

11) One of your COPD patients visits your office to consult you about the hazards of airplane traveling. You educate him and give few advices. Regarding airplane traveling for patients with COPD, which one is the *wrong* statement?

a) There is a risk of expansion of non-functional pulmonary bullae
b) There is a risk of producing excessive abdominal gases
c) There is high risk of dryness of the bronchial secretions
d) All patients with resting PaO_2 <9 kPa on air require supplementary O_2
e) Hypercapnia is an absolute contraindication to airplane traveling

12) A 44-year-old man presents with productive cough and sputum production over the past 3 years. He admits to smoking heavily, about three packets of cigarettes per day for the past 25 years. You diagnose chronic bronchitis. With respect to COPD, which one is the *incorrect* statement?

a) Central cyanosis may be observed
b) Peripheral edema may indicate the development of cor pulmonale

c) Weight loss is uncommon and is suggestive of malignancy
d) Only 15% of smokers are likely to develop clinically-significant COPD
e) Smoker's emphysema is usually more prominent in the upper zones of the lung

13) A 45-year-old man is referred to you because of progressive exercise intolerance. He smokes a lot, but never drinks alcohol, and there is no family history of note. You suspect obstructive airway disease. Which one of the following is *not* an obstructive lung disease?
a) Emphysema
b) Chronic asthma
c) Bronchiectasis
d) Farmer's lung
e) Acute asthma

14) You discuss the histological changes of asthma with one of the pathology department colleagues. Histopathological features of bronchi in asthma patients are all of the following, *except?*
a) There is smooth muscles hypertrophy and hyperplasia
b) b- Desquamation of the epithelium and edema of the submucosa
c) Thickening of the basement membrane is rarely seen
d) Mucus plugs are common, especially in severe asthmatic episodes
e) The changes are initially reversible

15) A 16-year-old athletic male develops mild intermittent asthma. He consults you today because he develops the symptoms of chest tightness and wheeze every time he exercises. You tell him that there are effective ways to prevent and lessen these complaints. Regarding the prophylaxis of exercise-induced asthma, all of the following measures may be used, *except?*
a) Adequate warm-up exercises
b) Pre-treatment with inhaled β_2-agonists
c) Pre-treatment with intravenous steroids
d) Nedochromil sodium
e) Montelukast

16) A 14-year-old boy visited the Emergency Room. He developed severe shortness of breath and wheeze. He was unable to give any history. His past records showed recurrent hospital admissions for asthmatic attacks. After admission, he received treatment and you arranged for chest X-ray. Regarding chest X-ray findings in asthma, which one is the *wrong* statement?

a) May be entirely normal
b) Proximal bronchiectasis may indicate the presence of allergic bronchopulmonary aspergillosis
c) Fluffy transient and patchy changes may reflect Churg-Strauss vasculitis
d) During acute attacks of asthma, bilateral hyperlucency is a rare finding
e) Pneumothorax should always be looked for in severe cases not responding to standard treatment

17) The mother of a 7-year-old boy visits the physician's office because her son has developed intermittent asthma. She inquires whether her son's asthma can be controlled or not using certain measures. The following are considered to have a high efficiency when applied to prevent asthma provocation, *except*?

a) Avoid contact with dogs, cats, and horse for animal dander-induced cases
b) Avoid all possible drugs that may induce asthma attack for drug-induced asthma
c) Avoid exposure to chemicals or change occupation if necessary for occupational asthma
d) For feathers in pillows, substitute latex foam pillows for feathers in pillows-induced asthma
e) For food, try to identify and eliminate it from diet for food-induced asthma

18) A 23-year-old college student visits the Emergency Room after developing dyspnea and wheeze. He gives a history of asthma, for which he takes regular beclomethasone inhaler daily. FEV_1 is 50% of his previous best value. Regarding the treatment of acute asthmatic attacks, all of the following are correct, *except*?

a) The objective is to maintain PaO_2 above 8-8.5 kPa with O_2 therapy
b) High flow, high concentration O_2 should be given even in the presence of hypercapnia
c) Systemic steroids should be given in all acute severe asthmatic attacks
d) Peak expiratory flow rate has no role in the initial assessment and management
e) Nebulized β_2-agonists are preferred over their intravenous preparations

19) A 32-year-old man presents with severe attack of asthma. Your intern says that the patient displays many signs of severity, even after receiving an optimal medical therapy, and he asks what to do next. Which one of the following steps is *not* applicable at this point?
   a) Ipratropium bromide 0.5 mg should be added to the nebulized β_2-agonists
   b) Magnesium infusion may be considered
   c) Loading dose intravenous aminophylline should be given to patients who are already on oral theophylline
   d) Intravenous β_2-agonists may be used
   e) Consider mechanical ventilation

20) You are consulting the respiratory care unit to get their opinion about using mechanical ventilation as mode of therapy for one of your patients, who has developed severe acute asthmatic attack. Which one of the following is *not* an indication for mechanical ventilation in acute severe asthma?
   a) Exhaustion, confusion, and drowsiness
   b) Coma
   c) Sudden respiratory arrest
   d) Peak expiratory flow rate <50% of the patient previous best value
   e) $PaCO_2$ >6 kPa and is rising

21) A 23-year-old man visits the pulmonary outpatients' clinic, stating that he has daily productive cough since 3 years. There is a history of severe childhood whooping cough. You examine the patient and you diagnose bronchiectasis after doing some tests. Which one of the following statements with regard to bronchiectasis is *wrong*?
   a) May be caused by pulmonary sequestration
   b) May result from bronchomalacia
   c) Hemoptysis is common and may be life-threatening
   d) Clubbing is rare
   e) Dry bronchiectasis indicates upper lobe involvement

22) A 21-year-old male was admitted to the hospital because of severe chest infection. He was diagnosed with cystic fibrosis several years ago. He thin and pale and you detect finger clubbing. With respect to cystic fibrosis, which on is the *incorrect* statement?
   a) The commonest mutation is the 508 in CFTR gene on chromosome 17

b) Staphylococcal infections tend to occur early in the course of bronchiectasis while pseudomonas infections tend to complicate advanced cases

c) Spontaneous pneumothorax should be suspected if sudden deterioration develops

d) Nasal polypi are common

e) Nephrotic range proteinuria may occur

23) You are taking care of a 12-year-old cystic fiborosis boy. He has bronchiectasis, nasal polypi, steatorrhea, and weight loss. Choose the *incorrect* statement about this boy's disease?

a) Treatment with recombinant human DNAase may decrease infective exacerbations in a subgroup of patients

b) Treatment with recombinant human DNAase has been shown to improve the pulmonary function and sense of wellbeing in some patients

c) Recombinant human DNAase is delivered to the bronchial tree by a nebulizer

d) Treatment with recombinant human DNAase is effective in cystic fibrosis as well as in other causes of bronchiectasis

e) Treatment with recombinant human DNAase is expensive

24) A 34-year-old alcoholic man was brought to the Acute and Emergency Department. There are high fever, dyspnea, cough, and rusty sputum. Chest X-ray reveals right upper lobe pneumonic opacity. Regarding community acquired pneumonias, all of the following statements are correct, *except?*

a) The commonest cause is *Streptococcus pneumoniae*

b) *Staphylococcus aureus* is an uncommon cause

c) Many organisms produce overlapping clinical pictures which are not pathognomonic of any single infectious agent

d) It is easy to differentiate between typical and atypical pneumonias on clinical grounds

e) In severe community acquired pneumonia, *Legionella* infection must be excluded

25) A 54-year-old man presents with fever, breathlessness, cough, and lung opacities on chest X ray, 6 days after recovering from a mild flu. Objectives in the management of pneumonias are all of the following, *except?*

a) Get a radiological confirmation of the diagnosis

b) Exclude other conditions which may mimic pneumonias

c) Obtain a microbiological diagnosis

d) Asses the severity of pneumonia

e) Looking for the development of complications is not indicated initially

26) A 17-year-old high school student presents with a 6-day history of fever, dry cough, and shortness of breath. Clinical examination is consistent with a moderate right-sided pleural effusion. The following microbiological investigations should be done routinely in all cases of community-acquired pneumonias, *except?*
a) Sputum for Gram stain and culture
b) Sputum for acid-fast bacilli and culture
c) Pleural fluid aspirate
d) Blood cultures
e) Serology with acute and convalescent titers to diagnose *Mycoplasma, Chlamydia, Legionella* and viral pneumonias

27) A 33-year-old male patient was referred from a rural hospital as a case of severe pneumonia. His accompaniment asks if the patient will live or die during the next few days. Features suggestive of high mortality in pneumonias are all of the following, *except?*
a) Respiratory rate >30 cycles/minute
b) Systolic blood pressure <90 mmHg
c) Blood urea >17 mmol/L
d) Hypoalbuminemia
e) Positive blood culture

28) A 46-year-old man presents with a 1-week history of fever, malaise, productive cough, and right-sided pleural chest pain. You think of pneumonia and you plan to exclude similar diseases. Which one of the following is *not* a differential diagnosis of acute pneumonia?
a) Pulmonary infarction
b) Pleural/pulmonary tuberculosis
c) Atypical pulmonary edema
d) Sub-phrenic abscess
e) Paracolic abscess

29) A 34-year-old female was treated in the medical ward for idiopathic thrombocytopenic purpura and was receiving high doses of systemic corticosteroids. Yesterday, she developed fever, chest tightness, and productive sputum. With regard to hospital-acquired pneumonia, all of the following statements are wrong, *except?*
a) Its definition does not include postoperative pneumonia
b) Does not include a new pneumonia while the patient is on a ventilator

c) Includes certain types of aspiration pneumonias
d) Occurs in up to 20% of in-hospital patients
e) Gram-negative organisms usually predominate

30) You attend a medical symposium in Ghana about tuberculosis and its subtypes and complications. They mention cryptic military tuberculosis and you have learned many things about this disease entity. Regarding cryptic military tuberculosis, all of the following statements are correct, *except?*
a) Usually occurs in elderly females
b) Chest signs are rare
c) Leukemoid reaction may occur
d) Pancytopenia may develop
e) Chest X-ray is usually abnormal

31) You work in a local tuberculosis health center and you are about to interpret the result of tuberculin test of one of your patients who has fever and cough. False negative tuberculin testing may be encountered in all of the following, *except?*
a) Miliary tuberculosis
b) Immune-suppressive therapy
c) Early in the course of tuberculous meningitis
d) Elderly patients
e) AIDS population

32) A 33-year-old Indian immigrant has developed fever, weight loss, night sweating, and cough over the past 3 weeks. His sputum is full of acid-fast bacilli. You start anti-tuberculosis medications. With respect to the treatment of pulmonary tuberculosis, which one is the *correct* statement?
a) Thiacetazone is not used for AIDS patients
b) Optic neuritis is usually caused by pyrazinamide
c) Hemolytic anemia with rifampicin is very common
d) Streptomycin mainly affects hearing rather than vestibular function
e) Pyridoxine should be given to all patients who receives isoniazid

33) A 41-year-old female, who has chronic persistent asthma, states that her disease control has become gradually difficult and the amount of sputum has increased. You examine the patient and order chest X-ray, which reveals proximal bronchiectasis. After conducting further tests, you diagnose allergic bronchopulmonary aspergillosis (ABPA). Regarding ABPA, which one is the *incorrect* statement?

a) Serum IgG precipitins testing against *Aspergillus* is usually strongly positive
b) Immune-complex disease, not an invasive one
c) Blood eosinophilia is very common
d) Not caused by *Aspergillus clavatus*
e) Usually there is preexistent lung disease

34) A 55-year-old heavy smoker man presents with global wasting, clubbing, and massive hemoptysis. Spiral chest CT scan reveals a large irregular right sided hilar mass with bilateral mediastinal lymph node enlargement. You think he has lung cancer. Regarding lung cancer, which one is the *incorrect* statement?
   a) Passive smoking is responsible for 5% of all lung cancers
   b) Exposure to naturally occurring radon may result in 5% of all lung cancers
   c) May be caused by chromium and cadmium exposure
   d) The incidence is slightly higher in rural than urban dwellers
   e) The incidence of adenocarcinoma is rising in the Western World

35) A 61-year-old man, who is a known case of COPD, states that his sputum is streaked with blood every day and he is feeling tired all the time. You run a battery of investigations and you diagnose lung cancer. With regard to bronchogenic carcinoma, choose the *wrong* statement?
   a) Responsible for 25% of all cancer deaths
   b) The cause of about 80% of cancer deaths in women and 40% cancer deaths in men
   c) The commonest cause of cancer death in men
   d) The most rapidly increasing cause of cancer death in women
   e) More than 3 folds increase in cancer deaths since 1950

36) A 62-year-old man presents with dry irritative cough, weight loss, and global wasting over a matter of 2 months. He is a passive smoker, as his wife smokes 3 packets per day. Which one of the following with respect to chest X-ray findings in lung cancer is the incorrect one?
   a) Normal film virtually excludes lung cancer
   b) There may be pleural effusion
   c) Lung, lobe, or segmental collapse
   d) Broadening of the mediastinum
   e) Rib destruction

37) A 54-year-old man will undergo thoracic surgery to remove a malignant lung mass. The surgeon thinks that surgery cannot be done. Which one of the following is *not* a contraindications to surgery in non-small cell lung cancer?
a) Esophageal invasion
b) FEV$_1$ less than 2.5 L
c) Malignant pleural effusion
d) Severe ischemic heart disease
e) Contralateral mediastinal lymph node involvement

38) A 55-year-old man presents with left-sided upper chest pain. He is a lifelong heavy smoker and drinks alcohol at weekends. There are malaise, dullness at the left lung apex, and wasting of left hand small muscles. The left pupil is smaller than the right one. What is the likely diagnosis?
a) Post-primary pulmonary tuberculosis
b) Apical aspergilloma
c) Pancoast tumor
d) Left apical lung collapse
e) Malignant secondary tumors in both lungs

39) A 34-year-old woman visits the physician's office because of persistent dry cough which has been increasing over the past few months. She neither smokes nor drinks alcohol. Chest X-ray discloses a mass in the posterior mediastinum. The mass is rounded and lies in front of the dorsal vertebrae. Which one of the following is *not* a posterior mediastinal mass?
a) Paravertebral abscess
b) Neurogenic tumor
c) Foregut duplication
d) Diaphragmatic hernia through foramen of Morgagni
e) Diaphragmatic hernia through foramen of Bockdaleck

40) A 32-year-old black male develops fever, arthralgia, and painful nodules on both shins over a matter of 2 weeks. Chest X-ray film shows bilateral hilar lymph node enlargement but there are no parenchymal lesions. You diagnose acute sarcoidosis. Regarding the presentation of sarcoidosis, all of the following are correct, *except*?
a) May be asymptomatic and is discovered incidentally through chest X-ray films in 30% of cases
b) Ocular symptoms are the presenting feature in 5-10% of patients
c) Skin sarcoid is the presenting complain in 5% of cases
d) About 30% of patients present with symptoms of hypercalcemia

e) Superficial lymph node enlargement is the presenting feature in 5% of cases

41) A 27-year-old woman is referred to you from the dermatology department after establishing the diagnosis of skin sarcoid. You consider starting medical treatment. Which one of the following is *not* an indications for treatment of sarcoidosis with oral steroids?
a) Hypercalcemia
b) Ocular sarcoidosis
c) CNS involvement
d) Stage I or II disease
e) Cardiac sarcoidosis

42) A 63-year-old female presents with a one-year history of progressive shortness of breath and dry cough. There is cyanosis and tachypnea. After conducting some tests, you believe the final diagnosis is cryptogenic fibrosing alveolitis. With respect to cryptogenic fibrosing alveolitis, choose the incorrect statement?
a) Twice as common among cigarette smokers than in nonsmokers
b) Bronchoalveolar lavage shows predominance of neutrophils
c) Clubbing is seen in 60% of cases
d) Query association with Epstein-Barr virus infection
e) Serum LDH is elevated in a minority of patients

43) A 25-year-old farm worker complains of chest tightness, cough, and fever about 8 hours after returning from the farm. This has been happening since 4 months. During these episodes, examination reveals crepitation in both mid and lower lung zones. Regarding extrinsic allergic alveolitis and other lung diseases due to exposure to organic dusts, which one of the following associations is *correct*?
a) Byssinosis-*Penicillium cassie*
b) Malt worker's lung-*Aspergillus fumigatus*
c) Maple barks stripper's lung-*Cryptostroma corticale*
d) Cheese worker's lung-thermophilic *Actinomycetes*
e) Bird fancier's lung-*Aspergillus clavatus*

44) A 60-year-old man presents with a one-year history of dyspnea and dry cough. There is a 30-year history of occupational asbestos exposure. Examination reveals bibasal crepitation. You do chest X-ray and you consider asbestosis. Regarding asbestosis, all of the following are correct, *except*?

a) The commonest type of asbestos fibers produced world-wide is the white one (chrysotile)
b) Exposure mainly occurs through mining and milling of the mineral
c) Greatly increases the risk of lung adenocarcinoma, especially in smokers
d) Remarkably increases the risk of pleural mesothelioma, especially in smokers
e) The overall risk of malignancy is higher in smokers than in non-smokers

45) A 44-year-old male presents with fever, cough, and malaise over few weeks. Chest X-ray shows bilateral fluffy migrating opacities. Blood count discloses an eosinophil count of 1900 cells/mm^3. Pulmonary eosinophilia may be seen in all of the following, *except*?
a) Nitrofurantoin
b) *Mansonella streptocerca*
c) c *Wuchereria bancrofti*
d) Phenylbutazone
e) Wegener granulomatosis

46) A 45-year-old man is referred to your office as a suspected case of alveolar proteinosis, after having had certain symptoms for some time. You do bronchoscopy to solidify the diagnosis. Regarding alveolar proteinosis, which one is the *incorrect* statement?
a) Spontaneous remission occurs in up to 30% of cases
b) Fever and hemoptysis are common
c) There is a 5% remission rate after doing whole lung lavage
d) Air bronchogram on chest X-ray is a common feature
e) Diffuse bilateral shadowing, mainly around the hili, is the usual chest X-ray finding

47) You read an article in a medical journal about a lung disease called lymphanigoleiomyomatosis. Which of the following you have *not* read?
a) Usually seen in females during their childbearing age
b) Pleural effusions tend to be chylous
c) Recurrent pneumothoraces may complicate the disease
d) d- Hormonal ablation and progestin therapy are highly effective
e) Patient may present with hemoptysis

48) A 56-year-old smoker presents with chest tightness. Examination reveals clubbing and signs of right-sided pleural effusion.

Pleural fluid cytology uncovers the presence of malignant-looking cells. Which one of the following is *not* a common cause of pleural effusion?
a) Yellow nail syndrome
b) Meig syndrome
c) Acute rheumatic fever
d) Subphrenic abscess
e) Myxedema

49) A 29-year-old tall man presents with sudden severe right-sided pleural pain and shortness of breath. You examine the patient and order chest imaging. You find large right-sided pneumothorax but there is no mediastinal shift. Which one of the following does *not* call for tube drainage in pneumothorax?
a) Tension type
b) More than 2-5 liters of air has been aspirated
c) If you face resistance during aspiration
d) Pneumothorax with underlying COPD
e) Any degree of dyspnea

50) A 64-year-old smoker male presents with chest pain and shortness of breath over 2 months. Examination reveals wasting, clubbing, and paradoxical respiration. Chest X-ray reveals elevated left hemi-diaphragm and left-sided hilar mass. You think he has diaphragmatic paralysis because of lung cancer. Elevation of a hemi-diaphragm does *not* result from which one of the following?
a) Severe pleuritic chest pain
b) Subphrenic abscess
c) Pulmonary infarction
d) Eventration of the diaphragm
e) Intercostal nerve palsy

51) A 22-year-old man visits the outpatients' clinic. A recent chest X-ray, which was done as part of pre-employment assessment, revealed a rounded opacity in the left upper lung zone. He denies symptoms, and there is no past-history of exposure to infectious patients. Which one of the following is *not* a common cause of solitary lung nodule?
a) Single metastasis
b) Bronchial carcinoma
c) Wegener granulomatosis
d) Tuberculoma
e) Lung abscess

52) A 15-year-old schoolboy presents with fever, chest tightness, and cough for 1 week. Examination reveals a tinge of jaundice. Chest X-ray shows bilateral lung infiltrates and hilar lymphadenopathy. Blood film reveals rouleaux formation. Cold agglutinin titer is positive. His presentation is caused by which one of the following?

a) *Streptococcus pneumoniae*
b) *Staph aureus*
c) *Chlamydia pneumoniae*
d) *Mycoplasma pneumoniae*
e) Influenza virus type B

53) A 72-year-old non-smoker man presents with a 2-day history of confusion. His past history is unremarkable, and he is on no medications. Chest X-ray reveals a lobar type of pneumonia. You tell his son that his confusion is a complication of his chest infection. Which one of the following is *not* an intrathoracic complication of bacterial pneumonia?

a) Adult respiratory distress syndrome
b) Pneumothorax
c) Empyema thoracis
d) Pericarditis
e) Dorsal spine subdural abscess

54) A 43-year-old HIV-positive homosexual male presents with shortness of breath and cough for some time. Chest examination is unremarkable, but his chest X-ray reveals diffuse and bilateral pulmonary infiltrates. Which one of the following is *not* a usual cause of this radiographic finding in AIDS patients?

a) Viral pneumonias
b) *Pneumocystis carinii*
c) Bacterial pneumonias
d) Tuberculosis
e) Fungal pneumonias

55) A 34-year-old AIDS patient was admitted to the Emergency Room. He developed fever, chest pain, exertional dyspnea, and dry cough over 1 week. Chest X-ray revealed bilateral interstitial infiltrates. Serum LDH was high. You considered *Pneumocystis carinii* pneumonia. Which one of the following is *not* a complication of this type of infection?

a) Development of lung cysts
b) Respiratory failure
c) Pneumothorax
d) Expiratory air flow obstruction

e) Lung abscess

56) A 35-year-old office secretary female presented with persistent dry cough for few months. She was diagnosed with cough-variant asthma but had not improved upon receiving anti-asthma medications since then. You re-examined the patient and did chest X-ray. There was a benign-looking mass in the right upper lung zone. Which one of the following is *not* a benign pulmonary tumor?
a) Carcinoid
b) Sclerosing hemangioma
c) Leiomyoma
d) Amyloidoma
e) Chemodectoma

57) A 58-year-old man underwent hemicolectomy to remove colonic cancer 3 months ago. Today, he presented with chest tightness and dry cough. Chest X-ray reveals multiple bilateral lung nodules that were highly consistent with secondary tumors. Which one of the following is *not* a cause of multiple lung nodules?
a) Synchronous primary bronchogenic carcinomas
b) Kaposi sarcoma.
c) Mycoplasma pneumonia
d) Pulmonary arteriovenous malformations
e) Broncholithiasis

58) A 31-year-old male visits the physician's office because of a 1-year history of exertional breathlessness. He has just finished his chemotherapeutic regimen for stage 3 testicular cancer. Chest X-ray is unremarkable. You suspect bleomycin-induced bronchiolitis obliterans organizing pneumonia. Which one of the following is *not* a well-recognized cause of this chest disease?
a) Gold
b) Amiodarone
c) Aspirin
d) Cocaine
e) Mitomycin-C

59) A 28-year-old female presents with a 2-month history of chest tightness and orthopnea. Her past medical history is unremarkable and there is no family history of note. She denies infectious or toxic exposure. Examination reveals signs of left-sided pleural effusion, an observation that is confirmed by doing chest X-ray film.

Pleural fluid aspiration and further examination disclose an exudative type of effusion. Which one of the following is a cause of exudative type of pleural effusion?

a) Acute atelectasis
b) Superior vena cava syndrome
c) Urinothorax
d) Constrictive pericarditis
e) Post-coronary artery bypass grafting

60) A 55-year-old man, who was diagnosed with seropositive rheumatoid arthritis 15 years ago, presents with cough and exertional breathlessness. He takes gold injections at regular intervals and he is doing well in terms of joint complaints. Open lung biopsy reveals bronchiolitis obliterans organizing pneumonia. Which one of the following is *not* a pulmonary complication of rheumatoid arthritis?

a) Thoracic cage immobility
b) Apical fibrobullous disease
c) Pneumothorax
d) Follicular bronchiolitis
e) Bronchospasm

Respiratory Medicine

Answers

1) c.

The visceral pleura is insensitive to pain, and the pleural pain is due to involvement of the parietal pleura.

2) e.

Acute uncomplicated pneumonia causes tachypnea (and CO_2 wash-out) and hypocapnia. The other stems are result in hypoventilation with CO_2 retention and cyanosis, which can be corrected by giving oxygen.

3) a.

Large pulmonary arteriovenous malformation would cause right to left shunting of blood; the hypoxemic blood cannot be corrected with oxygen. The other causes of this type of hypoxemia which is not corrected by O_2 therapy are any right to left shunt (at cardiac or lung level) as well as decreased O_2 carrying capacity of blood (anemia and inactivated hemoglobins).

4) c.

Lung ventilation/perfusion (V/Q) scan is contraindicated in atrial septal defects (ASD) because the macro-aggregates of albumin loge in the cerebral and renal circulations with devastating effects. Patients with high clinical probability and a high-probability V/Q scan had a 95% likelihood of pulmonary thromboembolism (PE); patients with low clinical probability and a low-probability V/Q scan had only a 4% likelihood of PE. A normal V/Q scan virtually excludes PE. MRI is usually used in mediastinal lesions, apical lung lesions, and vertebral lesions; CT scan is mainly used for parenchyma lung lesions. Bronchoscope is useful in central lung tumors. Ultrasound examination is very sensitive at detecting small pleural effusions.

5) e.

Fibrosing alveolitis causes reduction in both DLCO and KCO, because it is an interstitial lung disease. Causes of *low* DLCO with *normal* spirometry (i.e., disorders to consider when there is isolated decrease in DLCO) are:

1. Pulmonary vascular disease (mild to severe decrease in DLCO): such as chronic recurrent pulmonary emboli, idiopathic pulmonary arterial hypertension, and pulmonary vascular involvement with connective tissue diseases and vasculitides (systemic sclerosis, systemic lupus erythematosus, and rheumatoid arthritis).

2. Early interstitial lung disease (mild to moderate decrease in DLCO) before the vital capacity falls below the lower limit of the normal range.

3. Anemia.

4. Increased carboxyhemoglobin level in blood.

Causes of isolated *increased* DLCO (DLCO is considered to be elevated when it is greater than 140% of predicted):

1. Severe obesity.

2. Asthma (during attacks; an uncommon findings in clinical practice).

3. Polycythemia.

4. Pulmonary hemorrhage.

5. Left-to-right intracardiac shunting.

6. Mild left heart failure (due to increased pulmonary capillary blood volume).

7. Exercise just prior to the test session (due to increased cardiac output).

Another important theme is the *reduced* DLCO with obstructive spirometry. In middle-aged and elderly people, think of emphysema-predominant COPD, while α1-antitrypsin deficiency, cystic fibrosis, bronchiectasis, and lymphangioleiomyomatosis are the usual causes in young people.

6) e.

The indications of LTOT are:

1. PaO_2 <7.3 kPa (with *any* level of $PaCO_2$) and FEV_1<1.5 liters.

2. PaO_2 between 7.3 to 8 kPa *PLUS* either pulmonary hypertension, peripheral edema, or nocturnal hypoxemia.

I've been told that there was an MRCP question about the complications of LTOT. There is no evidence that LTOT has adverse effects in patients with COPD when administered correctly. Although there are theoretical concerns about potential toxicities in patients administered oxygen in high concentrations (above 50%) for extended time periods (e.g., absorptive atelectasis, increased oxidative stress, and inflammation), clinical experience has provided little support for these concerns in the setting of LTOT.

7) e.

Fever may be due to many causes; do not assume that the patient has develops an exacerbation simply because of fever. The first 4 stems are well-established modes of presentations. Some patients may be managed out- of-hospital; however, others need hospital admission and aggressive treatment. Indications of admission are: high risk co-morbidities (including pneumonia, cardiac arrhythmia, congestive heart failure, diabetes mellitus, renal failure, or liver failure); inadequate response of symptoms to outpatient management; marked increase in dyspnea; inability to eat or sleep due to symptoms; worsening hypoxemia; worsening hypercapnia; changes in mental status; inability to care for oneself (i.e., lack of home support); and uncertain diagnosis. In stable chronic bronchitis, the sputum is mucoid and the predominant cell is the macrophage. During exacerbations, sputum usually becomes purulent with an influx of neutrophils. The Gram stain usually shows a mixture of organisms. The most frequent pathogens cultured from the sputum are *Streptococcus pneumoniae* and *Haemophilus influenzae*. Other oropharyngeal flora such as *Moraxella catarrhalis* have been shown to cause exacerbations. However, cultures and even Gram stains are rarely necessary before instituting antimicrobial therapy in the outpatient setting.

8) c.

Systemic/inhaled steroids have no effect on the progression of the disease. The other indications are:
1. If the exacerbation is the presenting feature.
2. If there is a poor response to bronchodilator therapy.
Inhaled low-dose steroids *may* be used in severe COPD with history of severe recurrent exacerbations requiring hospitalizations, but they do not affect the rate of decline of FEV₁ over time.

9) d.

Atopy patients have a better survival, and to date, no drug treatment (aside from LTOT and smoking cessation) has been shown to affect the disease outcome. Patients with COPD have usually been smoking at least 20 cigarettes per day for 20 or more years before symptoms develop. Chronic productive cough, sometimes with wheezing, often begins when patients are in their forties, although the patients are frequently less aware of these symptoms than the persons they live with.

10) d.

Diuretics should be strongly considered if there is edema or raised JVP. The other stems are applicable and true.

11) e.

Hypercapnia is a *relative* contraindication to airplane travel, as well as a $PaO_2 <$ 6.7 kPa on air. All patients with resting $PaO_2 < 9$ kPa on air require supplementary O_2; those patients would develop profound reduction in their PaO_2 upon airplane travelling.

12) c.

Approximately 20% of patients with moderate and severe emphysema experience weight loss (and loss of body fat). There is increased muscle protein breakdown that is believed to be due to systemic factors, such as systemic effects of lung inflammation, oxidative stress, and an excess of circulating cytokines. These abnormalities of muscles contribute to decreased exercise performance. This weight loss may call for investigating an occult malignancy.

13) d.

Farmer's lung is extrinsic allergic alveolitis (hypersensitivity pneumonitis, HP) affecting the pulmonary interstitium. Pathologically, acute HP is characterized by poorly formed, non-caseating interstitial granulomas or mononuclear cell infiltration in a peribronchial distribution, often with prominent giant cells. Well-formed granulomas are not commonly seen, in contrast to their frequent identification in patients with sarcoidosis. The frequency of various pathologic findings has been described in farmer's lung and is probably similar in other types of HP.

14) c.

Thickening of the basement membrane contributes to chronic disability and poor response in chronic asthma cases. All of the changes, in the long-term, become irreversible.

Differential diagnosis of asthma includes: upper airway narrowing, vocal cord polyp (or tumors), vocal cord dysfunction (paralysis or functional), tracheal stenosis or compression, lower airway narrowing, chronic bronchitis and emphysema, airway tumors, airway compression (e.g., by tumor, mass, enlarged vessel or lymph nodes), airway stricture (stenosis), airway foreign bodies, airway inflammation (e.g., sarcoidosis, bronchiectasis), airway edema (e.g., heart failure), bronchoconstriction (e.g., due to pulmonary embolus, anaphylaxis, post-viral, toxic/irritant inhalation), and low lung volume (e.g., obesity).

15) c.

Common asthma triggers are allergens, respiratory infections (including chronic sinusitis), inhaled irritants (such as tobacco smoke, strong perfume, chlorine-based cleaning products, and air pollutants), physical activity (exercise, especially in cold air, although regular exercise can improve general fitness), emotional stress, gastroesophageal reflux, and hormonal fluctuations. Intravenous steroids have no role in prophylaxis against exercise- induced asthma. The others stems can be applied effectively and these have a well-established efficacy.

16) d.

Hyperlucency is common during acute asthmatic attacks (due to air trapping) and actually may be the only finding. In chronic cases, the chest X-ray may resemble that of emphysema. The chest radiograph is almost always normal in patients with asthma, however. Many clinicians favor obtaining a chest radiograph for new-onset asthma in adults, for the purpose of excluding the occasional alternative diagnosis (e.g., the mediastinal mass with tracheal compression or congestive heart failure). In contrast, chest radiographs are definitely recommended in the evaluation of severe or "difficult-to -control" asthma, for the detection of co-morbid conditions (e.g., allergic bronchopulmonary aspergillosis, eosinophilic pneumonia, or atelectasis due to mucus plugging).

17) e.
Stems a, b, c, and d are highly efficacious measures to prevent asthma attacks. Stem "e" has low efficacy, and preventive measures against mites and pollens have low to uncertain efficacy, unlike the popular belief. Stem "e" is a very effective measure in eczema, not in asthma.

18) d.

In the management of acute asthmatic attacks, patients' PaO_2 should be kept >8-8.5 kPa while their O_2 saturation should be maintained above 93% using high flow, high concentration O_2. Oral steroids are used, but if patients have vomiting or are unable to swallow, then the intravenous route is used. Peak expiratory flow rates should be obtained in all cases admitted to the Emergency Room in order to assess asthma severity and for subsequent and follow-up management. Viral, but not bacterial, respiratory infections are common precipitants of asthma exacerbations. It is therefore not surprising that antibiotics have been shown not to affect the course of acute asthmatic attacks. Antibiotics are now reserved for patients with fever, leukocytosis (>15000 cells/mm^3), and a pulmonary infiltrate on the chest radiograph.

19) c.

Aminophylline should be avoided in patients who are already on oral theophylline, as clinical toxicity may ensue rapidly, besides enhancing an increased mortality.

20) d.

Such flow rate's value (peak expiratory flow rate below 50% of the patient previous best value) indicates a severe attack, but it is not necessarily an indication for mechanical ventilation. Other indications for mechanical ventilation in acute asthma are PaO_2<8 kPa (and falling), and blood pH that is low (and falling), i.e. deteriorating blood gases despite optimal therapy. The most direct assessment of airflow obstruction is measurement of spirometry or peak expiratory flow rate (PEFR). Peak flow is easier for most patients to perform, because it does not require a sustained expiratory effort. However, some patients are too dyspnic to perform this test until bronchodilator therapy has been given. Whenever possible, the PEFR should be measured initially to provide a baseline and at successive intervals during treatment. Predicted values differ with size and age, but a peak flow below 120 L/min or an FEV_1 below 1.0 liters indicates severe obstruction for all but unusually small adults.

21) d.

Clubbing and hemoptysis are both common, and absence of large amount of sputum is not against the diagnosis (so-called bronchiectasis sicca).

22) a.

The mutant gene lies on chromosome 7. Many mutations have been detected. Sinusitis and bronchiectasis are the major respiratory manifestations of cystic fibrosis (CF), and the latter may be the sole feature of CF in adults. Clues suggesting the presence of this disorder are upper lobe radiographic involvement and sputum cultures showing mucoid *Pseudomonas aeruginosa*. Adults with CF diagnosed after the age of 20 years are less likely than children to have homozygous CFTR mutation, pancreatic insufficiency, diabetes mellitus, and infection with *Pseudomonas aeruginosa*. They also have higher FEV_1 values and increased colonization with non-tuberculous *Mycobacteria*. Patients with Young syndrome exhibit clinical features similar to individuals with cystic fibrosis, including bronchiectasis, sinusitis, and obstructive azospermia. However, they do not have increased sweat chloride values, pancreatic insufficiency, abnormal nasal potential differences, or abnormal CF mutations; affected individuals are often middle-aged males identified during the evaluation for infertility. Long-term bronchiectasis results in secondary AA amyloidosis and nephrotic syndrome.

23) d.

Treating other causes of bronchiectasis with recombinant human DNAase (rDNAase) has been shown to produce deleterious effects; it is only used in bronchiectasis of cystic fibrosis. The other modes of treatment in cystic fibrosis are continuous high dose oral ibuprofen and gene therapy. rDNAase is administered via nebulizers (during nebulization, it should not be diluted or mixed with any other drugs, as this may inactivate the drug). Common adverse effects are chest pain, pharyngitis, and voice alteration while conjunctivitis, rhinitis, hemoptysis, and wheezes are uncommon side effects. It is contraindicated when there are hypersensitivity to dornase alfa, Chinese hamster ovary cell products (e.g., epoetin alfa), or any component of the formulation.

24) d.

Many patients have an overlapping picture, and it is usually difficult to differentiate between typical and atypical pneumonias on clinical grounds only.

25) e.

The development of complications adversely affects the outcome and should be looked for in all cases. The other stems are definitely applicable.

26) c.

In general, parapneumonic effusion should be sampled if it meets *any* of the following criteria:
1. Free-flowing but layers >10 mm on a lateral decubitus film.
2. Loculated.
3. Associated with thickened parietal pleura on a contrast enhanced CT scan (a finding that is suggestive of empyema).
In general, pH <7.20 or a glucose <60 mg/dL is an indication for drainage of the effusion. The other lab tests mentioned in the question should be ordered to diagnose the causative agent. In 20% of cases, no infectious agent can be isolated.

27) c.

Risk factors for mortality or a complicated course from community-acquired pneumonia are:
1. Historical factors: age >65 years, suspicion of aspiration, congestive heart failure, chronic obstructive pulmonary disease, diabetes mellitus, chronic alcohol abuse, chronic renal failure, chronic liver disease of any etiology, previous splenectomy, hospitalization during the prior 12 months, and altered mental status.
2. Physical findings: temperature >38.3°C, respiratory rate >30 cycles/minutes, systolic blood pressure <90 mm Hg, and evidence of extra-pulmonary sites of infection.
3. Laboratory abnormalities: white blood cell count <4000 or >30000/mm^3, hematocrit <30%, PaO_2 <60 mm Hg or $PaCO_2$ >50 mm Hg on room air, blood urea >7 mmol/L, serum creatinine >1.2 mg/dL, multi-lobar or rapidly progressive radiographic infiltrates, positive blood culture, and hypoalbuminemia.

28) e.

Atypical pulmonary edema may be unilateral and/or localized and may be extremely difficult to differentiate from pneumonia.

However, the absence of fever and the presence of underlying heart disease favor pulmonary edema over chest infection.

29) c.

The definition of hospital acquired pneumonia is pneumonia that occurs 48 hours or more after hospital admission (which was not incubating at the time of admission); the term also encompasses postoperative pneumonia and new pneumonia in patients on ventilators. It occurs in 2-5% of hospital admissions and is predominantly of Gram-negative type. Aggressive treatment is indicated, as the majority of patients are already debilitated by another illness, e.g. immune suppression, malignancy,...etc.

30) e.

Chest X-ray is usually normal in cryptic military tuberculosis, which usually presents as pyrexia of unknown origin, hepatosplenomegaly, unexplained weight loss, and malaise.

31) c.

Usually, tuberculin skin testing becomes falsely negative lately in the course of tuberculous meningitis in 25 -50 % of cases. The other causes of false negative tuberculin skin testing are sarcoidosis, childhood exanthems, uremia, and systemic malignancy. A person infected with *M. tuberculosis* will react to intra-dermal injection of tuberculin with a delayed type hypersensitivity response mediated by T-lymphocytes. Cellular infiltration by T-cells (in combination with other recruited inflammatory cells) results in maximal induration at 48 to 72 hours after inoculation with intra-dermal antigen. A period of 2 to 12 weeks after primary infection is generally required for skin test conversion to occur. The ability to mount such a response is usually maintained for many years, although reactivity may wane among the elderly. Children who have received the BCG vaccine generally demonstrate tuberculin skin test reactions of 3 to 19 mm several months after vaccination. The size of the skin test does not correlate with the degree of protection conferred by the vaccine. Most of these reactions wane significantly over a ten year period of time. Tuberculin skin tests are *not* contraindicated in BCG-vaccinated persons; in fact, skin tests should always be done several months following vaccination to document baseline tuberculin reactivity.

32) a.

Because it is a bacteriostatic medication (not cidal), thiacetazone is contraindicated in AIDS patients who have active tuberculosis. Ethambutol may result in optic neuritis; rarely, isoniazid can cause this optic neuropathy. Rifampicin-induced hemolytic anemia is a rare complication; the resultant pre-hepatic jaundice may be easily confused with that caused by drug-induced hepatitis. Streptomycin usually damages the peripheral vestibular apparatus rather than causing hearing impairment. The *routine* prescription of oral vitamin B6 for patients who take (or are planned to take) long-term isoniazid is no longer recommended; only high-risk groups should be offered this therapy, e.g. HIV patients, malnutrition, chronic diarrhea, and alcoholics.

33) a.

Serum IgG precipitins testing against *Aspergillus* is weakly positive in ABPA; ABPA is an immune-complex disease and is not an invasive one. The total (and specific) serum IgE level is greatly raised. Pulmonary and peripheral blood eosinophilia occur; the latter may be very high. *Aspergillus clavatus* is a cause of malt worker's lung and cheese worker's lung. The disease commonly develops in those with underlying lung disease, such as long-standing asthma (1-2% of asthmatics) and cystic fibrosis (1-15% of patients).

34) d.

About 90% of lung cancers are associated with cigarette smoking; therefore, prevention is easier than treatment. The reverse is true in stem "d". Exposures to asbestos, radon, arsenic, ionizing radiation, haloethers, polycyclic aromatic hydrocarbons, and nickel are also associated with lung cancer development.

35) b.

Lung cancer is responsible of about 4% of cancer deaths in women and 8% cancer deaths in men.

36) a.

Some tumors are truly endobronchial and may not produce any changes on the plain X- ray films, especially early in the course; hence, bronchoscopy is required in high-risk clinical setting.

37) b.

Contraindications to surgical treatment of non-small cell lung cancer are: T4 (stems a, c), N3 (stem e), M1 (metastatic disease), and FEV_1 less than 0.8 L. The presence of severe or unstable cardiac or other medical conditions would preclude this form of treatment.

38) c.

The scenario fits Pancoast tumor of the left lung apex. Pancoast tumors are much less common than other lung cancers; the rate of Pancoast tumors varies from 1-3% of all lung cancers. A major issue with Pancoast tumors is the delay in diagnosis. Most Pancoast tumors are squamous cell carcinomas or adenocarcinomas; only 3-5% are small cell carcinomas. Adenocarcinoma is sometimes found in this location and can even be metastatic. Involvement of the phrenic or recurrent laryngeal nerve or superior vena cava obstruction is not representative of the classic Pancoast tumor. The location of the tumor, rather than its pathology or histology of origin, is significant in producing its characteristic clinical pattern. Most tumors are extra-thoracic. Differential diagnoses of masses in the apical chest are primary tumors of the thyroid, larynx, and pleura. Other causes may include infectious disorders of the lung, aneurysms of the subclavian vessels, amyloid of pleura, and multiple myeloma.

39) d.

Stem "d" is an anterior mediastinal mass.

40) d.

Although hypercalcemia occurs in 20- 30% of sarcoidosis, but as a presenting complaint, it accounts for 1% of cases or even less.

41) d.

The other indications of starting glucocorticoids in sarcoidosis are rapidly worsening stage II/III with deteriorating lung function test, hypercalcemia/hypercalciuria, lupus pernio, and major organ involvement (heart, brain, eye,...etc.).

Radiographic Stages of Sarcoidosis

Stage	Radiographic abnormalities
0	None
I	Bilateral hilar and/or mediastinal adenopathy without pulmonary parenchymal abnormalities
II	Hilar and/or mediastinal adenopathy with pulmonary parenchymal abnormalities (generally a diffuse interstitial pattern)
II	Diffuse pulmonary parenchymal disease without nodal enlargement
IV	Pulmonary fibrosis with evidence of honeycomb change (end-stage lung disease)

Guidelines for Starting Glucocorticoids in Sarcoidosis

Clinical Status	Recommendation
Asymptomatic pulmonary disease, or mild dry cough	Observation usually suffices; consider oral steroids if worsening occurs
Stage II or III pulmonary disease with symptoms and/or abnormal function	Systemic therapy with oral steroids
Severe extra-pulmonary disease, including heart, central nervous system, eyes, and kidneys	Systemic therapy with oral steroids should always be considered whenever there is vital organ involvement
Hypercalcemia/hypercalciuria	There is a risk of nephrocalcinosis and renal stone formation; start oral steroids
Lupus pernio	High likelihood of being associated with chronic fibrosing sarcoidosis; start oral steroids

42) e.

Serum LDH is elevated in the majority of patients. There is a query association with antidepressants and EBV infection. Excess neutrophils in bronchoalveolar lavage (BAL) portends a bad prognosis (neutrophilic BAL is also seen in pneumoconiosis and sarcoidosis, while it is lymphocytic in extrinsic allergic alveolitis). Serum ANA and rheumatoid factor are positive in up to 50% of cases; not necessarily indicate an association with an underlying collagen vascular disease.

43) c.

The following are correctly paired associations:
1. Byssinosis-textile industries with cotton, flax, and hemp dust.
2. Malt worker's lung-*Aspergillus clavatus*.
3. Cheese worker's lung-*Aspergillus clavatus* and *Penicillium cassie*.
4. Bird's fancier lung-avian serum proteins.

44) d.

The commonest type of asbestos fibers produced world-wide is the white one (chrysotile; 90%). Exposure mainly occurs through mining and milling of the mineral, as well as demolition, ship breaking, break-pads, pipe and boiler lagging. A strong association exists between asbestosis and lung cancer development, especially in smokers. The risk is multiplicative rather than additive, which may reach 50 folds. However, smoking in patients who are/were exposed to asbestos has no adverse effects in asbestos-related mesotheliomas.

45) e.

Churg-Strauss vasculitis may be the cause, but not Wegener granulomatosis. Pulmonary eosinophilia occurs in acute and chronic eosinophilic pneumonia, hyper-eosinophilic syndrome, fungal and parasitic infections, and drugs (PAS and sulfasalazine). Differential diagnosis of *acute* eosinophilic pneumonia:
1. Acute interstitial pneumonitis: there is rapid onset of acute respiratory failure of unknown causation. The case fatality rate is 50–80%. It can be distinguished from acute eosinophilic pneumonia by doing bronchoalveolar lavage (BAL); there is predominance of neutrophils the BAL fluid of acute interstitial pneumonitis.
2. Acute respiratory distress syndrome: this is typically caused by an underlying condition (e.g., acute pancreatitis, multiple blood transfusions, and massive trauma), which is readily identified.
3. Cryptogenic organizing pneumonia: this disease can have a subacute to fulminant onset. The fulminant form is associated with a toxic exposure (such as chemotherapy or radiation therapy involving the chest) or connective tissue diseases. In the absence of an exposure history, biopsy is necessary.
4. Diffuse alveolar hemorrhage: it presents with shortness of breath with or without hemoptysis. It can be distinguished from acute eosinophilic pneumonia by doing bronchoalveolar lavage, which returns sanguineous fluid. Lung biopsy may be necessary for diagnosis of the underlying cause.
5. Drug-induced lung disease: clinically, it may be differentiated from idiopathic acute eosinophilic pneumonia only by identification and removal of the causative agent.
6. Hypersensitivity pneumonitis: typically, there is an exposure history that precedes the onset of symptoms. Eosinophilia on lavage is generally < 20% of total leukocyte count.

Differential diagnosis of *chronic* eosinophilic pneumonia:

1. Allergic bronchopulmonary aspergillosis: this is an allergic disorder of the airways producing asthma-type symptoms. It is suspected in patients with hard-to-control asthma. Serum IgE level is a good initial screen but a normal level in a patient on corticosteroids does not exclude the diagnosis.

2. Churg–Strauss vasculitis: a form of systemic vasculitis of the respiratory tract, with the triad of paranasal sinus disease, bronchial asthma, and peripheral eosinophilia. Biopsy demonstrating eosinophilic infiltration of blood vessels is required for definitive diagnosis.

3. Cryptogenic organizing pneumonia: radiographically, the subacute and chronic forms may mimic chronic eosinophilic pneumonia. The absence of both peripheral and bronchoalveolar eosinophilia essentially rules out other chronic eosinophilic pneumonia. Lung biopsy may be necessary to establish the diagnosis.

4. Drug-induced lung disease: In the differential diagnosis of chronic eosinophilic pneumonia, drug-induced lung disease must always be ruled out by careful review of health products including over-the-counter, herbal, and vitamin supplements.

5. Helminthic infestation: typically caused by *Ascaris lumbricoides* or *Strongyloides stercoralis*.

Patients complain of episodic cough and wheezing. It is more common in residents of rural, socioeconomically depressed regions. Identification of the parasite in feces, bronchoalveolar lavage fluid, or lung tissue establishes the diagnosis.

6. Idiopathic hypereosinophilic syndrome: a progressive disorder characterized by peripheral blood eosinophilia and eosinophilic infiltration of tissues. It is distinguished from chronic eosinophilic pneumonia by the presence of extra-pulmonary disease and the relatively *poor* response to corticosteroids.

7. Löeffler syndrome: this disease is characterized by migratory pulmonary infiltrates and peripheral eosinophilia. Most reported cases likely represent helminthic infections.

46) c.

Pulmonary alveolar proteinosis (PAP), which is also known as pulmonary alveolar phospho-lipoproteinosis, is a diffuse lung disease characterized by the accumulation of amorphous, periodic acid-Schiff (PAS) -positive lipoproteinaceous material in the distal air spaces. There is little or no lung inflammation, and the underlying lung architecture is preserved.

This lipoproteinaceous material is composed principally of the phospholipid surfactant and surfactant apoproteins. Whole lung lavage may result in remission in 50% of patients. Cough and dyspnea are common.

47) d.

The primary pathological abnormality in this disease is the proliferation of atypical smooth muscle cells around bronchovascular structures and into the interstitium. Hormonal ablation therapy and progestins are of doubtful value. The only established treatment is lung transplantation.

48) d.

Subphrenic abscess is considered a common cause (the resulting pleural effusion is reactionary). The other uncommon causes are uremia, asbestos-related effusion, and Dressler syndrome.

49) e.

Chest tightness is very common in pneumothorax and does not necessarily indicate massive air collection; yet, severe dyspnea is an indication for tube drainage. Excessive coughing during simple aspiration is another indication for tube drainage.

50) e.

Phrenic nerve palsy results elevation of the hemi-diaphragm. The other causes are: excess gas in the stomach or colon; large tumors or cysts in the liver; any cause of reduction in the volume of the lung (lobectomy or lung fibrosis).

51) c.
Wegener granulomatosis, as well as bronchogenic cysts, pulmonary sequestration, lymphoma, benign tumors, aspergilloma, and rheumatoid nodule are considered uncommon causes of solitary pulmonary nodule (or the so-called pulmonary coin lesion).

52) d.

Mycoplasma pneumoniae is one of the causes of bilateral hilar lymphadenopathy. It also causes hemolytic anemia and hepatitis (both can result in jaundice), and positive cold agglutinin (in 50% of cases), which is responsible for the rouleaux formation.

53) e.

Uncomplicated para-pneumonic effusions are common but frank empyemas are not. Stem "e" can be an extra-thoracic manifestation of metastatic abscesses, e.g., in *Staph aureus* lung infection.

54) c.

Bacterial pneumonias in AIDS patients tend to produce focal consolidations or cavities. A normal chest X- ray in such context is more likely to be due to tuberculosis or *Pneumocystis carinii*. Radiographic appearance of pulmonary diseases in HIV infected individuals:
1. Diffuse interstitial infiltrates: *Pneumocystis carinii* pneumonia, tuberculosis, fungal pneumonias, and viral pneumonia.
2. Nodules or masses: tuberculosis, Kaposi's sarcoma, pulmonary lymphoma, cryptococcosis, Aspergillosis, and *Mycobacterium avium* complex.
3. Cavities: bacterial pneumonia, tuberculosis, and Aspergillosis.
4. Focal consolidation: bacterial pneumonia and tuberculosis.
5. Normal chest radiograph: *Pneumocystis carinii* pneumonia and tuberculosis.

55) e.

Lung abscess formation is not a well- known complication of *Pneumocystis carinii* pneumonia (PCP). Infection with PCP, and to some extent bacterial pneumonias, may be associated with expiratory airflow obstruction that persists after resolution of the acute infection, as well as long-term decline in pulmonary function (including permanent decreases in FEV_1, FVC, FEV_1/FVC, and the diffusing capacity of carbon monoxide). Survivors of PCP can also develop lung cysts that may cause spontaneous pneumothorax.

56) a.

Carcinoid tumors are low-grade malignant tumors. The other benign pulmonary neoplasms are neurofibroma, myoepithelioma, granular cell tumor, and clear cell tumor.

57) c.

Primary bronchogenic carcinoma result in multiple lung nodules (cannon balls); synchronous tumors or local lung extension can produce this picture. Causes of multiple pulmonary nodules:
1. Benign: infections (granulomas, septic emboli, parasites), noninfectious granulomas (Wegener granulomatosis, sarcoidosis, rheumatoid arthritis), pulmonary arteriovenous malformations, silicosis, vasculitis, and broncholithiasis.
2. Malignant: metastatic cancer, lymphoma, metastatic bronchogenic carcinoma, Kaposi sarcoma, and synchronous primary bronchogenic carcinomas.

58) c.

Aspirin can cause bronchospasm and pulmonary edema. Clinical presentations of drug-induced lung disease:
1. Barotrauma: amphetamines, cocaine, and marijuana.
2. Bronchiolitis obliterans: gold and penicillamine.
3. Bronchiolitis obliterans with organizing pneumonia: amiodarone, bleomycin, cocaine, cytosine arabinoside, gold, methotrexate, mitomycin C, penicillamine, and radiation.
4. Bronchospasm: acetylcysteine, amphetamines, aspirin, beta blockers, cocaine, cytosine arabinoside, dipyridamole, intravenous contrast, methacholine, non-steroidal anti-inflammatory drugs, pentamidine (inhaled), paclitaxel, propellants, and protamine,
5. Drug-induced lupus: chlorpromazine, hydralazine, isoniazid, penicillamine, phenytoin, and procainamide.
6. Eosinophilic lung disease: beta lactam antibiotics, carbamazepine, phenytoin, isoniazid, L-tryptophan, minocycline, nitrofurantoin, NSAIDs, pentamidine (inhaled), sulfonamides, and tetracycline.
7. Pleural effusion: amiodarone, bromocriptine, drugs associated with drug-induced lupus, interleukin-2, methysergide, nitrofurantoin, sclerotherapy, and tocolytics.

8. Pulmonary edema: aspirin, amphetamines, chlordiazepoxide, cocaine, cytosine arabinoside, hydrochlorothiazide, interleukin-2, opiates, protamine, tocolytics, and tricyclic antidepressant overdose.

9. Pulmonary fibrosis: amiodarone, azathioprine, bleomycin, busulfan, chlorambucil, combination chemotherapy, cyclosphosphamide, melphalan, mercaptopurine, mitomycin C, nitrofurantoin, nitrosoureas (BCNU, CCNU, methyl-CCNU), procarbazine, and radiation.

10. Pulmonary hemorrhage: anticoagulants, cocaine, high-dose combination chemotherapy, mitomycin C, penicillamine, and platelet IIb/IIIa inhibitors.

11. Pulmonary hypertension: anorectic agents and cocaine.

12. Pneumonitis: amiodarone, azathioprine, bleomycin, cocaine, cyclohosphamide, gold, methotrexate, nitrofurantoin, penicillamine, paclitaxel, and radiation.

13. Talc granulomatosis: amphetamines, cocaine, opiates.

59) e.

Causes of transudative pleural effusion are congestive heart failure, nephrotic syndrome, liver cirrhosis (with ascites), myxedema, sarcoidosis, Fontan procedure, and peritoneal dialysis. Chylothorax is a cause of exudative effusion, while urinothorax is a cause of trasudative one. Fontan procedure is an operation that is done for single ventricle or tricuspid atresia (with pulmonary stenosis).

60) e.

Thoracic cage immobility and apical fibrobullous disease are also seen in ankylosing spondylitis. Pulmonary manifestations in rheumatoid arthritis are:

1. Airways disease: cricoarytenoid arthritis, airflow limitation (obstruction), follicular bronchiolitis, bronchiolitis obliterans, and bronchiectasis.

2. Pleural disease: pleuritis, empyema (aseptic or septic), pneumothorax secondary to a ruptured necrobiotic nodule, and bronchopleural fistula.

3. Rheumatoid nodules (necrobiotic nodules).

4. Interstitial lung disease: interstitial pulmonary fibrosis, bronchiolitis obliterans organizing pneumonia, Caplan syndrome, apical fibrobullous disease, and amyloidosis.

5. Drug-induced lung disease: methotrexate pneumonitis, penicillamine, and gold.

6. Thoracic cage immobility.

7. Pulmonary vascular disease: pulmonary arterial hypertension and vasculitis.

8. Infections: reactivation of tuberculosis.

This page was intentionally left blank

Chapter 12
Rheumatology

Questions

1) A 56-year-old man presents with symmetrical deforming arthropathy. His rheumatoid factor is positive. With respect to rheumatoid factor, which one is the *incorrect* statement?
   a) Antibody directed against Fc portion of IgG
   b) IgG, IgM, or IgA class
   c) Diagnostic of rheumatoid arthritis
   d) Can be detected by many lab methods
   e) Found in almost 100% of cases of secondary Sjögren syndrome and Felty syndrome

2) A 45-year-old woman, who is a known case of long-standing rheumatoid arthritis, presents with an exacerbation of her disease. Her C-reactive protein is raised and there are high ESR, normochromic normocytic anemia, and thrombocytosis. Regarding C-reactive protein (CRP) and ESR, all of the following statements are correct, *except?*
   a) CRP, as an acute phase reactant, closely mirrors the degree of inflammatory processes
   b) CRP is the single most useful direct measure of acute phase responses
   c) ESR is an indirect measure of acute phase responses
   d) ESR is characteristically very low in heart failure and sickle cell disease
   e) In systemic lupus erythematosus, the ESR and CRP are elevated during relapses

3) A 51-year-old man, who was diagnosed with psoriatic arthropathy, looks pale. Blood counts reveal normochromic normocytic anemia and hemoglobin of 8.9 g/dl. All of the following statements regarding peripheral blood counts in inflammatory disorders are true, *except?*
   a) Normochromic normocytic anemia is much more common than the hypochromic microcytic one
   b) The platelets are usually elevated and reflect disease activity
   c) The differential white cell count is highly variable
   d) In systemic necrotizing vasculitis, usually there is leukocytosis with relative neutropenia
   e) In systemic inflammatory diseases, neutrophilia may be induced by corticosteroids

4) A 25-year-old female presents with arthralgia, fever, and malar rash. Her urine sediment is active. Blood tests show neutropenia and positive Coombs test. The ESR is moderately raised. Serum ANA is positive and displays a speckled pattern on staining. Regarding anti-nuclear antibodies (ANA), choose the *wrong* statement?

a) The routine detecting test is indirect immune-fluorescent technique
b) The higher the titer, the greater the significance
c) ANA is directed against one or more components of the cell nucleus
d) In drug-induced lupus, it is positive in 30% of cases
e) The sensitivity and specificity of the detecting test vary widely

5) A 54-year-old woman developed right wrist fracture after an apparently trivial fall. Her dorsal spine is kyphotic. Blood biochemical profile is normal. You consider osteoporosis and plan to do DEXA scan. Which one of the following is *not* an indications for bone mineral density measurement?
   a) Previous low trauma fracture
   b) Family history of osteoporotic fracture
   c) Systemic diseases associated with high risk of osteoporosis
   d) Patient on long-term prednisolone therapy
   e) Body mass index >3 Kg/m^2

6) A 65-year-old male was diagnosed with small-cell lung cancer recently. Today, he presents with diffuse bone pains, which are responsive to diclofenac. He denies falls or skeletal trauma. Whole-body radionuclide bone scan is consistent with wide-spread skeletal metastases. Regarding radionuclide bone scanning, which one is the *incorrect* statement?
   a) Utilizes an intravenous injection of ^{99}mTc-biphsphonate and this is detected by γ-camera
   b) Early post-injection images reflect increased vascularity
   c) Delayed images indicate increased bone remodeling processes
   d) Particularly useful in the assessment of local painful skeletal areas with normal or inconclusive plain radiographs
   e) Extremely useful in the assessment of multiple myeloma extent

7) You attend a symposium about the clinical implication of radionuclide bone scanning. You have learned that this investigation is of value in solving the differential diagnosis of bone pain. The main indications for bone scintigraphy are all of following, *except?*
   a) Bone metastasis
   b) Bone or joint sepsis
   c) Early osteonecrosis
   d) Stress fractures
   e) Osteomalacia

8) A 43-year-old male complained of low back pain for the last 2 months. His physician diagnosed simple mechanical back pain. However, you examined him and find "red flags" against that initial diagnosis. All of the following are considered to be red flags for a possible spinal pathology in patients with back pain, *except?*

a) Presentation between the ages of 20-50 years
b) The pain is constant, progressive, or unremitting
c) Dorsal spine pain
d) Presence of systemic symptoms, such as fever, weight loss, and sweating
e) Past history of tuberculosis

9) A 31-year-old soldier presents with low back pain. The pain improves with rest and is exacerbated by activity. He denies back trauma. Features of simple mechanical back pain are all of the following, *except?*

a) Sudden onset, precipitated by sudden lifting or bending
b) The pain is variable but improves with rest
c) The age of the patients is usually between 20-50 years
d) No systemic symptoms
e) About 50% of cases will improve after 1 year

10) A 54-year-old female has severe spinal stenosis and she asks about medications that could relieve her low back pain and left sciatica. Which one of the following is *not* consistent with radicular pain in spinal stenosis?

a) Unilateral leg pain that is worse than the associated back pain
b) Pain radiates below the knee
c) Paresthesia within the same area of pain distribution
d) Reflex changes in lower limbs
e) Does not respond to an nti-epileptic medication

11) Your junior house officer attends a lecture about bone diseases and heard of a disease called DISH. He asks about this disease and what are the radiological findings. Regarding diffuse idiopathic skeletal hyperostoses (DISH), all of the following statements are correct, *except?*

a) Affects 10% of males and 8% of females by the age of 65 years
b) In most cases, there is an associated obesity, hypertension, and type II diabetes mellitus
c) Defined as florid new bone formation along the anteriolateral aspects of 4 contiguous vertebral bodies
d) There is disc space narrowing and marginal vertebral body sclerosis
e) Usually an asymptomatic radiological finding

12) A 63-year-old obese male complains of knee pains and stiffness. He denies fever, joint trauma, or hand symptoms. There is effusion in both knees with crepitus upon movement. You tell him that this is osteoarthrosis. Regarding osteoarthrosis, all of the following statements are true, *except?*
   a) Restriction of joint motion may be due to pain and muscle spasm
   b) The pain is usually variable and intermittent
   c) The pain is mainly seen in weight-bearing joints and is relieved by rest
   d) The pain is insidious over many months and years
   e) Joint deformities usually result in prominent joint instability

13) A 30-year-old Marfan syndrome man complains of knee pains. Plain knee X-ray films are consistent with osteoarthrosis. The list of causes of premature (<45 years of age) osteoarthrosis does *not* include which one of the following?
   a) Localized long-term joint instability
   b) Prior joint disease
   c) Late avascular necrosis
   d) Ochronosis
   e) Growth hormone deficiency

14) A 69-year-old woman has left-sided knee pain and swelling. Otherwise, she is well and takes no medications. Patellar tap sign is positive, and you are about to aspirate the joint. Regarding osteoarthrosis (OA) in elderly people, all of the following statements are correct, *except?*
   a) OA is the major musculoskeletal cause of pain and disability
   b) Aging is not a contraindication to strengthening and aerobic exercises
   c) Total hip replacement with rehabilitation is an excellent cost-effective treatment for severe disabling knee or hip OA
   d) Oral paracetamol and topical NSAIDS are safe
   e) Coexistent calcium pyrophosphate crystal deposition is rarely found

15) A 49-year-old man is referred to you for further evaluation of his rheumatoid arthritis. He is doing well on ibuprofen and methotrexate. He reports little morning stiffness but complains of dyspepsia. Gastroscopy reveals multiple superficial gastric erosions. Regarding rheumatoid arthritis (RA), all of the following statements are correct, *except?*
   a) The commonest cause of inflammatory arthritis
   b) Seen worldwide and in all ethnic groups
   c) The incidence is lowest in African Americans and highest among Pima Indians
   d) In Caucasians, the prevalence is about 10-15% with female to male ratio of 9:1

e) No single factor as an etiology has been identified to date

16) A 32-year-old woman has been recently diagnosed with seropositive rheumatoid arthritis and she asks you about the cause of her joint problem and extra-articular features. With respect to the etiologies of rheumatoid arthritis (RA), choose the *wrong* statement?
   a) HLA DR1 is the major susceptibility haplotype in most ethnic groups
   b) The concordance rate in monozygotic twins is 15% versus 3% in the dizygotic ones
   c) Female gender is considered a risk factor and this susceptibility is increased post-partum and by breast-feeding
   d) Cigarette smoking is a risk factor for RA and for positivity for rheumatoid factor in non-RA subjects
   e) Up to 50% of genetic contribution to susceptibility is due to genes in the HLA region

17) A 57-year-old man presents with progressive right knee pain, stiffness, and swelling. Rheumatoid factor is negative and knee synovial biopsy is consistent with active rheumatoid arthritis. The patient is due to receive disease-modifying medications. Regarding the pathology of rheumatoid arthritis, which one of the following is *incorrect*?
   a) The earliest change is swelling and congestion of the synovial membrane and the underlying connective tissue
   b) The infiltrating T-lymphocytes are mainly CD8+ cells
   c) Subcutaneous nodules consist of central fibrinoid necrosis surrounded by palisading mononuclear cells
   d) The muscles adjacent to the inflamed joints atrophy and there may be low-grade lymphocytic myositis
   e) Effusions into the joints are mainly seen in active disease

18) A 29-year-old female complained of morning stiffness and painful movements of her hands, wrists, and ankles over the last 2 months. You find spindling of the fingers, doughy swollen wrists, and widening of the forefeet. The ESR is raised and there are normochromic normocytic anemia and thrombocytosis. The final diagnosis turns out to be rheumatoid arthritis. Regarding rheumatoid arthritis (RA), all of the following are correct, *except*?
   a) Popliteal cysts usually develop in combination with knee synovitis
   b) Rheumatoid nodules reflect seropositivity and severe erosive disease
   c) Dry eye is the commonest ocular complication of RA
   d) Asymptomatic pericarditis is the commonest cardiac manifestation
   e) Brain vasculitis has serious devastating consequences

19) A 57-year-old female was diagnosed with rheumatoid arthritis 14 years ago. Routine bloods show neutropenia. The spleen is palpable and you palpate cervical lymph nodes. She reports recurrent infections. Regarding Felty syndrome in Rheumatoid arthritis (RA), which one of the following is the *incorrect* statement?
   a)   Afflicts 50% of RA patients
   b)   Develops in the context of long-standing deforming but inactive RA
   c)   Leg ulceration and skin hyperpigmentation are well-recognized features
   d)   There are thrombocytopenia and abnormal liver function tests
   e)   More common in elderly Caucasian females

20) A 58-year-old man, who was diagnosed with rheumatoid arthritis, had been recently prescribed D-penicillamine. He declined this form of therapy, because he surfed the internet and found that this medication might result in many serious side effects and complications. Regarding side effects of D-penicillamine, all of the following statements are correct, *except?*
   a)   Metallic taste in the mouth is reversible
   b)   Myasthenia gravis is a serious but rare complication
   c)   Thrombocytopenia and pancytopenia can develop at any time and are the major concerns
   d)   Febrile reactions are common and benign conditions
   e)   Membranous nephropathy may develop

21) A 34-year-old male has developed rheumatoid arthritis. He prefers oral gold over the injectable form. With respect to the side effects of gold therapy, choose the *incorrect* statement?
   a)   The resulting pruritic skin rash may respond to antihistamines
   b)   Corticosteroids treat severe exfoliative skin rash
   c)   Unlike D-penicillamine, pancytopenia and aplastic anemia are not well-known adverse effects
   d)   Membranous type of nephropathy may develop
   e)   Alopecia can occur

22) One of your rheumatology department colleagues asks you to attend a symposium about seronegative spondyloarthritides, recent advances in their managements, and current research results. The following features of seronegative spondyloarthritides are correct, *except?*
   a)   There are sacroiliitis and spondylitis
   b)   The peripheral symmetrical polyarthritis affects the upper limbs more than the lower ones

c) No association with rheumatoid factor
d) Tendency for familial aggregation
e) Characteristic overlapping extra-articular manifestations

23) A 34-year-old man complains of low back pain and morning stiffness which persist for several hours after awaking from sleep. The pain is relieved by activity and is worsened by rest. He denies back trauma. The right eye is sore and he has heel pain. Your work-up is directed towards ankylosing spondylitis. Regarding the epidemiology of ankylosing spondylitis, which one is the *incorrect* statement?
a) HLA B27 is positive in 90% of affected patients
b) Its peak age of onset is between 6th and 7th decades
c) Male to female ratio is 3:1
d) The overall incidence is 0.5% of the general population but may be much higher in Pima and Haida Indians
e) There is an association with fecal *Klebsiella aerogenes* carriage and chronic non-infective prostatitis

24) A 51-year-old man was diagnosed with ankylosing spondylitis before 15 years. Recently, he has noticed exertional shortness of breath. You hear an early diastolic murmur down the left sternal border. This is aortic regurgitation. All of the following statements with respect to ankylosing spondylitis are correct, *except?*
a) Anterior uveitis is the commonest extra-articular disease
b) Peripheral arthropathy is responsible for 10% of the disease's presentation
c) Apical upper lobe pulmonary fibrocystic disease occurs in 90% of patients
d) Bilateral hip involvement portends poor prognosis
e) Osteoporosis may complicate the spinal involvement and adds more to the disability

25) A 29-year-old male presents with acute painful right hip swelling and fever. He reports a diarrheal illness 2 weeks ago, but denies local trauma or unprotected sex. You detect a rash on his penis, and find planter hyperkeratosis and oral ulcers. HLA B27 is positive. You diagnose Reiter syndrome. Regarding the epidemiology of this syndrome, which one of the following is *wrong?*
a) The commonest cause of inflammatory arthritis in men aged 16-35 years

b) About 1-2% of young males with non-specific urethritis attending sexually transmitted diseases clinics have Reiter syndrome
c) During *Shigella* epidemics, up to 20% of HLA B27 positive young men will develop Reiter syndrome
d) Male to female ratio is 1:5
e) May be preceded by a *Chlamydia* urethritis

26) A 27-year-old man has developed Reiter syndrome after an apparently mild urethritis. He has the fully-fledged syndrome, and he is afraid that he will spend the rest of his life with this disease. He is desperate for help. All of the following statements about Reiter syndrome are correct, *except?*
a) Nail dystrophy may be indistinguishable from that of psoriasis
b) Circinate balanitis occurs in 1% of cases and is extremely painful
c) Buccal erosions are observed in 10% of cases, usually painless, and last only for few days
d) Keratoderma blennorrhagicum is seen in 15% of cases and is clinically and histologically indistinguishable from pustular psoriasis
e) Spondylitis, chronic erosive arthritis, recurrent acute arthritis, and uveitis are the commonest causes of long-term morbidity

27) A 37-year-old male has a long-standing skin disease manifested by red oval plaques with silvery white scales on extensor services. His nails display coarse pitting and there is onycholysis. Recently, has been experiencing hand and wrist pain and stiffness. Rheumatoid factor is negative. You consider psoriatic arthropathy. Regarding the epidemiology of psoriatic arthropathy, all of the following statements are correct, *except?*
a) In 20% of cases, it precedes the onset of skin disease
b) Simultaneous occurrence of skin plaques and joint involvement is very common
c) Approximately, 20% of cases of seronegative spondyloarthritides are due to psoriatic arthropathy
d) Develops 0.1% of the general population
e) The usual age of onset is between 25 and 40 years

28) A 31-year-old female, who is a known case of psoriatic arthropathy, asks you about the treatment options and the long-term prognosis. She has also surfed the internet and found that there are many subtypes of this disease. With respect to psoriatic arthropathy, choose the incorrect statement?
a) The commonest clinical pattern is the asymmetrical inflammatory oligoarthritis, which is responsible for 40% of cases
b) The characteristic psoriatic nail changes are highly uncommon

c) Absence of skin lesions may be due to poor clinical examination
d) Uveitis mainly occurs in patients who have spondylitis and HLA B27 positivity
e) The X-ray findings of psoriatic spondylitis may exactly resemble those of reactive arthritis

29) A 58-year-old obese male presents with acute painful swelling of the left 1st metatarsophalangeal joint. The overlying skin is red and he is feverish. He reports two previous attacks in the same area, and admits to drinking alcohol a lot. This is gout. Regarding the epidemiology of gout, choose the *incorrect* statement?
a) There is a strong male predominance, with male to female ratio of 10:1
b) The overall prevalence is 1% of the general population
c) Serum levels of uric acid are normally higher in females than in males
d) Probably, 95% of hyperuricemic individuals never develop clinical gout
e) Serum uric acid levels are re-distributed in the community as a continuous variable

30) A 54-year-old male develops gout. You tell him that he should avoid certain medications which may worsen his condition. All of the following medications increase serum uric acid, *except?*
a) Cyclosporine
b) Bendroflumethiazide
c) Low dose aspirin
d) Frusemide
e) Azapropazone

31) A 54-year-old man has developed gout. You consider the long-term use of a hypouricemic agent in order to prevent further attacks. Which one of the following is *not* an indications for drug prophylaxis in hyperuricemia and gout?
a) Evidence of tophi
b) Recurrent attacks of acute gouty arthritis
c) Evidence of joint or bone damage
d) Associated renal disease
e) Concomitant treatment with high-dose aspirin

32) Gout is one of the diseases that can result in considerable morbidity in old people. Regarding gout in elderly people, all of the following statements are correct, *except?*

a) Usually secondary to chronic diuretic therapy or chronic renal failure
b) In contrast to primary gout, secondary gout in elderly people usually presents as painful tophi rather than acute attacks
c) There is an increased risk of allopurinol toxicity
d) Acute attacks in elderly people are better treated with joint aspiration and intra-articular injection of steroids
e) Acute attacks can safely be treated with high-dose colchicine

33) You assess a patient, who was referred by his general practitioner as a suspected case of pseudogout. Calcium pyrophosphate dihydrate crystal deposition in joints does *not* result from which one of the following?
a) Sporadic age-related process
b) Pre-existent joint damage and osteoarthrosis
c) Hypoparathyroidism
d) Hypothyroidism
e) Wilson disease

34) You have read an article, which was published recently in a medical journal, about the clinical features of pseudogout and how to screen certain at-risk patients. Which one of the following is *not* an indications for screening metabolic or familial predisposition of calcium pyrophosphate dihydrate (CPPD) deposition and pseudogout?
a) Early onset disease, <55 years of age
b) Florid polyarticular chondrocalcinosis
c) Recurrent acute attacks of pseudogout without chronic arthropathy
d) Presence of clinical or radiological evidence of a predisposing disease
e) Presence of tophi

35) A 26-year-old man presents with acute painful left knee joint. The left knee is warm, swollen, and red and you find restricted active and passive movements. There are fever, lassitude, and leukocytosis. The ESR is raised. You suspect septic arthritis. With respect to this form of join infection, which one of the following statements is *incorrect*?
a) It is a medical emergency
b) Has a mortality rate approaching 10%
c) The objectives of treatment are pain relief, parenteral antibiotics, adequate drainage, and early active rehabilitation
d) Synovial fluid culture is positive in 90% of cases while initial Gram-stain is positivity is 50%
e) The commonest infective agent in adults is *Staphylococcus albus*

36) A 46-year-old female is referred to you as a suspected case of fibromyalgia. Initially, she was given a diagnosis of endogenous depression, but did not respond to drug therapy. Regarding fibromyalgia, all of the following statements are correct, *except?*

a) Its prevalence is 2-3% of the general population and occurs in 7% of woman above the age of 70 years
b) There is a strong female predominance, with female to male ratio of 10:1
c) Despite intensive and invasive investigations, no structural, inflammatory, or endocrine abnormality has been found
d) The ESR is characteristically high
e) The mainstay treatment is low-dose amitriptyline with graded aerobic exercises

37) A 48-year-old male presents with hip fracture after a minor fall. He is using any medications for his chronic severe asthma. DEXA scan reveals a T-score of -3.1 at both hips. The list of medications that increase the risk of osteoporosis does *not* include which one of the following?

a) Corticosteroids
b) b Unfractionated heparin
c) GnRH analogues
d) Sedatives
e) Tibolone

38) A 65-year-old female has developed progressive loss of height and protuberant abdomen. Her dorsal spine is kyphotic. Lateral x-ray film of the dorsal spine reveals severe osteopenia and vertebral compression fractures. Which one of the following medications does *not* reduce the risk of vertebral osteoporotic fractures?

a) Bisphosphonates
b) Hormonal replacement therapy
c) Raloxifene
d) Calcitonin
e) Vitamin-D with calcium

39) A 55-year-old woman has been referred to you after developing right-sided Colle's fracture. Which one of the following medications decreases the risk of non-vertebral osteoporotic fractures?

a) Corticosteroids
b) Calcium
c) Tibolone

d) Calcitonin
e) Bisphosphonates

40) A 25-year-old epileptic female presents with proximal weakness, waddling gait, and bone pain. She is on long-term phenytoin. X-ray films reveal the presence of Looser zones. Regarding osteomalacia and rickets, all of the following statements are correct, *except?*
   a) Osteomalacia secondary to aluminum toxicity requires bone biopsy for diagnosis and may respond to desferrioxamine
   b) Hypophosphatasia-associated cases are resistant to treatment due to pyrophosphate accumulation
   c) Hypophosphatemic rickets responds well to life-long phosphate replacement *plus* vitamin D metabolites
   d) Type II vitamin D-dependent rickets responds favorably to active vitamin D metabolites
   e) Bisphosphonate-induced osteomalacia is usually transient and asymptomatic

41) A 76-year-old man complains of continuous pain in his right thigh and leg. You find prominent deformity of that limb. He says that his hearing is not good. You notice that his skull is enlarged. Apart from prominently elevated serum alkaline phosphatase level, his blood tests are unremarkable. Regarding Paget disease of the bone, all of the following statements are correct, *except?*
   a) Rarely spreads to new bones once the diagnosis is made
   b) The primary abnormality is increased bone osteoclastic resorption accompanied by marrow fibrosis, increased vascularity, and osteoblastic repair
   c) The classic presentation is bone pain, bone deformity, deafness, and pathological fractures
   d) High output cardiac failure is rare and usually develops in patients with extensive disease
   e) Osteosarcoma eventually develops in all patients

42) A 28-year-old athlete presents with polyarthralgia, fever, and facial rash. She has right-sided pleural pain, dark urine, and reticular rash on her thighs. There are positive ANA testing, raised ESR, positive Coombs test, low hemoglobin, and leucopenia. Regarding systemic lupus erythematosus (SLE), all of the following statements are correct, *except?*
   a) There is monoclonal B-cell and T-cell activation
   b) None of the diverse clinical manifestations of SLE can be attributed to a single antigenic stimulus

c) Exacerbations can be induced by sun light, infection, and pregnancy
d) Although anti-dsDNA antibodies are highly specific of SLE, they are positive only in 30-50% of active disease
e) The renal lesions are due to immune-complex disease, while the brain damaging effect is mainly due to the antibody-mediated cell cytotoxicity

43) A 31-year-old SLE female patient complains of shortness of breath upon exertion. You find bipedal leg edema. Examination reveals S_3 gallop and mitral regurgitation. Echocardiography shows ejection fraction of 33%. You consider lupus cardiomyopathy. All of the following statements with respect to SLE are correct, *except?*
a) The classic butterfly rash is seen in 20-30% of cases
b) Lupus arthritis is non-deforming and non-erosive
c) The commonest renal lesion is diffuse proliferative glomerulonephritis
d) Chest pain is usually due to pleuritis and/or pericarditis
e) The commonest CNS manifestation is chorea

44) A 37-year-old female presents with tight smooth facial skin, Reynaud phenomenon, and shortness of breath. The skin of the fingers looks tight and shiny with non-pitting edema. Serum anti-Scl70 is positive, and the DLCO is low. You find conduction defects on 12-lead ECG. Regarding scleroderma, all of the following statements are correct, *except?*
a) The etiology is still unknown, and there is no consistent genetic, geographical, or racial association
b) Erosive arthropathy is very common
c) Restricted hand movement is mainly due to skin disease rather than joint disease
d) Pulmonary involvement is the major cause of morbidity and mortality
e) One of the main causes of death is hypertensive renal crisis

45) A 37-year-old female, who is a known case of scleroderma, presents with rapidly progressive renal impairment and severe hypertension. You suspect scleroderma renal crisis. Which one of the following statements regarding scleroderma is *incorrect?*
a) The 5-year survival is approximately 70%
b) Poor prognostic factors at the time of diagnosis are old age and proteinuria
c) Recurrent occult upper gastrointestinal hemorrhage may indicate watermelon stomach

d) Pulmonary hypertension is more common in the diffuse variety than in the limited one

e) In 20-30% of the diffuse type, anti-Scl70 antibody is positive while anti-centromere antibody is positive in 70% of the limited type cases

46) A 51-year-old female complains of eye grittiness and eye mucosal threads. She is intolerant to wind and has to drink water frequently when eating to facilitate swallowing. You find bilateral parotid gland enlargement and urticarial skin lesions. Anti-Ro and anti-La antibodies are positive. All of the following statements about Sjögren syndrome are correct, *except?*

a) More common in females between 40-60 years of age and HLA B8/DR3

b) Parotid sialogram may show sialectasis, which may be diffuse, punctate, or cavitary

c) ANA is positive in almost 100% of primary Sjögren syndrome patients

d) Extensive dental carries and oral candidiasis are common.

e) Previous history of head and neck irradiation should be excluded beforehand

47) A 56-year-old male presents with painful and tender proximal symmetrical myopathy, shortness of breath, and palpitations. Serum muscle enzymes are raised, EMG is myopathic, and inflammatory myositis on quadriceps muscle biopsy is found. Regarding polymyositis and dermatomyositis, all of the following statements are correct, *except?*

a) Proximal muscle weakness is very common while facial weakness occurs in 5% of patients

b) Dermatomyositis muscle biopsy shows characteristic peri-fascicular atrophy due to capillaritis

c) Serum creatine phosphokinase level is usually normal

d) The childhood type carries a poor prognosis, because of associated visceral vasculitis

e) EMG may show pseudo-myotonic pattern

48) You attend a symposium about recent medical advances in vasculitides. Your interns ask you what you have learned from this symposium. Which one of the following you have *not* learned?

a) Fever, malaise, weight loss, anorexia, and sweating are common non-specific manifestations

b) There are neutrophil leukocytosis and raised ESR

c) Hepatitis B surface antigen is positive in 5-50% of polyarteritis nodosa patients

d) Oral co-trimoxazole may prevent relapses in localized Wegener granulomatosis

e) cANCA is found in to 90% of active Churg-Strauss vasculitis

49) A 61-year-old man presents with fever, lassitude, painful urticaria, weight loss, and rapidly progressive azotemia. The ESR is raised and there are neutrophilic leukocytosis and negative testing for ANA and rheumatoid factor. You suspect a systemic vasculitis process. All of the following statements about systemic vasculitides are correct, *except?*

a) Hypertension occurs in 30% of cases of polyarteritis nodosa

b) Paranasal sinus disease may indicate Churg-Strauss or Wegener granulomatosis

c) Pulmonary artery involvement is seen in Takayasu arteritis

d) Giant cell arteritis is treated by 10-20 mg daily prednisolone

e) In Kawasaki disease, the mortality is 1% and disease relapse is rare

50) A 46-year-old male, who is a known case of rheumatoid arthritis, takes long-term prednisolone. Today, he presents with painful right hip. X-ray films are suggestive of avascular necrosis of the right femoral head. All of the following medications are correctly matched with their musculoskeletal effect, *except?*

a) Glucocorticoids-osteomalacia

b) Thiazide diuretics-secondary gout

c) Procainamide-drug induced lupus

d) Penicillamine-myasthenia gravis

e) Amphetamines-vasculitis

Rheumatology

Answers

1) c.

The slide latex agglutination, SCAT, and Rose-Waaler tests can be used to detect rheumatoid factor. The slide (latex) agglutination test detects rheumatoid factor of the class IgM; IgG and IgA are not picked up by this test. Rheumatoid factor is not diagnostic of rheumatoid arthritis; however, high serum titer of this factor portends poor prognosis in rheumatoid arthritis. In terms of sensitivity, it is present in the majority of patients with erosive rheumatoid disease but may only appear after months or years of disease, once the diagnosis is beyond dispute. It is therefore neither sufficient nor necessary for the diagnosis. IgG rheumatoid factor has greater specificity for major rheumatic disease.

2) e.

C- reactive protein (CRP) rises rapidly and falls quickly; therefore, it is a direct measure of the degree of the acute phase reactants and mirrors the magnitude of inflammatory reactions. The ESR changes relatively slowly; in cases of clinical improvement, CRP normalizes rapidly while the ESR lags behind for a variable period of time (which may wrongly suggest non-effective management). Polycythemia rubra vera is another cause of low ESR in spite of the florid acute phase reactant changes (in these cases, measurement of the plasma viscosity reflects the degree of acute phase reactants). SLE usually does not raise serum CRP level; high values indicate associated infection, trauma, or serositis in SLE.

3) d.

The RBCs are hypochromic microcytic in quarter of cases of anemia of chronic diseases; this anemia does not respond to iron therapy. It is not an iron deficiency stats, but there is a defect in the kinetics and utilization of iron. However, it may be accompanied by iron deficiency anemia; e.g., due to long-term treatment with NSAIDs. Thrombocytosis is a useful marker of active inflammatory diseases, as are leukocytosis and neutrophilia. Leukocytosis and neutrophilia are found in systemic vasculitis.

4) d.

The routine detecting test is *indirect* immune-fluorescent technique (using rodent organs or human cell lines). Like rheumatoid factor, low tires of ANA may be encountered in healthy people.

The sensitivity and specificity of the detecting test vary widely; it depends on the antigen preparation used in the test and whether we detect IgG or IgM type. ANA is directed against one or more components of the nucleus (hence the designation anti-nuclear). Serum ANA (mainly of anti-histone H_2, A and B subtypes) is positive in almost 100% of drug-induced lupus; a negative titer virtually excludes drug-induced lupus in the appropriate clinical setting.

5) e.

A fracture (e.g., Colle's or femoral neck) is said to be of a low threshold when patients fall from standing height or less. The other indications for doing DEXA are radiological evidence of osteoporosis (on X-ray films, e.g. severe osteopenia of vertebral bodies) and premature gonadal failure, clinical features of osteoporosis (loss of height and progressive kyphosis), body mass index <19 Kg/m^2 (obese people are protected against osteoporosis), and follow-up of treatment with anti-osteoporosis medications. Bone mineral density (BMD) in patients receiving osteoporosis treatments changes slowly; therefore, repeat measurements of DEXA scan T-scores should be done after at least 1-2 years of starting therapy. Assessing BMD at the spine is better than the hips for assessing the overall treatment response, because precision is better and changes in BMD at the spine occur more quickly (higher content of trabecular bone).

6) e.

^{99}mTc-biphsphonate is an isotope that is taken up by bones; therefore, it is utilized in bone scintigraphy. Early films (post-injection) are useful to detect increased bone/joint vascularity, for example, in primary and secondary bone tumors, as well as Paget disease of the bone and inflammatory synovitis. After few hours, the isotope localizes itself to areas of increased bone remodeling. Although there are many indications for doing bone scintigraphy, but it is mainly useful in the assessment of local painful skeletal areas which have normal or inconclusive plain radiographs. Bone scanning is abnormal only in the presence of fractures in multiple myeloma; otherwise, it is useless because the increased myeloma osteoclastic activity is not coupled by augmentation in bone osteoblastic activity.

7) e.

The other indications for bone scintigraphy are reflex sympathetic dystrophy and hypertrophic osteoarthropathy.

It is also useful in the assessment of the extent of Paget disease of the bone. Bone scanning is abnormal in the presence of fractures in multiple myeloma. Bone scintigraphy is not required to diagnose/follow-up osteomalacia patients.

8) a.

Simple mechanical low back pain usually targets people between the ages of 20-50 years; the occurrence of this pain in patients younger than 20 or older than 50 years should be taken seriously (a red flag). The other red flags are the presence of painful spinal deformity, severe symmetrical spinal deformity, saddle anesthesia, progressive neurological signs in lower limbs, sphincter dysfunction, and a sensory level on the trunk. Past histories of back trauma, HIV, steroid use, and systemic malignancy should always prompt a search for causes other than the simple mechanical one.

9) e.

About 90% of cases of simple mechanical back pain improve after 6 weeks; therefore, it has an excellent prognosis. The other features of this type of back pain are: tendency for recurrent episodes; pain that is limited in the back or thigh but never radiates below the knee; and no clear-cut nerve root signs.

10) e.

The prognosis is reasonable, with 50% recovery at 6 weeks in patients who have developed radicular pain of spinal stenosis. Neuropathic pains respond favorably to anti-epileptics; NSAIDs are weak agents against neuropathic pain.

11) d.

Diffuse idiopathic skeletal hyperostoses is a relatively common disease of the spine, that is usually asymptomatic (rarely, patients report low back pain). Hyperinsulinemia is another association. The disease's definition does not encompass spondylotic changes (disc space narrowing, marginal vertebral body sclerosis, and posterior apophyseal joint involvement).

12) e.

The constellation of joint pain, muscle spasm, capsular fibrosis, and intervening large osteophytes results in joint movement limitation. There is palpable, sometimes audible, coarse crepitus.

Most patients report pains that are variable and intermittent; so-called "good days and bad days." The pain is usually augmented by activity (joint movement) and is usually relived by rest. In spite of being a wide-spread disease of the skeleton, many patients insist that the pain is in few joints only; rarely, the pain is wide-spread in many joints at the same time. Although the affected joint is painful and deformities may occur, but joint instability is a rare occurrence.

13) e.

Pre-mature (occurrence <45 years of age) osteoarthrosis is almost always mono-articular. The commonest causes are prior joint disease and localized joint trauma. Arthropathy afflicts 50% of acromegaly patients (not growth hormone deficient patients).

14) e.

Osteoarthrosis is a painful disease that you should treat appropriately to minimize disability. Oral paracetamol and topical NSAIDs have no significant drug-drug interaction and are often effective at relieving pain. Calcium pyrophosphate dihydrate crystal deposition is a common age-associated phenomenon (and accompaniment) and may precipitate acute pseudo-gout attacks in patients with osteoarthrosis.

15) d.

In Caucasians, the prevalence is about 1-1.5 %, with female to male ratio of 3:1. Whatever the initiating stimulus is, rheumatoid arthritis is characterized by persistent cellular activation, autoimmunity, and the presence of immune complexes at sites of articular and extra- articular lesions. This leads to chronic inflammation, granuloma formation, and joint destruction.

16) a.

HLA DR4 is the major susceptibility haplotype in most ethnic groups, e.g. found in 75% of Caucasians with RA. HLA DR4 positivity is more common in patients with severe disease. Indian and Israeli patients are usually HLA DR1 positive, while HLA DW15 is more common in Japanese cases. The following factors, when found at the time of presentation, are associated with a poor outlook: higher baseline disability, female sex, involvement of the feet's metatarsophalangeal joints, positive rheumatoid factor (especially when high titers are present), and disease duration greater than 3 months.

Prominent extra-articular manifestations and disease that is active for 1 year without any intervening remissions are the other poor prognostic indicators. Peri-articular osteopenia is usually found on joint X-ray films in the first 6 months; erosions are usually uncommon within the first year. Strictly relying on marginal erosions to establish the diagnosis should therefore be discouraged. It is not appropriate to await radiographic changes before making the diagnosis.

17) b.

The inflamed synovium is infiltrated by CD4+ lymphocytes, plasma cells, and macrophages. The disease's granulomas may also be found in the sclera, pericardium, pleura, and lungs. The chronically inflamed joints gradually develop fibrous or bony ankylosis. The regional lymph nodes draining the actively inflamed joints are frequently hyperplasic.

18) e.

Baker cysts may rupture and this event may simulate calf deep venous thrombosis; both may coexist together. Rheumatoid nodules are usually found in extensor body surfaces, sclera, and lung. Episcleritis is a benign complication but sclertitis is serious (the underlying uveal tract may also be inflamed). Acute pericarditis occurs in 30% of nodular seropositive patients, but the constrictive type is rare. The CNS is surprisingly spared in rheumatoid vasculitis, unlike the situation in systemic vasculitides.

19) a.

Felty syndrome complicates <1% of long- standing rheumatoid arthritis patients. The joint disease itself is seropositive, nodular, and deforming, but is *inactive*. Weight loss, recurrent infections (from defective T and B cell function and neutropenia), and sicca syndrome are common accompaniments. The resulting anemia is normochromic normocytic. Most patients are between 50-70 years of age.

20) d.

Metallic taste in the mouth is a common side effect, but frank mouth ulceration and pemphigus are rare adverse events. Penicillamine can also result in drug-induced lupus and Goodpasture syndrome (both are rare). Rapidly falling platelet count, mild thrombocytopenia, and proteinuria call for stopping penicillamine therapy for some time.

It can be re-introduced slowly later (if these adverse effects recur, then stop it for good). The occurrence of febrile reactions and pancytopenia are mandatory indications to stop this medication immediately and forever.

21) c.

Marrow suppression and aplastic anemia may occur during therapy with gold compounds, and both carry a significant mortality.

22) b.

In ankylosing spondylitis for example, the peripheral *asymmetrical oligoarthritis* affects lower limbs more than the upper limbs. Enthesopathy is a common occurrence. Iritis develops in 20- 25% of patients (and 20% have conjunctivitis), while ascending aortitis affects 4% of the patients (usually after many years); both are the commonest extra-articular manifestations. Apical fibro-cystic disease (which resembles post-primary tuberculosis) is rare (about 1% of patients).

23) b.

The peak incidence of ankylosing spondylitis is in the 2nd and 3rd decades. The incidence is higher in Pima and Haida Indians (because of higher prevalence of HLA B27 in these populations). The *Klebsiella* carriage may be responsible for joint and eyes disease flares-ups.

24) c.

Anterior uveitis occasionally precedes joint disease. Conjunctivitis occurs in 20% of cases. Extra-spinal (peripheral joints) involvement is observed in 40% of cases. This is the sole presenting feature in 10% of patients only (i.e., precedes the spinal symptoms); usually asymmetrical at first and may cause inflammatory symptoms that mainly affect the hips, knees, ankles or shoulders. In a further 10% of patients, symptoms begin in childhood as one variety of pauciarticular juvenile idiopathic arthritis. Apical upper lobe pulmonary fibrocystic disease occurs in 1% of patients. Aortitis, aortic and mitral regurgitations, and cardiac conduction defects may also occur and increase the disease morbidity. Bilateral hip involvement portends poor outlook because of the resulting loss of hip flexion; patients would be unable to compensate for the spinal rigidity.

25) d.

The male to female ratio is 15:1 (there is a strong male predominance). Reiter syndrome may be preceded by *Chlamydia* urethritis, intestinal *Salmonella*, *Shigella*, *Campylobacter*, and *Yersinia* infections.

26) b.

Circinate balanitis develops in 20-50% of cases and are usually *painless* and easily escape notice. The first attack of arthritis is usually self-limiting; spontaneous remission of symptoms is the usual outcome and occurs within 2-4 months of onset. However, recurrent or chronic arthritis develops in more than 60% of patients and is not necessarily related to further infections. Extra-articular features are uncommon but include cardiac abnormalities (aortic incompetence, conduction defects, and pericarditis), peripheral neuropathy (usually in the form of foot drop and ulnar neuropathy), and CNS (especially seizures and meningoencephalitis).

27) b.

The synchronous onset of skin disease and joint disease is uncommon and is observed in 5% of cases only. Psoriatic arthritis occurs in about 1 in 1000 of the general population and in 7% of patients with psoriasis. Approximately 20% of all patients with seronegative polyarthritis have psoriasis, while the prevalence of psoriasis in seropositive rheumatoid arthritis is no higher than that in the general population (suggesting that the association of the skin disease with seronegative arthritis does not arise by chance alone). The onset is usually between 25 and 40 years of age.

28) b.

Arthritis mutilans complicates 5% of cases only. The nail changes of psoriasis occur in 85% of psoriatic arthritis, while they are found in 1/3 [rd] of uncomplicated psoriasis patients. The typical psoriasis skin lesions may only develop in hidden areas, such as genital areas, natal cleft, and scalp. Although anterior uveitis occurs, but conjunctivitis is much more common. There are asymmetrical, coarse, and non-marginal syndesmophytes as well as asymmetrical sacroiliitis; the same changes are found in reactive Reiter arthritis.

29) c.

Serum levels of uric acid are higher in males than females. Gout is a disease of post-pubertal males and post-menopausal females. About 1% of patients with primary gout have a specific inherited enzyme defect of purine synthesis. Such abnormal enzymatic activity should be suspected if gout develops before the age of 25 years, if uric acid stones are the presenting feature, or if there is a strong family history of early-onset gout. Apart from hyperuricemia, other risk factors and inter- related associations for primary gout include obesity, high alcohol ingestion, type IV hyperlipidemia, hypertension, and coronary artery disease. Hyperuricemia is common while clinical gout is not.

30) e.

Azapropazone is a uricosuric agent. The other agents that result in increased serum uric acid are pyrazinamide, alcohol, and lead (saturnine gout). High-dose aspirin promotes uric acid excretion by the kidneys, while low-dose aspirin enhances hyperuricemia.

31) e.

The other indications for chronic hypouricemic therapy are gout with a very high serum uric acid level as well as hyperuricemia due to congenital enzyme defects (such as HGPRTase deficiency). Stem "e" is a powerful uricosuric measure; however, the mere concomitant use of low-dose aspirin does not call for hypouricemic therapy.

32) e.

Chronic diuretic therapy of more than 18 months duration may result in hyperuricemia and secondary gout; the hands, not the feet, are the usual targets in those patients. A daily dose of 100 mg of allopurinol is better to be prescribed in elderly people to lessen allopurinol toxicity. NSAIDs and colchicine are best avoided in the elderly because of increased incidence and severity of toxicities. There is no conclusive evidence that asymptomatic hyperuricemia itself is damaging; treatment is therefore unnecessary unless there is a strong family history of gout, uric acid renal stone formation, or persistently very high serum levels (>0.6 mmol/L). Causes of secondary hyperuricemia should always be considered and treated.

Renal uricosuric agents should avoided in uric acid over-producers (gross uricosuria is already present), in renal impairment (obviously, they are ineffective), and in patients with urolithiasis (this would definitely increase uric acid stone formation).

33) c.

Calcium pyrophosphate dihydrate crystal deposition in hyaline and fibrocartilage of joints (so-called chondrocalcinosis) is a common age-associated phenomenon (in people >55 years) that particularly targets the knee. Hyperparathyroidism, not hypoparathyroidism, can result in chondrocalcinosis. The other causes of chondrocalcinosis are hemochromatosis and hypophosphatasia. Chondrocalcinosis can be sporadic, familial, and metabolic disease- associated. It is often clinically silent but may result in acute self-limiting attacks of synovitis (termed pseudogout) or simply may present as a chronic form of arthritis (that shows a strong association and overlap with osteoarthrosis; both are age-related phenomena).

34) e.

Tophi are the hallmark of prolonged hyperuricemia, not pseudogout; however, both may coexist. Metabolic screening for chondrocalcinosis includes serum calcium (for hypercalcemia), alkaline phosphatase (very low serum levels may point out towards hypophosphatasia), magnesium, and ferritin and liver function testing (for hemochromatosis).

35) e.

The mortality rate of septic arthritis is generally about 10%, and there is an appreciable morbidity. Inpatient management is a must (patients should not be treated on an outpatient basis). About 70% of cases are due to *Staphylococcus aureus*; the synovial fluid culture is positive in 90% of cases, while initial Gram stain is positivity is 50%. Synovial culture is positive in 25% of gonococcal cases, while about 70% positivity rate is obtained when the genital tract swabs are cultured. Joint aspiration is essential whenever septic arthritis is suspected. Fever with peripheral neutrophilic leukocytosis and raised ESR occur in most patients; however, they may be normal in elderly, immune-compromised patients, or early in the course of the disease.

36) d.

The 2 well-recognized risk factors for fibromyalgia are age and female sex; however, no age is exempt, but the disease is rare in young people. Only 2 consistent associations have been shown and these are sleep abnormality and abnormal pain perception. The ESR should be within its normal reference range (for age and sex). The mainstay treatment is low-dose amitriptyline with graded aerobic exercises; this is an evidence-based intervention. Recognized risk factors (other than ageing and female gender) include a wide variety of life events (all of which are linked with psychosocial distress) and these are: divorce, marital disharmony, family history of alcoholism, violent traumatic injury (or assault), and self-reported childhood abuse.

37) e.

Tibolone is a hormone receptor modulator, which acts as a partial agonist at estrogen, progesterone, and androgen receptors. It is used in the prevention of postmenopausal osteoporosis, symptomatic treatment of hot flushes and associated sweating resulting from menopause (surgical or natural), improvement of bone mineral density in patients with established postmenopausal osteoporosis, and is used in the treatment of osteoporosis.

38) e.

Vitamin-D and calcium increase the bone mineral density and decrease the risk of non-vertebral osteoporotic fractures; data are still not available regarding the reduction in vertebral fractures.

39) e.

Only three medications have been consistently shown to decrease the risk of *non*-vertebral osteoporotic fractures and these are bisphosphonates, hormonal replacement therapy, and calcium *plus* vitamin D. The response of osteoporosis to treatment should ideally be monitored by doing spinal bone mineral density measurements 1-2 years after starting this form of therapy. An increase in this bone mineral density of between 5-8% is expected while using bisphosphonates and hormonal replacement therapy; the increase in bone mineral density with the use of other agents has been reported to be 1-3%.

Some patients still develop fractures while receiving anti-osteoporosis therapy and this is not necessarily a reason for stopping this form of treatment; even the most potent anti-osteoporosis agent would reduce the risk of such fractures by 50%, i.e. there is no 100% protection.

40) d.

Aluminum toxicity-associated osteomalacia is encountered in chronic renal failure; very rare nowadays because of the elimination of aluminum from dialysate fluids. Congenital defects in bone-specific alkaline phosphatase result in hypophosphatasia (and low serum alkaline phosphatase), which results in a severe resistant form of osteomalacia and rickets. Hypophosphatemic rickets responds well to life -long phosphate replacement *plus* vitamin D metabolites; no need for calcium at all. Type I vitamin D resistant rickets responds well to active vitamin D metabolites; type II disease is very resistant to calcium and vitamin D metabolites. Bisphosphonate-induced osteomalacia is usually transient and asymptomatic, and is mainly seen in Paget's disease patients who are given high doses of etidronate

41) e.

Paget disease of the bone rarely spreads to new bones once the diagnosis is made, indicating that the bone targeting insult occurs early in life. The disease causes bone thickening with abnormal architecture and reduced mechanical strength. Those who have limited cardiac reserve may develop high output heart failure when the disease is wide-spread (because of increased bone vascularity and blood shunting). Osteosarcoma development is a rare event and portends a very poor prognosis; it should be suspected in any rapid increase in local pain or swelling of an affected bone.

42) a.

There is polyclonal B and T cell activation and hypergammaglobulinemia. Fever, weight loss, and mild lymphadenopathy commonly accompany active disease. Gastrointestinal involvement is rare and other causes of abdominal pain should always be considered, i.e. appendicitis, perforation secondary to drugs, or infection.

43) a.

The other skin manifestations are discoid rash, photosensitive rash, livedo reticularis, lupus panniculitis and lupus profundus, purpura, skin ulceration, nail-fold infarcts, and bullous lesions. Contractures around joints may occur due to tendon inflammation and may be seen as bone deformity (Jaccoud's arthropathy). Diffuse glomerulonephritis portends a poor prognosis and should be treated aggressively. Pleural effusions of SLE are exudative; rarely the chest pain is due to myocardial infarction. Mild neuropsychiatric manifestations (such as depression or psychosis) and seizures are the usual neurologic features. Antibody-mediated destruction of peripheral blood cells may cause neutropenia, lymphopenia, thrombocytopenia, or Coombs positive hemolytic anemia. The degree of leucopenia, most commonly lymphopenia, is often a good guide to disease activity. Overall, the 5-year survival is >90%.

44) b.

Erosive arthropathy is highly uncommon in scleroderma. Muscle weakness and wasting is usually due to low-grade myositis. One of the main causes of death is hypertensive renal crisis, which is characterized by rapidly-evolving malignant hypertension and renal failure. Treatment is by angiotensin-converting enzyme inhibition, even if renal impairment is present.

45) d.

Pulmonary hypertension is 6 times more common in the limited type of scleroderma than in the diffuse one. Steroids and cytotoxic medications are indicated in patients with myositis or alveolitis. No agent has demonstrated efficacy in arresting or improving skin changes. Poor prognostic factors at the time of diagnosis are old age and proteinuria, diffuse skin disease, high ESR, low DLCO, and pulmonary hypertension. Watermelon stomach occurs in 20% of patients.

46) c.

ANA is positive in almost 100% of cases of secondary Sjögren syndrome of rheumatoid arthritis. In the primary one, ANA is positive in about 50-70% of patients. Anti-Ro and anti-La antibodies are positive in 50-70% of cases and both are more suggestive of the primary syndrome. Most patients have raised ESR secondary to hypergammaglobulinemia and one or more antibodies (of which, antinuclear and rheumatoid factors are the most common).

This sicca syndrome can be diagnosed by Schirmer's tear test (which measures the tear flow over 5 minutes using absorbent paper strips placed in the lower lachrymal sac; a normal result is greater than 6 mm of wetting). If diagnosis is still questionable, it can be secured by demonstrating the presence of focal lymphocytic infiltrate in the minor salivary glands on lip biopsy; parotid gland biopsy is not required. Previous history of HIV, hepatitis C infection, sarcoidosis, and head and neck irradiation should be excluded before diagnosing the primary type of Sjögren syndrome.

47) c.

The serum level of creatine phosphokinase is usually elevated and is a guide to disease activity. However, a normal value does not exclude the diagnosis, particularly in juvenile cases, where $2/3^{rd}$ of patients only have a raised creatine phosphokinase the time of diagnosis. Anti-Jo1 antibodies correlate with pulmonary parenchymal involvement (and interstitial fibrosis). Extra-ocular muscle involvement is seen in 2% of cases only. Occasionally, a biopsy may be normal, particularly if the myositis is patchy; in such cases, MRI is a useful means of identifying areas of abnormal muscle that are amenable to biopsy. The weakness in inclusion-body myositis is predominantly distal, although proximal weakness does occur; serum creatine phosphokinase is usually marginally raised and EMG studies may show neuropathic changes besides the myopathic ones

48) e.

Neutrophilic leukocytosis is encountered; prominent peripheral blood eosinophilia points out towards Churg-Strauss vasculitis. Hepatitis B surface antigen is positive in 5-50 % of polyarteritis nodosa; the treatment is mainly targeted against the hepatitis B infection (with interferon gamma or lamivudine). Oral co-trimoxazole has been shown to prevent relapses in localized Wegener granulomatosis, and sometimes methotrexate is used. Serum cANCA is found in 90% of active Wegener granulomatosis and reflects disease activity. Anti-neutrophil cytoplasmic antibodies (ANCA) are directed against enzymes present in neutrophil granules.

49) d.

Giant cell arteritis should be treated by high doses of steroids, such as 80 mg/day of prednisolone, as the risk of visual loss is high.

Early detection of Kawasaki disease and prompt initiation of therapy with aspirin and intravenous immunoglobulin have reduced the death rate to well below 1% and the prevalence of coronary artery aneurysms to approximately 5%. The pulmonary arteries are involved pathologically in up to 50% of Takayasu arteritis; however, symptoms related to pulmonary arteritis are less common.

50) a.

Glucocorticoids cause proximal myopathy with type II fiber atrophy, osteonecrosis, and osteoporosis (not osteomalacia). Femoral neck fracture, dislocation/fracture dislocation, and minor bone trauma may result in "traumatic" osteonecrosis. Non-traumatic causes of osteonecrosis are SLE, alcoholism, sickle cell disease, glucocorticoids administration (it is rare with non-iatrogenic Cushing syndrome), pancreatitis, pregnancy, chronic renal failure or hemodialysis, hyperlipidemia, radiation, smoking, hyperuricemia/gout, HIV, DIC, organ transplantation, and last but not least, idiopathic. Approximately 1% of patients treated with penicillamine develop autoimmune myasthenia gravis; penicillamine-induced disease shares many of the characteristics of primary myasthenia gravis. Myasthenia induced by penicillamine resolves when the drug is withdrawn, in the majority of cases.

This page was intentionally left blank

You may also try other MRCP self-assessment books, which were written by Osama S. M. Amin:

1. Get Through MRCP;BOFs:

This was published initially by the Royal Society of Medicine (RSM) Press Ltd, London in 2008. Now the book is published by CRC Press of Taylor & and Francis. *Get Through MRCP Part 1: BOFs* provides over 600 questions and answers, allowing the reader to test their knowledge in preparation for the MRCP Part 1 examination. Questions are presented in the style used in the real examination, and answers are supplemented with useful additional explanatory material to help the reader understand why their answer was right, or wrong. The book offers a useful review of all elements of the syllabus, so the reader can feel fully prepared when they enter the examination room.

2. Mock Papers for MRCPI, 2nd Edition:

This was published in December 2016 by Lulu Press Inc. In this book, you will find 3 mock papers. Each one contains 100 questions in a Best of Five (BOF) format. Self-assess, try to complete each paper within 3 hours, check out your answers, read the explanation, and re-read accredited medicine textbooks to fill in the gap in your knowledge. Each question has an "objective"; try to review what the objective is about. This is the only self-assessment book specifically written to imitate the MRCPI part I examination.

3. Neurology: Self-Assessment for MRCP(UK) and MRCP(I):

This was published in September 2016 by Lulu Press Inc. You will find 792 questions of different formats, distributed into 3 chapters. Chapter 3 has many photographic materials.

This page was intentionally left blank

Lightning Source UK Ltd.
Milton Keynes UK
UKHW03f1612250418
321647UK00001B/32/P